Race and Photography

Race and Photography

RACIAL PHOTOGRAPHY AS
SCIENTIFIC EVIDENCE,
1876-1980

Amos Morris-Reich

The University of Chicago Press   CHICAGO & LONDON

AMOS MORRIS-REICH is an associate professor in the Department of Jewish History and the director of the Bucerius Institute for Research of Contemporary German History and Society at the University of Haifa. He is the author of *The Quest for Jewish Assimilation in Modern Social Science* and the editor of collected essays by Georg Simmel and Sander Gilman.

The University of Chicago Press, Chicago 60637
The University of Chicago Press, Ltd., London
© 2016 by The University of Chicago
All rights reserved. Published 2016.
Printed in the United States of America

25 24 23 22 21 20 19 18 17 16   1 2 3 4 5

ISBN-13: 978-0-226-32074-8 (cloth)
ISBN-13: 978-0-226-32088-5 (paper)
ISBN-13: 978-0-226-32091-5 (e-book)
DOI: 10.7208/chicago/9780226320915.001.0001

Library of Congress Cataloging-in-Publication Data
Morris-Reich, Amos, author.
Race and photography : racial photography as scientific evidence,
1876–1980 / Amos Morris-Reich.
pages : illustrations ; cm
Includes bibliographical references and index.
ISBN 978-0-226-32074-8 (cloth : alk. paper) — ISBN 978-0-226-32088-5 (pbk. : alk. paper) —
ISBN 978-0-226-32091-5 (ebook) 1. Photography in ethnology—History. 2. Photography—
Scientific applications—History. 3. Photography in ethnology—Europe—History.
4. Photography—Scientific applications—Europe—History. 5. Photography in
ethnology—Palestine—History. 6. Jews—Europe, Eastern—Identity. 7. Jews—Europe,
Central—Identity. 8. Germans—Race identity. 9. Günther, Hans F. K., 1891–1968.
10. Clauss, Ludwig Ferdinand, 1892–1974.   I. Title.
GN347.M67 2016
305.80022'2—dc23

∞ This paper meets the requirements of ANSI/NISO Z39.48-1992
(Permanence of Paper).

CONTENTS

ACKNOWLEDGMENTS

The question of what constitutes a critical position vis-à-vis the history of racial photography has been an ongoing concern for me while writing and researching this book. A couple of years ago, at a conference in London, a young colleague criticized my presentation as "uncritical." Her evaluation was valid in so far as that my work departs from what is commonly regarded as a "critical" position in the study of racial photography. In fact, I believe that this traditional criticism constitutes a superficial, weak, and inadequate stance. Thus, already early on in the project, I felt reluctant to approach my subject matter from an external perspective based on historically anachronistic theories of photography and positions dramatically independent from the scientific and political beliefs of the authors in question. In my view, this perspective—frequently based on ready-made categories and shaped by a safe, external point of view—fails to alter our already fixed values and historical understanding of the subjects and contexts under study.

In my quest for a different methodology for this work, I decided to apply the same standards of reading, interpretation, and contextualization to *all* versions of science, to all concepts of race, and to all photographic practices and images included in this book. This approach does not shrink from the study of individuals who today are regarded as dubious and racist pseudoscientists (and whose work it is sometimes truly painful to read). In fact, although I never agree with them, I read those authors—just like any other individuals that I examine—from within their respective discourses, assumptions, beliefs, use of scientific evidence, and argumentation. In short, I read them with a degree of what philosophers call "charity." My investigation of racial photography therefore reestablishes how—according to these authors—photographs functioned within the study and construction of categories of race. In other words, my main focus is not on documenting these authors' inconsistencies, contradictions, and weaknesses, but on analyzing the interconnections and interdependencies, tensions, and mutual flows between racial photography and other chapters in the history of science and culture. In this vein, I uncover how deeply flawed ideas, theories, beliefs, and practices can be internally coherent and consistent. In taking those photographic texts seriously (within the parameters outlined above), we can therefore reconstruct how exactly they appealed to and attracted (certain) contemporaries. Naturally, this strategy does not mean to rehabilitate such authors, ideas, or beliefs: on the contrary, by reintegrating racial photography as scientific evidence into the history of science, this methodology attempts to push the boundaries of a critical history of photography and science at large.

I thus perceive my methodology—which grants a certain amount of consideration to the study of even highly contested authors—as an "internal" perspective. At times, however—like on a Möbius strip—I found myself resorting to a more external mode of analysis that, initially, I had been determined to overcome. The reason for this is that applying the same strategies of reading, contextualization, and interpretation to different kinds of authors did not work in the same way. At certain moments, upholding this strategy meant to expend more and more energy, and more and more charity, to a point at which this effort in itself had to be acknowledged. I sought to recognize the significance of such moments for the historical picture at large. At least for me, these moments mark the boundaries of the critical position I have developed in this study.

Researching and writing this book took over eight years and took shape through the exchange of ideas of various kinds with numerous friends and colleagues on general as well as particular aspects of the history of science, Jewish history, the history of antisemitism, the history of photography, German history, and sociology of the kinds of ideas dealt with in this book. I would like to thank Eugene Sheppard, Scott Ury, Pawel Macjeko, Zur Shalev, Arnon Keren, Gur Alroey, Eli Lederhendler, Gerhard Wolf, Anne Harrington, Paul Mendes-Flohr, Tom Weber, Ezra Mendelsohn, Shulamit Volkov, Steven Aschheim, Eli Dresner, Ariel Meirav, Meirav Almog, Sharon Livne, Veronika Lipphardt, Snait Gissis, Shai Lavi, Shai Abady, Yossi Ben-Artzi, Guy Raz, Felix Wiedemann, Anja Laukötter, Marcelo Dascal, Paul Weindling, Marius Turda, Stefanie Schüler-Springorum, Shoshana Blum-Kulka, Yehuda Elkana, Jochai Rosen, Irene Tucker, Martin Ritter, Danny Shrire, Daniel Greene, Tania

Munz, Orit Abuhav, Dimitry Shumsky, Michael Miller, Menachem Brinker, Eva Illouz, Julia Adeney Thomas, Elizabeth Edwards, Ann Stoler, Jonathan Judaken, Francois Guesnet, David Feldman, Yossi Ziegler, Uwe Hoßfeld, Malachi HaCohen, Ofer Ashkenazi, Ayelet Ben-Yishai, Arik Dubnow, Baruch Eitam, and Cedric Cohen Skally.

In different parts of the book, I had the privilege of working with three different superb language editors: Mical Raz, Simon Cook, and Marie Deer. Their contribution to the development of the ideas, their clarification, and their exposition is far greater than merely editing my language.

Several people read major parts of the manuscript at different stages and gave me feedback that was critical for shaping the form and the content of the book. I am deeply indebted in this respect to Sander Gilman, Vered Maimon, Margaret Olin, Danny Trom, and Mitchell Ash. I would also like to thank the two external reviewers of the book for their careful reading and for their detailed and constructive remarks.

Between 2005 and 2013, I had the repeated privilege of staying at Department II of the Max Planck Institute for the History of Science (Max-Planck-Institut für Wissenschaftsgeschichte [MPIWG]) in Berlin. To a great extent, the subject of this study, its main complex of questions, and its historical epistemological methodology originated there. I would like to thank Department II of the MPIWG for its hospitality and access to its outstanding library and research facilities. In my first extended stay at the institute—still in the old building near Friedrichstraße—Kelley Wilder introduced me to the history of scientific photography in conversations that had a long-lasting effect on me. I also greatly benefited from discussions with Fernando Vidal as well as other visiting scholars at the institute. I am particularly indebted to Lorraine Daston for her generosity in allowing me to test the direction of my ideas in conversations with her and for her guidance and support.

I would like to thank the Simon Dubnow Institute for Jewish History and Culture in Leipzig for a very helpful research stay in the summer of 2007, which allowed me to spend a month perusing the endless Nazi racial journals in the Nationalbibliothek. I would like to express my gratitude to Dan Diner for this invitation and for my conversations with him. I also greatly benefited from my exchange with Nicolas Berg and Suzanne Zepp.

Some of the archival work for this book was carried out in several trips to archives in Vienna. I am deeply indebted to Margit Berner of the Anthropological Department at the Naturhistorisches Museum in Vienna, who shared with me her knowledge and time and guided me through the archives of her museum as well as those of the University of Vienna's Anthropological

Institute and other museum archives in Vienna. In Germany, I would like to thank the staff of Eugen Fischer's estate—now situated in Dahlem—for their assistance. In Israel, I wish to express my gratitude to Anat Benin of the Central Zionist Archives and Revital Hovav, the director of the photography collection of the Jewish Art and Culture Collection at the Israel Museum—both located in Jerusalem. I also wish to thank Yigal Feliks for his patience with scanning materials for this book.

In 2003, after I submitted my PhD dissertation, I started to discover my interest in the subject of race and photography. But only after several postdoc positions—and after I was awarded a five-year Polonsky Fellowship at the Jerusalem Van Leer Institute in 2007—did I find the courage to embark on a completely new study based on new materials and different methodologies; a project that, I knew, would at best yield fruit after several years of research. I am profoundly grateful to Gabi Motzkin for this opportunity, for his ongoing support, and for the exchange of ideas in conversations with him. I only enjoyed one year of the outstanding conditions of this fellowship because I was then offered a tenure-track position at the Department of Jewish History and Thought at the University of Haifa. But by then I was already committed to the new project. I am aware that with the deep crisis in the humanities in Israel (as well as around the world), a tenure-track position is nowadays—not only statistically—almost like winning the lottery. I am deeply obliged to my academic and administrative colleagues in my department, in the Faculty of Humanities, and in the Research Authority.

The administrative and academic staff of the Bucerius Institute for the Research of Contemporary German History and Society at the University of Haifa was crucial to the preparation of the manuscript for publication. I would like to thank Lea Dror, the administrative director of the institute, for her commitment and energy. I also wish to express my appreciation to Sharon Livne for our ongoing work together as well as to Franziska Tsufim and my PhD students—Julia Werner and Tally Gur—for their diligent assistance as well as for their support with technical aspects of the preparation of the book for print. I am particularly indebted to Katharina Konarek, who—with the assistance of Nina Schröter—handled the clarification of the status of photographs and the at times rather complex process of acquiring the respective reprint permissions. (Her documentation of that process, which brings together history and law, could serve as the basis for an interesting article.) Finally, I would like to thank the ZEIT-Stiftung Ebelin und Gerd Bucerius in Hamburg for their ongoing support of the Bucerius Institute.

Researching and writing this book were made possible by several generous grants. I would like to express my gratitude to the German Israel "Young" Grant Fund (2009), the Shpilman Institute for Photography Research Grant (2011), and the Israel Science Foundation (2012–2016). I am further obliged to the Minerva Foundation for funding that allowed me to carry out several research stays in Germany between 2005 and 2008. This book was published with the support of the Israel Science Foundation Humanities Program (2014).

I am enormously proud that David Brent chose to work with me, and, at the University of Chicago Press, I also had the privilege of working closely with Priya Nelson, Ellen Garnett Kladky, Ryo Yamaguchi, and Erik Carlson. I am grateful for their extraordinarily professional handling of the manuscript. I thank Bonny McLaughlin for the detailed index.

Finally, I want to thank my mother, Eva Morris, and my brother, Michael Morris-Reich, for their support. My wife Orit Siman-Tov and our children now know much more about racial photography than they wish to. Orit (who is a photographer) and I quickly realized how big a gap exists between the perspectives of a photographer and a historian of photography. Particularly, her observations about technical, practical, and pragmatic aspects of photography and photographs—which to her seemed obvious and self-evident—were quite often crucial and of a kind that to a nonpractitioner are hardest to perceive and grasp. I am deeply grateful for the support, interest, and (almost) endless patience I received from her and our children. The presence of Abigail, Yael, and Shaul Kenneth (who entered this world not long ago) made the effort enjoyable.

NOTE ABOUT SCANNING AND
REPRODUCTION OF PHOTOGRAPHS

This book features numerous reproductions of photographs from various kinds of sources. The particular context of reproductions of photographs concerning the scientific study of "race" accentuates some of the practical questions that are involved in scanning and reproducing photographs in a study of this kind. I do not know of simple or good solutions to these questions. At the outset, I wish to indicate one set of concerns with regard to scanning and reproducing photographs that pervaded this study.

There is a tension between photographs as material objects and the mechanical potential of photographs to be endlessly reproduced. Some photographs can be turned over and the notes on their back be read. But photographs can also exist only as negatives from which any number of reproductions can be made. The questions that pertain to the scanning and reproduction of photographs, then, touch on fundamental features of photography as a medium. Preparing a book manuscript necessitates standardization. But this standardization in the current context implicitly reifies the conception of photography as a nonmaterial, "transparent" medium, a conception that the text attempts to historically deconstruct. This tension is only sharpened by the fact that the great majority of the authors studied in this book believed photography to be a mechanical, neutral form for the representation or reproduction of reality.

The recognition of the heterogeneous nature of photographs in terms of their material form, their condition, and the way they appear to the historian (in a publication, unpublished in an archive, etc.) is essential for a history of racial photography as scientific evidence. Differences in material form and in the ways that their copies can be obtained greatly influence the form and the quality of the reproductions in this book. Most of the photographs were printed in books or in articles in journals. Some of these books were published eighty, ninety, or a hundred years ago (using different techniques and on different kinds of paper). In some cases, even in less than ten years, the photographic reproductions in the copies that I came across faded significantly. Many of these photographs can only be found today in the books from which they were scanned. Book exemplars of the same edition sometimes vary in the condition of the photographic reproductions. Some of the reproductions in books can be found in archives as well. It is common that these photographs have no "originals" and exist only as reproductions held by multiple archives. Others exist in more than one form: on glass, film, or paper, or a photocopy of a photograph (often a photograph can be found in multiple forms). Even the policies of archives, such as the Museum of Natural History in Vienna or the Central Zionist Archive in Jerusalem, which for instance do not allow researchers to scan photographs or negatives on their own but provide them with professional scans, affect the quality of the reproductions. The quality of these later reproductions is better than scans of reproductions made from books printed eighty years ago.

Scanning and reproducing photographs necessarily involves technical decisions and various forms of digital manipulation, such as framing, cutting details out of larger layouts, and the determination of contrast, brightness, color, or focus to mention just a few. The scanning and reproduction practices that were used in this study do not attempt to emphasize dramatically these material aspects of photographs, but they also do not attempt to even them out or conceal them. The ontological ambivalence or even duality of the photographic medium as one that possesses a material aspect but that is irreducible to that aspect and not fully determined by empirical or material appearance in a given instance is built into not only the photographic medium as such but also the attempt to write chapters of its history.

INTRODUCTION

Photography played a notable role in the vast body of "racial" literature produced in the Weimar and Nazi periods. Its realistic nature did not halt its demise in racial studies after World War II, which seems to have been both immediate and faster than was the case with other scientific mediums, such as statistics. In the American and European cultural imagination, a dark shadow has blotted out this period in photography's history, which epitomized all that is spurious, venal, and nefarious in subordinating human individuality to racial categories. The idea that photography could be used for the study of "race" never rose again.

Although this collective amnesia is perfectly understandable, the post-1945 historical memory conceals the fact that photography, once intimately linked to dominant trends in science and culture, was long a major medium in the scientific study of race. This book attempts to bring this connection back to light. Its chapters track the trajectory of racial photography from 1876 through transformations in the early decades of the twentieth century, up to Weimar and Nazi Germany, and beyond. German and Jewish subjects are the primary focus, although on occasion other authors and contexts are examined in order to fully develop the historical picture. In a certain sense, this book is intended as a provocation in the face of a history that has become culturally and politically fixed. Overall, the book traces some of the major facets of racial photography's use in science and scholarship by attempting to answer the following questions. What were the most prominent features and main purposes of scientific racial photography? How were these purposes linked to "racial" theories and to other scientific fields? What were the epistemological statuses of photographic practices? What forms of "visual perception" were embedded or implied in racial photography? And last but not least, what was photography's influence on racial studies?

To address these questions requires delving into a complex web of people, ideas, scientific beliefs, technologies, instruments, and practices. In the search for answers, I thus probe several genealogies of scientific photography, race, and visual perception from the last third of the nineteenth century to the Nazi period and its aftermath.

The resultant investigation introduces us to a number of writers who employed photography for their research and scientific agendas. These were writers of very diverse backgrounds, ethnicity, and social status who held to a variety of "racial" views and pursued divergent research goals and whose works express, explicitly or cryptically, a number of different approaches toward visual perception. Their careers began in the last third of the nineteenth century (terminus a quo) and continued through the establishment of West Germany. Some were major figures in the development of scientific photography, whose developments bore—theoretically or practically, directly or indirectly—on the study of race in the early twentieth century: Alphonse Bertillon, the founder of forensic photography in criminal identification; Sir Francis Galton, the founding father of eugenics and composite photography; and Rudolf Martin, who contributed perhaps more than any other scholar in the German-speaking world to the standardization of anthropometry. Also discussed are Carl Heinrich Stratz, Redcliffe Salaman, and Eugen Fischer, the last of whom, in the same period, developed photography within a type of racial science that used observations based on the principles of modern genetics rather than statistics and measurement. Some of the other scholars I discuss are Arthur Ruppin, a leading figure in the Zionist movement in the early twentieth century, founding member of the "Covenant of Peace" (Brit Shalom), and the father of Jewish demography and modern Jewish sociology; Hans F. K. Günther, generally considered the most important racial theorist and researcher of Nazi Germany but who was famous in Germany already in the 1920s and, in effect, remained a writer on racial issues in West Germany even after the war; Ludwig Ferdinand Clauß, a pupil of Edmund Husserl, one of the leading racial theorists in Nazi Germany who, shortly after his death, was recognized by Yad Vashem as one of the "righteous of the nations" (the title was revoked a few years later). I also address many other researchers and photographers, but this list is already sufficient to show that many issues in modern Jewish history—such as various racial and antisemitic ideas, the relationship between Jewish and non-Jewish scientists and researchers, and Jews as "objects"—are interwoven, nonreductively, into this study.

Analysis of the work of writers of such diverse backgrounds reveals a crowded, tension-riddled, and even contradictory paradigm, yet one with patently common epistemological features. These shared features relate to the connection between concepts of race and the use of photography, a connection that itself results in complex forms of visual perception (and also embodies desperately important political dimensions, which, for the moment, we leave to one side). These forms of visual perception are hard to understand or conceptually reconstruct unless a number of ideological features of the period's culture are explained, and this is what the main part of the introduction attempts to do. In six sequential but also parallel sections, I excavate core methodological and historical and sociological cross sections, which themselves underwrite and are interwoven into the individual cases dealt with in the chapters that follow and indeed also the trajectory of racial photography as a whole. In these sections I elaborate, first of all, my use of the notion of practical epistemology, which is employed in this book in order to recover the assumptions that accompanied uses of photography in the study of race. Second, I identify the "reactionary logic" that is to be found as the basis of much of the writing on race that we shall subsequently encounter. Third, I identify the tension, which is at the basis of the development of racial photography, between belief in "race" as stable and observable on the one hand and as increasingly fluid and concealed on the other. Fourth, I reconstruct the shift to a subjective paradigm of visual perception that occurred over the course of the nineteenth century—a paradigm shift from which racial photography emerged and to which it responded. Fifth, I turn to the history of mutual stigmatization that surrounded the politicization of the sciences of "race" in the 1920s, one product of which was that the status of the literature and the photographic practices studied in this book were rendered as essentially pseudoscientific. Finally, in the sixth section I attempt to argue why, for the history of racial photography, the category of imagination is as essential as is that of sense perception. I conclude this introductory chapter with a justification of the decision to resituate racial photography in the context of science rather than of politics.

PRACTICAL EPISTEMOLOGY

The modernist and realistic-critical writings on visual perception and photography developed in the 1920s by the likes of Theodor Adorno, Siegfried Kracauer, and Rudolf Arnheim have become, justifiably, objects of research in their own right and are now considered classics in modern cultural studies. Most of the writers dealt with here, however, viewed photography as a realist medium, that is, as a reproduction of reality. But the starting point of my analysis is not the content of the photography—the "this is how it was"—but rather the "this is how somebody observed a particular object at a particular point of time." From my own perspective the history of racial photography thus intersects with the history of observation. Or, put another way, from the point of view of the analysis of this study, the realist tradition in photography cannot be understood without drawing on some of the insights developed by the modernist and critical traditions. This book, then, describes the uses of photography, their incorporation of and transformative effects on forms of visual perception. To these ends my analysis is based on the idea of "practical epistemology."

While the book deals with many aspects of practical epistemology, a schematic survey will suffice at this point.[1] Practical epistemology refers to reasoning that develops from practice to theory, that is, from the effects of a scientific practice to its underlying epistemological assumptions. I have experienced the force of certain practical considerations on more than one occasion. During my postdoctorate year in Chicago, for example, I had planned to undertake a relatively small comparative study of the relationship between cultural anthropology and racial sciences, with special reference to the work of Günther. Interested primarily in the methodological aspects of his anthropology, I skipped over the pages of photographs that appear in many of Günther's books. But when I walked along the streets of Chicago, surrounded by its living mosaic of social types, I suddenly became aware of the fact that I was examining minute physiognomic details of other passersby. This ability to take in even subtle distinctions caught me by surprise. I realized in a flash that even though I had passed over the photographs in Günther's books, nevertheless they had somehow affected my visual perception. If more than eighty years after their publication and extensive circulation the photographs could still have such a powerful effect on my consciousness, then one can only imagine the intensity of the effects produced nearer the time of their publication. A closer look at these photographs—their technical quality, their diversity, their arrangement on the paper, their correlation with the written text, the time and energy that went into their collection, and so on, reveals that they not only presumed but also created complex forms of knowledge and that they served as conduits of forms of visual perception with immediate and socially effective consequences. One of my goals in this book has been to interpret these forms and bring them to our awareness.

Practical epistemology refers to a form of analysis somewhat different from classical epistemology—from the bottom up, from praxis to the social—and different also in the epistemological, metaphysical, or ontological knowledge that it assumes. In the context of racial photography, this is the meeting point of knowledge and assumptions of knowledge, of praxis and belief, of individuality and intentionality, concepts, ideas, and social and cultural constraints. But "questions of epistemology are also questions of social order,"[2] and in the case of racial authors, the practical perspective may be more interesting than the "classical," although the two are obviously inseparable. The importance of an analysis based on photographic angles, for example, is a case in point. Modernist photographer-designer Moholy-Nagy and many others introduced specific angles into their photographic work, such as the now well-known elevated point above a street scene. Consequently, modernist photographs could be identified by their angles rather than by their subject matter or format. Similarly, racial photographic matrices could also be identified by a particular set of angles, such as the frontal and profile angles. The photograph, therefore, is a powerful starting point for a practical epistemological analysis. Contrary to the assumption that "photographs speak for themselves," photographic authors must make numerous (visual and other) assumptions as to their audience's knowledge in order to get their main ideas across.

What are the visual practices of photographic presentation in racial books and their sources? How are they connected to the ideas in the written text? In a different context, Hike Thode-Arora showed that the advisers to the producers of German adventure films in the 1920s had earlier been directors of amusement parks in which exotic animals and human types from remote parts of the globe had been displayed.[3] The knowledge and experience of these consultants, who had accumulated "special types" (*charakteristische Typen*) at the turn of the century did not go to waste when the parks fell into a slump during the First World War; the consultants of the "strikingly unusual" simply shifted their activity to a new cultural field—exotic travel films.

Even though some overlap exists in dealing with the racial "other" (the film advisers and theme park owners were members of the German Anthropological Society), nevertheless, the practical use of photography in books on race comes from other sources and disciplines and had its sights set on other goals. For example, the managers of the amusement parks and the producers of exotic travel films both realized the importance of creating "authenticity." They did this inter alia by means of the use of identifiable ethnic objects (e.g., the "Indian" turban) or the reconstruction of authentic local flora and fauna, which was considered de rigueur for the visitors' or viewers' experience.[4] The authors of race studies, by contrast, used photography to create a different kind of authentic context, a sort of antithetical one. That is, they frequently employed photography to isolate a race in order to show that it was not a product of the environment. Their goals, however, underwent changes in accordance with shifts in underlying concepts.

I am not sure if I am willing to go as far as claiming that photography was indispensable for the emergence of the racial sciences, as Georges Didi-Huberman claimed vis-à-vis Charcot and the invention of hysteria. But, as I show, photography was deeply part of their emergence.[5] One example, which I explain in detail, is the Mendelian theory that penetrated racial discourse at the beginning of the twentieth century and generated specific forms of racial photography with particular goals, forms of layout, and subtitles. Another example deals with forms of visual proof that were closely tied to psychological "thought experiments." The practical approach also discerns the difference among writers regarding writing about photography and using photography. Kracauer and Walter Benjamin *wrote about* photography, but they *used* photography much less. Benjamin's *A Short History of Photography* contains only a handful of photographs. I discuss writers who almost completely refrained from theorizing about photography but who used it profusely, presenting thousands of photos methodically, deliberately, and diversely. In attempting to bring out the intimate connection between race, photography, and visual perception, the practical epistemological perspective proves itself particularly powerful.

"REACTIONARY LOGIC"

All of the writers discussed in the book shared, in one way or another, what may be called a "reactionary logic." Since my use of *reactionary* is somewhat idiosyncratic, a brief elaboration is called for. *Reactionary* does not here refer to the authors' political views (regardless of their importance). The second half of the nineteenth century witnessed widespread processes of modernization throughout Europe and especially in Germany. Two basic forms of intellectual response to this modernization can be distinguished, historically and analytically: "modernist" and "reactionary."[6] As a matter of fact, history never neatly fits a model or scheme, but focusing on these two basic types of intellectual response allows us to place the subjects of this book within a broad historical context. In reality, of course, alongside obvious cases are others that are difficult to classify: border cases, areas of interpretive decision, cases that belong to the two abovementioned types, and dynamic cases that change over the course of time. Nevertheless, our basic distinction allows us to identify the various intellectual mechanisms of response, to map out the connections between the forms of discourse and argument on the one hand and the ideological positions on the other, and to comprehend the various deviations, alternatives, and permutations.

The two responses—"modernist" and "reactionary"—embodied different views of the modern age, which nevertheless concurred that the era was unlike anything that preceded it, a veritable sui generis. While modernist responses tended to approve of modernization as a necessary process, they did so reservedly, aware of inherent problems and dangers. Reactionary responses, however, regarded modernization with far more misgiving and negativity. Even though the racial writers cannot all be placed within the same camp, there is a close relationship between the idea of "race" and the reactionary logic as a form of response to modernity. That is, there is a structural relationship between the belief in the natural division of the human species into races and the belief that modern conditions increasingly undermine and erode that boundary. As directly pertaining to humans, and therefore inevitably political, "race" is arguably the most important category of reactionary response to modernity.

While the writers discussed in this book do not all share the same concept of race, they do share a more general epistemological dimension of their thinking. The best way to conceive of the "racial" dimension in Weimar and Nazi literature is by way of the Kantian term *regulative ideal*. This is because, while it is impossible to identify any common denominator across various discourses with regard to definitions of race or ways to study it, on a higher plane diverse discourses were based on an implicit agreement that "race" is essential for understanding such general concepts as man, society, nature, history, and life itself. In the literature that this book deals with, "race" is not an isolated concept but a broad category, a language that defines forms of thought and expression, connotations and images, expressions of desire and meaning (and is also the key to them), and responses to specific historical, cultural, social, and political conditions. The exchange of ideas of race between the biologically oriented sciences (physical anthropology, human genetics) and branches of the humanities (philosophy, art history) is methodologically and terminologically complex and open to various historical interpretations.[7] Despite this heterogeneous point of departure, however, all writers under discussion adhere to the paradigm of "race." One of the problems we face—the sign of a genuine historical question—is the contradiction rooted in this point. On the one hand, it is hard not to regard the centrality of "race" as a snowball effect in which, to a certain degree, terminological differences among the writers regarding race are, in retrospect, only trifling nuances compared with the tremendous unfolding catastrophe. On the other hand, the centrality of the category and its diversification means that there was no single verdict that applied to the writers who conceived of phenomena differently or developed notions of race for contradictory purposes.[8]

The racial writers cannot be understood unless they are placed in the wider context of the crisis of modernity; indeed, their responses cannot be considered other than as specifically modern. Racial writers developed concepts, tools, and technologies and constituted subjects and objects that can all be dated to a particular point in time and to specific historical, social, and political circumstances. Nevertheless, these writers experienced their principles and discoveries as if they had existed from time immemorial, as if they were eternally valid rules of nature. Time-bound, constructed categories are projected backward in time as though they had always been there.

Many of these writers were responding to what they felt was the disintegration of traditional forms of social, community, family, and natural life in a spate of serious, immediate processes with irreversible implications. Seen in this light, "race" is both the form and content of their response. In the research conducted in the last third of the nineteenth century, race increasingly appeared as a natural, anthropological, or biological category that defined collective and personal identity and strongly resisted a reality perceived as problematic, degenerate, and corrupt. This description applies to prewar German, to French, and British writers and certainly also to a vanquished Germany after World War I, to which last we can add the disintegration of the political order, economic crisis, and heavy opposition to the new democratic framework that the victors had imposed.

A paradox exists in the relationship between the crisis of modernity and the rise of the term *race*, a paradox whose significance is insufficiently understood. Louis Dumont spoke of a similar condition regarding the connection between democratization, equality before the law, and the rise of racism as a social phenomenon when he claimed that racism was a social response to the rescission of social differences that had been anchored in law. But I refer to something different: scientific racism's establishment of race as a natural category that was disintegrating. Like the discovery or invention of "scenery," as Danny Trom has noted, race appeared when the European cultural arena was viewed as rapidly eroding because of accelerated modernization, industrialization, and urbanization. Race is "discovered" as being on the verge of collapse. The laws of nature come apart at the seams. A good example, and one that scholars in the late nineteenth and early twentieth century viewed as apolitical, is the strong belief that primitive races are crushed by modern civilization and done away with in a brutal, tragic, quick, inevitable process. If this assumption was the ethical and scientific basis for anthropology's quest to document forms of life about to become extinct, then the racial writers observed the changes in the natural order of things as resulting from the decline of the superior races (which were usually those to which they believed they belonged). This reactionary logic is necessary, then, for analyzing the connection between race, photography, and visual perception.

SCIENTIFIC PROGRESS AND THE FLUIDITY OF RACIAL DIFFERENCES

Racial scholars of the late nineteenth and early twentieth century claimed that the essence of racial differences is intuitive and obvious as well as elusive, fluid, and often practically indiscernible. The dialectic in the development of racial photography is directly related to the tension between the belief in the legitimacy of racial difference and the sciences that study it and the slippery content of racial photography and the increasingly difficult way of substantiating it. I do not share the growing trend in recent years that situates the Holocaust and persecution of the Jews in the historical context of European colonialism, for the social and philosophical roots of colonial racism and antisemitism are profoundly different. Nevertheless, the following chapter shows that the scientific photographic racialization of Jews is inseparable from the colonialist context, or to be more precise, from the European roots of this context. Some postcolonial research findings are undoubtedly useful here; others, especially those dealing with power struggles between European colonialists and indigenous Africans or Asians, are less helpful. The most relevant findings, however, deal with the fluidity of racial differences.

Ann Stoler, a pioneer in postcolonial "race" studies, notes that colonial authority was based on two false assumptions: first, that the Europeans and natives of the colonies were two biologically and socially disparate entities that could be immediately separated; second, that the boundaries between them had already been somehow demarcated and were naturally and unequivocally self-understood.[9] Stoler links this distinction to Ian Hacking's assertion concerning the "looping effect" of classification systems, that is, their ability to force distinctions on what they merely describe.[10] The racial writer's basic dilemma lies in the gap between his belief that races are separate entities and that racial differences are unalterable, important, intuitively identifiable, and the fluidity of social reality. In our context, this dilemma is characterized by the writer's ceaseless attempts to comprehend and scientifically represent elusive and fluid racial differences by way of an effective, valid, and definitive method. The result is an endless series of "conclusive" attempts.

There is a technological dimension to the internal progression of scientific photography that is not specific to the field. The introduction of a more advanced technology or method normally drives out the older one, underpinned by the assertion that the new is fault free, more precise, cheaper to employ on a large scale, and easier to operate. In other words, it relativizes the former's efficiency. After DNA checking was introduced, for example, fingerprinting came to be considered insufficiently precise. Or, another example, chronologically closer to the people discussed in this study, Galton's fingerprint identification method increasingly pushed aside Bertillon's complex statistical anthropometric method, although the latter's was not completely invalidated.

From the end of the 1880s in the Americas, parts of Europe, and certain colonial outposts, historians discern a gradual weakening of racial divisions as under changing social and economical conditions, class and ethnic background no longer invariably coincided. On some occasions, members of established racial groups could no longer be racially distinguished. Examples include educated African Americans in North American urban centers (e.g., as reflected, in William DuBois's early twentieth century "Negro type" photographs), acculturated bourgeois Jews in central Europe, or, in the opposite direction, the emergence of a racialized class of poor urban white Dutch in Jakarta. Social possibilities of "looking like" were further facilitated by breakthroughs in medical technologies, such as plastic surgery (which was initially illegal).[11] As we shall see, transformations in scientific photography may lag behind such social processes, but they are nevertheless intimately connected to the internalization of perceived racial differences. One consequence of all this is that in the last years of the nineteenth century, the discernment of actual differences was increasingly replaced with the discernment of potential differences. If Cesare Lombroso, the Italian Jewish founder of the Italian school of positivist criminology, still believed that with the aid of photography he could identify criminals, now only the potential for criminality could be identified. Scientific photography at the turn of the nineteenth century occupied itself with the study of the invisible in various fields and contexts (from galloping horses to the substantiation of spirits)—racial phenomena being just one of these contexts.[12]

Photography is a powerful tool for studying race. In the first chapter of this book, I attempt to show the dialectic inherent in the development of racial photography. I attempt to determine the nature of race as a scientific object and the role of scientific photography as a key to processes of racialization. But I also show how racial photography is symptomatic of the tension between the belief in self-evident racial difference and its actual social elusiveness. In other words, racial photography assumes that race is permanent and the scientific method flawed. The development of scientific photography was part of the attempt to surmount this tension by developing better instruments and theories that supposedly stood on firmer epistemological ground. This is true for compositional photography, which was designed to abstract the "type" from a sample group of individuals (chap. 1). It is also true for photographic observation based on Mendelian principles, according to which, and following specific patterns, offspring manifest characteristics of their parents (chap. 1, 3). And it is also true for phenomenological photography, which was designed to capture the aesthetic essence of race, conceived mainly as a spiritual feature (chap. 4). I examine the labyrinthine dynamic racial asymmetries involved in attempts to employ photography in the study of race and show that photography was interwoven into racial concepts that were far less deterministic than writers believed them to be. The study of minor fluctuations as well as major changes in the use of photography sharpens our understanding of the dynamic relationship between the conceptual transformations of "race" and the advances made in scientific photography.

THE DISINTEGRATION OF VISUAL PERCEPTION IN THE NINETEENTH CENTURY

Because we assume our own perceptual experience to be in some way universal, we tend to resist the idea that visual perception is culturally dependent or that scientific models of visual perception might influence the way that people actually see. Nevertheless, to fully understand the paradigms of visual perception with which this book is concerned, it is necessary to take note of the transformation in the conceptualization of visual perception that historians have recently traced through the nineteenth and early twentieth centuries—a transformation directly linked to the modernization and rationalization of European social life and the sources of some of the "threads" of visual perception discussed in this book. Fundamental to this conceptual revolution was the displacement of an older geometrical optics by the emergence of a new paradigm of visual perception based on physiology and ultimately also psychology, and hence the replacement of an early modern objective model of vision with a new, modern, subjective model.

Jonathan Crary is the pioneer of the study of this nineteenth-century transformation of vision. Crary juxtaposes the objective model of vision that dominated early modern theories of art and sight with the modernist model that evolved over the course of the nineteenth century as a consequence of the physiological investigation of the nervous system.[13] Within a relatively short period, according to Crary, the older model of visual perception as determined by the passive reception of geometrically analyzable optical phenomena was replaced by a modern understanding that experience in general, and vision in particular, was the product of a living, embodied subject. Crucially, because it was now recognized that visual perception rested on the inherently subjective contingent cognitive activity, so the visual experience of an observer came to be seen as inherently unreliable and even arbitrary.

While the roots of this new subjective understanding of vision can be traced to the physiological research conducted in the first decades of the nineteenth century, the philosophical implications of the new paradigm seem only to have been worked out in the second half of the century. Between 1856 and 1866, Hermann von Helmholtz published a three-volume work on vision. What was clear from his work was that the eye is not a transparent organism but an amazing instrument marred by aberrations, capable of error, and imbued with inconsistencies in processing visual information.[14] By the late 1860s the studies of Helmholtz and others, such as Gustav Fechner, had convinced many not only of the inherently subjective basis of sensory perception, including vision, but consequently also of the epistemological unreliability of visual observation. The new theory of vision, in other words, had cast a shadow over the credibility of the visual observations that since the early modern era had served as the supposedly secure basis of scientific knowledge.

By the end of the century, the earlier physiological investigation of the role of the human body in the construction of vision had begun to give way to a growing fascination with the role of the human mind. In the 1880s Wilhelm Wundt distinguished between the field of vision (*Blickfeld*) and the angle of vision (*Blickpunkt*). This differentiation stood in marked contrast to the fixed, mathematical, "bodiless" perspective established in the Renaissance.[15] Wundt also discovered the two forms in which a particular nervous stimulation is able to penetrate the consciousness. He termed the entrance of physical stimulation into the arena of consciousness "absorption" or "perception" (*Perzeption*) and its transformation into the focus of attention "apperception" (*Apperzeption*). Wundt, whose attempts to measure sensation continued those pioneered by Gustav Fechner, found that stimulation can be absorbed only if it crosses the border of the consciousness and becomes the focus of attention.[16] By the 1890s physiological studies of the eye further demonstrated that the field of vision is not, as hitherto imagined, the immediate result of the absorption of an object but rather the product of a string of complex operations connected to eye movements that, by way of the active intervention of the perceiving subject, result in a stable image.[17] Thus the physiological research of the 1890s now brought into prominence the psychological role of the subject in the construction of vision. This necessarily pointed the way toward a subsequent emphasis on nonphysical mental factors in the construction of vision (i.e., the role of consciousness and apperception).

Crary relates these nineteenth-century changes in the understanding of visual perception to the rise of those philosophical theories (of Nietzsche, Peirce, Heidegger, and Wittgenstein) that emphasized the way in which the subject is a product of language and social and cultural systems. In this context, the relevant, organic faculties (e.g., the eye itself) come to lose their centrality in the understanding of visual perception.[18] By the 1890s, even before the growth of popular cinema, understanding of visual perception had been gradually broken down within certain discourses into an ongoing, temporal, dynamic act. The paradox here is that the scientific study of visual perception established at the heart of the visual experience an active, unreliable subject, one that is the agent of optical activity, a locus in which the synthesis of vision takes place. It is, in certain respects, to this general shift in the understanding of vision that writers such as Günther and Clauß responded when they separated visual perception as *relative* (to race) from visual perception as *subjective*, lest the two be confused. In this book I clarify these issues.

It is still unclear whether a direct connection exists between the crisis of modernity and the changes that took place in the scientific view of visual perception in the latter half of the nineteenth century. Nevertheless, a growing number of accounts provide ample evidence of a feeling that the tempo of life was accelerating, and the "disintegration of form" that was felt in many areas of life—in art, for example—paralleled the processes of destruction. Many felt at the time that the number of sensory stimulations was dramatically increasing and that as a result, the human subject was unable to keep pace and was finding itself on the verge of collapse. Walter Benjamin, who could be taken here as a cultural seismograph, complained that the amount of stimulation was drastically climbing and that absorption was taking place in a "state of perception distraction."[19] Many found that rapid technological development, such as the development of fast trains and the invention of the telephone and the telegram, which revolutionized the movement of people and of information, or the illumination of large cities with electric lights as well as the modern factory that dissected and concentrated the senses of the laborer, all transformed urban experience and overflooded the senses of subjects. This aspect of "the crisis of modernity" was perceived as necessarily bringing about the fragmentation of the world as previously known. This feeling of disintegration was reflected in many fields of research, from nascent science of sociology to the psychologists' study of perception.[20]

The upsurge in the use of signs, numbers, newspaper photographs, and advertisements resulted in the eye's liberation from the strictures of horizontal (left to right) and vertical (top to bottom) movement and echoed the clamor of modern urban life. It also indicated a break in the conditions of sensory absorption as the horizon of optical experience expanded and demanded new types of visual perception. Even before the invention of the cinema in the 1890s, sensory absorption conditions had decomposed into units. This decomposition brought about a philosophical endeavor to go beyond the units and rediscover the unbroken intuition. Impressionist painters and philosophers such as Henri Bergson in France and Georg Simmel in Germany asked similar questions and developed a philosophy of "life" as nondecomposable. Cultural phenomena such as impressionist art or philosophies of life were thus tools for studying the disintegration of form that, at the same time, created those very phenomena.[21]

Photography is an integral part of these changes: the close-up expands space; a photograph of movement illustrates movement; the enlargement of part of a photo not only magnifies a particular detail but also reveals new, primitive structures.[22] Walter Benjamin developed the idea of the "optical unconscious" and related photography's capabilities (which are relevant to its uses in racial literature) to its "ability to reassemble fragments 'under a different law.' "[23] Merging, in a certain sense, the human eye and the camera lens, Moholy-Nagy theorized "photography as the new objective form of seeing." He emphasized the specificity of photography in revolutionizing the way the world is seen, theoretically and practically, the way it widens the capability of optical demonstration (*Darstellungsmöglichkeiten*), and its accurateness in describing (*Beschreibung*) a thing under observation.[24]

These processes were accompanied by the question, did the world of phenomena lack permanent, basic form? Thus, for example, in the late 1870s, by using multiple cameras Edward Muybridge separated into discrete images the composite movement of a galloping horse, the symbol of movement in many societies. These "solid" yet lifeless units of movement, each with the potential to be quantified, showed a world that stood beyond the realm of human sensory perception, and Muybridge was able to rearrange them in an order different from the temporal succession of subjective experience. As for theories of visual perception, the racial writers often expressed ideas that had once been considered revolutionary or avant-garde. In this respect they may be compared to Hans Sedlmayr, the Austrian art historian, as described by American-British historian Fredrick Schwartz: "Sedlmayr found himself in a position of more than one scholar at the time, in a very grey area between charlantry and the most advanced work in the human sciences."[25]

These ideas are especially relevant for radical notions of vision, such as the extrapolation of the whole from the fragmentary, physiognomic view; the ability to distinguish the "internal" from the "external" and the general from the particular; and the inverted relationship between reality and the creation of art.[26] At these junctions, the modernist avant-garde and the antimodernist are found in reciprocal relationships. A famous anecdote is told of an observer of Picasso's 1906 portrait of Gertrude Stein, who commented to the great artist that the painting bore no resemblance to the real person. Picasso replied, "she will." Picasso's answer can be interpreted in various ways. One is that Stein will gradually look like the image in the painting, that is, the referent will begin to resemble its artistic rendition. Another, perhaps stronger explanation, and one that is closely linked to the trends that took root in the use of racial photography soon after the comment was made, is that painting does not represent reality but is in partnership with it. The portrait does not represent Stein but rather elevates the act of viewing until the real object appears via its art form. There are parallels between the uses of photography by racial writers and other type of writers, such as Belá Balázs, who developed a theory of photography's ability to regenerate senses that had atrophied.[27] For Sedlmayr, at the other pole (politically, no doubt), "the organ for the grasp of the visible character has atrophied in most people."[28] In the next chapter I discuss how Sedlmayr, along with Wölfflin and Simmel, elevated the historicity and specificity of seeing—seeing here incorporates visual perception but also transcends it—as lost knowledge.[29]

The reactionary and the avant-garde writers often approach one another and even overlap regardless of their overt political differences, as is the case with Ludwig Klages and Walter Benjamin, certain racial writers, Rudolf Arnheim, or the cinematic theorist Belá Balázs. This last, a Hungarian-Jewish communist, shared with many of the racial writers a longing and hope for a new visual culture in an age of modern technology and the commercialization of capitalistic pleasure. Balázs nurtured an imaginary return to idyllic, preindustrial-age folklore.[30] The racial writers agreed on race but not on the medium of photography or cinema. Their uses of photography are not far from Balázs's view of cinema as a modern, popular "sensory-organ."[31] His book *The Visible Man* (*Der sichtbare Mensch*) (1924) comes close to physiognomic theories and concepts of expression (*Ausdruck*). Balázs tried to clarify the ramifications of racial difference.[32] He hoped that a democratic culture and a new order of visual perception would be created under the new conditions. In this light, Balázs may be termed a *modernist*. His conclusions are the opposite of those of the writers on race, yet some of their ideas and epistemological assumptions coincide.

One facet of the history of visual perception in this period is linked to the sense that it and the world of phenomena are disintegrating, that the frameworks of traditional knowledge are no longer satisfactory, that a radically new framework capable of capturing the changing situation is needed. What many of the writers in this period share is frustration with the cul-de-sac of historicism as well as disillusionment with a dominant neo-Kantianism that declares all knowledge to be mediated, and as such it blocks the fervently longed-for direct experience of the "thing in itself." Benjamin, for example, believed that photography shattered the prison walls of Kantian experience. On the other hand, the racial writers of the 1920s and 1930s were fettered by the reactionary moment. They did not look for this kind of total release. In contrast to the liberation that cinema offers from any spatially fixed point (which Erwin Panofsky noted) or the scientific view of a-perspectivism (the idea that one could be free of any specific perspective), they stated that photography proved that visual perception was relative to race, and they tried to determine visual perception and its limitations by claiming that they had discovered a racial perspective.[33]

The racial writers proposed another configuration for the dialectical relationship, described by Crary, between the perceived disintegration of reality, accelerated modernization and rationalization of society, and the attempt to discover, constitute, or anchor some kind of permanent, stable basis.[34] The more fluid the reality, the greater the urge was to establish it on a sturdy basis. Husserl will be treated in some detail in the discussion of racial imagination that appears in the chapter on his pupil, Ludwig Ferdinand Clauß. At this point, suffice it to say that Husserl's phenomenology (as with the impressionistic branches of art) exemplifies the profound dualism inherent in the attempt to grasp the permanent and essential—the "pure" in Husserl's terms—and distinguish it from the fleeting, secondary, and socially constructed. This was the case in conditions seen as the "disintegration of the world" and the "explosion" of vision into endless stimuli. Husserl's research project is one of many that testify to (and are symptomatic of) the crisis in perception. From this perspective, Husserl's yearning for universal structure is profoundly dualistic. His position—which influenced Gestalt views—states that rather than reality being a synthesis that is created by a perceiving subject, the perceived objects themselves—reality—have structure and order, and so they appear to us. The dualism consists in the fact that the search and yearning for order that is rooted in Husserl's phenomenology are the direct result of the intolerable possibility that underneath the raging, fragmentary flow of modern life, there is no structure or order, no fixed anchor—in other words, we live in an impermanent, foundationless, and perhaps meaningless world of floating symbols and fluctuating social relations. By the same token, the writers with whom I deal considered race the locus of their search for structure, stability, and meaning.

PSEUDOSCIENCE AND SCIENCE

The work of the writers dealt with in this book is today designated as *pseudoscience*, a term that became culturally fixed after 1945. The history of this designation, however, unearths a web of historical considerations that are crucial for the history of racial photography within science and its later legacy. Why did it come to be called *pseudoscience*, by whom, and in what context? And why is this terminology important for our interpretation?

Already by the 1920s the increasing politicization of "race" and its polarization within science had led to a stormy debate in both the United States and Germany. The writers whom I analyze stood in the center of this polemic. In the 1930s their opponents labeled them *pseudoscientists* and, to a large degree, succeeded in making this pejorative epithet stick. The establishment of this stigma was part of the intellectual struggle over the very nature of science, the highest domain of knowledge and rationality.

In one camp stood natural and social scientists, some with great scientific authority, who understood man as a creature within the natural world and believed that the study of human and social phenomena must conform to the concepts and categories of the natural sciences, first and foremost to the Darwinian theories of evolution and Mendelian genetics. This group was not homogenous, but after 1900 a growing number favored converting anthropology into a racial science.

In the other camp were scientists who harbored reservations over the gradual bioligization of society. The members of this camp differed over many issues, and this camp, too, defies easy generalization. Two common threads linking the majority of these scientists, however, were their opposition to both racial determinism and to the attempt to establish "race" as the main concept of the humanities and social sciences. On occasion, members of this group also displayed sensitivity to the political and social implications of racial concepts.

The struggle between these two camps contained epistemological, institutional, ideological, and political facets. Key differences were already discernable by the end of the nineteenth century, existing tendencies were intensified during World War I, but positions became far more politicized and polarized after 1918. A similar debate took place in Germany's Sociological Society before 1914, when "race" as an explanatory category was formally expelled from the sociological discipline.[35] In anthropology, by contrast, Germany's devastating defeat formed the immediate political background to the deepening of deterministic and increasingly overtly politicized and antisemitic notions of race. In contrast to Germany, where the whole spectrum of views with regard to race was present in the 1920s, in the United States the debate between proponents and opponents of racial determinism was institutionalized in the struggle between opposing camps, led by American German-Jewish Franz Boas and his students, on the one hand, and the antisemite and American proponent of racial determinism Madison Grant and his supporters on the other. When in 1918 Boas was elected president of the American Anthropological Association, Grant and his followers resigned from the association and established "The Galton Society." The American anthropological field was now institutionally split into two rival camps.[36]

In both the United States and Germany this controversy was accompanied by mutual stigmatization. The proracial criteria group labeled its opponents "Jews" who, for ideological motives, advocated interracial equality as an a priori position rather than basing their views on scientific findings (according to which Jews are racially inferior). The label "Jewish science" was intended to discredit their scholarly work and, in effect, hamstring the ability of a writer of Jewish background to voice scientific notions. The "Jewish" camp labeled its opponents *pseudoscientists*—a term that discredited the scientific legitimacy and validity of proracial science as a whole. The stigmas "Jewish science" and "pseudoscience" were, to a significant degree, weapons of mutual delegitimization in the internal scientific discourse, "floating signifiers" from a semiotic point of view, as opposed to established and permanent theories of different scientific approaches. It may be said, albeit cautiously, that the semiotic intentions of the two groups of stigmatizers were close. Both groups sought either to categorically deny the value of the other's scientific work because of its background (which automatically delegitimized the other camp's research) or to unconditionally invalidate their opponents' scientific work because of its conceptual basis (which utterly discredited their thinking). It should be noted that while I focus here on the sciences and the works of scholars who were branded pseudoscientists after World War II, the Holocaust, and decolonization, the polemical history of this branding has, to a large degree, been forgotten. In other words, today we simply take it for granted that by common consensus these intellectuals were pseudoscientists.[37]

In the twentieth century this racial controversy was arguably the single most important case in which the term *pseudoscience* was employed in an attempt to denigrate the scientific standing of opponents.[38] Initially the term was associated with the status of religion (Christianity) in relation to science; but since the early twentieth century it has been associated with the attempt to separate authentic science from what only masquerades as science. Indeed, in terms of the history of *philosophical* attempts to demarcate the notion of science, Karl Popper's principle of "demarcation," historically separate from the racial debate although temporally parallel, is absolutely indispensable.

In Vienna a spirited debate developed over the demarcation question. Primarily involving Karl Popper and members of the Vienna Circle who wanted to purify scientific language. Popper set forth a philosophy of falsifiability as a basic feature of those fields of science whose advances far outpaced development in other fields. In his 1935 *Logik der Forschung* (*The Logic of Scientific Discovery*), Popper coined the term *line of demarcation* to refer to the supposed border between genuine and bogus science. Surprisingly, given the date of the polemic and Popper's ethnic background (Jewish), he made no mention of "racial" concepts, although he must have encountered them. Instead, he focused on psychoanalysis and Marxism. In this period, however, Popper refrained from using the label *pseudoscience*.

After the Second World War, Popper used the term *pseudoscience* in his work *The Open Society and Its Enemies* (1945), and in his own translation of *Logik der Forschung* (major parts of which he rewrote at this time). Following his experience of racial persecution, which had transformed him into a displaced person living in a foreign land, and given the recent devastation caused by World War II with its crimes against humanity, fascism and antisemitism naturally appear in the subtext of these works; but even here, when he refers to pseudoscience Popper actually means astrology. In the translation of *Logic* he refers to Plato and Hegel as the fathers of fascism and racism, but he is silent about the scientists of race, some of whom were still alive and working in Berlin and Vienna. The twentieth century's great effort to establish an internal scientific conceptual system capable of differentiating between science and pseudoscience overlaps the period in which the racial scientists were branded "pseudoscientists," to a great extent by those who themselves were persecuted because of their origins. The two controversies, the racial and the philosophical, therefore, were historically independent of one another. In fact, if we apply

Popper's principle of demarcation to the racial debate, we must conclude that both critics of racism as well as their racially committed opponents failed the acid test of falsifiability. As an individual of Austrian Jewish descent, however, Popper (who was politically conservative) would be automatically excluded from the racially committed camp.

All of the evidence from the history of science regarding the distinction between science and pseudoscience testifies to specific changes in substance from case to case, scientific field to scientific field, period to period, and context to context. The term *pseudoscience* is extremely elusive and problematic, both epistemically and historically. It can be an indicator of internal scientific debates in respective historical contexts—a kind of historical, cultural seismograph, or a signpost for mapping the forces operating in a particular scientific controversy; but it is far less useful as a stable historical or analytical concept for categorization. Contemporary historians of science agree that the definition of science is normative and context specific. One would search in vain to find a transcendental definition that cuts across time and the concrete social contexts in which scientific activity is generated. Even radical conceptions of Jews as an "antirace" or ideas that Jews have undergone a "counterselection" or "negative selection," assertions that appear to move swiftly from the biological to the metaphysical, may be grounded in the biological notions of the period. What people who perceive themselves to be scientists, and whom the scientific community regards as scientists, produce in specific circumstances or contexts—is science. The label *pseudoscience* neither emerged from nor corresponds to a transcendental definition of science but is rather the result of a fierce controversy within science. The asymmetry in the use of the label is important for a semiotic analysis: only one camp tended to use it (the other used the other label); *pseudoscience* was recognized as a pejorative term at the time, and no scientist would identify his own work as pseudoscientific. The term was born in a political struggle in which one side depicted itself as void of self-interest, the untainted bearer of the torch of scientific knowledge.

In retrospect it is difficult to discuss the racial paradigm without employing the "pseudoscience" stigma, but we should keep in mind that the label's usefulness comes at a price: an implicit distinction between science and pseudoscience as opposites and a corresponding belief in their irreconcilability. The cultural reception of the racial literature as essentially pseudoscientific now determines the experience of its reading: the later reader cannot read these texts but as expressions of racist prejudice, void of a scientific base. It is certainly not my intention to rehabilitate this literature, but I do want to pose the question of whether it may have been just as close to the sciences and culture of the period as was the literature that opposed it. In this light, racial literature does not become less menacing or violent, but the discussion of it addresses the sweeping rejection of entire branches of knowledge-claims and science by the labeling of them as pseudosciences.

The question concerning the scientific nature of the racial paradigm is essential for the investigation undertaken in this book. But whether or not the paradigm is scientific or pseudoscientific interests me much less than does the question of how our historical understanding would change were we to read the various aspects of the paradigm not as pseudoscience but as genuine expressions of the science and culture of the time. My intention here is twofold: first, what happens to our understanding of this body of literature, and second, what happens to our understanding of fields of knowledge and disciplines that were not stigmatized as pseudosciences. This is why I deal extensively with the writers' research methodologies, practices, and respective scholarly traditions. At this point it should be clear that the Jewish context plays a key role in clarifying these issues, because the label "Jewish scientist" is the antithesis of "pseudoscience," and can be directly linked to the status of antisemitism in this discourse.

RACIAL IMAGINATION

Analyzing the photographs discussed in this book has led me to realize that beyond physiognomic differences, sensory perception, or the connection between perception and cognition, photography for these racial writers is linked to something deeper and more essential: imagination. By *racial imagination* I mean a number of joined but to a certain degree separated elements: imagination as a racial *characteristic* and as the *unifier* of the social fabric.

Since the inception of photography, it has been linked to vision and imagination in complex ways. In this study I show how imagination is deeply ingrained in the history of racial photography. In fact, a direct link exists between the two. Racial authors employed photography to capture what they considered to be both the visible and invisible features of race. In some ways they were like the mystics who try to apprehend imaginary worlds and objects that do not or cannot exist. If photography of the invisible is the attempt to embody imagination and make it look real (within historical, social, or cultural constraints), then, given its historically, scientifically, and culturally convoluted status, "race" is far more important in the annals of dubious photographic practices than are UFOs or mermaids.[39] In the following chapters I will return to the connection between photography, imagination, and forms of the invisible. At this point I would like to outline certain aspects of the history of imagination that are essential for the history of racial photography.

In an Andersonian vein, photography has been a central mode in which communities have imagined themselves.[40] But in contradistinction to Anderson's argument as to the emergence of nationalism, race theory, to the extent that it was shaped by notions of linguistic community and by legal notions of citizenship and state formation, offered a powerful critique of European nationalism. For race theorists, European states were inauthentic collectivities, racial ties being deeper and stronger; racial photography was for them connected to the creation of a deeper, more authentic and stable imagined community. Walter Benjamin's description of photography as the paradigmatic form of mechanically reproduced art could be coupled with, and in a certain sense made parallel to, Anderson's argument concerning the serialization involved in modern nationalism.[41] Whereas mechanically reproduced photography is often thought of as pertaining to the surface level, racial photography, on the contrary, pertains to the deeper, more permanent structure of what truly ties (and divides) individuals and communities.

While racial imagination has not, to the best of my knowledge, received detailed treatment in academic literature, Nazi racial literature has on more than one occasion been analyzed as bordering on mysticism, pseudomysticism, or irrationality. In general terms there is a kernel of truth in this. The writers on race, however, did not conceive themselves as irrational. But some of them conceived their own form of rationality as different from and deeper than the materialistic, limited, and superficial form of rationality. The debates over the link between imagination and reason—from Aristotle to Heidegger—become interwoven in the history of philosophy and science and serve as the necessary background for a historical discussion of the link between photography, race, and visual perception in the first half of the twentieth century.

The basic status of imagination as intermediate between perception and discursive thinking was established already by Aristotle and subsequently fascinated generations of philosophers.[42] Scholastic writers regarded the imaginative ability as conscious but prerational.[43] Renaissance thinkers viewed imagination as a tool that tied the senses to ideas, but rationalist philosophers of the early modern era saw it as a threat to the supremacy and clarity of reason.[44] Immanuel Kant subordinated imagination to the sublime, removing it from the realm of miracles and transforming it into "an instrument of reason."[45]

The upheavals that Kant introduced into the concept of the imagination are directly and intimately related to its status in the philosophy of the Weimar and Nazi periods. In his second major work, *Kant and the Problem of*

Metaphysics (1929), Martin Heidegger tried to show that in the first edition of *The Critique of Pure Reason*, Kant had discovered that imagination preceded sensibility and understanding. The power of imagination is so great in Heidegger's analysis of Kant that it operates behind our backs, unconsciously as it were, to the extent that it has remained hidden from us for two thousand years of philosophy.[46] Imagination has undermined the hierarchy of Western metaphysics: reason and sensation are simultaneously rational and irrational because they stem from the imagination, which is both receptive and spontaneous, that is, rational.[47]

According to Heidegger, Kant himself was so terrified by what he found that in the second edition of his book he excised his statements on imagination. Kant's retreat proved to Heidegger that he (Heidegger) was the first philosopher since the pre-Socratic philosophers who had faced the abyss of reason—and who had not turned back.

The a priori function of the imagination is to represent our actual experience in a particular form of pure concepts. Kant distinguished between *Sinn* (meaning) and imagination. Meaning serves intuition when the object is present, whereas imagination serves intuition when it is absent (absence is divided into two: an object that no longer exists or one that will exist in the future).[48] Equally important is the productive aspect of imagination. Imagination is coupled to the subject's activity, that is, to activity that precedes the experience of an object. In other words, imagination creates original representations that do not originate from an earlier experience but that provide representation with its experiential preconditions.

The picture gets even more complicated because it is difficult to discuss Heidegger's interpretation of Kant without first understanding the revolution with regard to imagination generated by Edmund Husserl, Heidegger's teacher. Husserl shifted the concept of imagination from the mental reproduction of an object in consciousness to visual perception as an *activity* of the consciousness, a mode of intentionality. Imagination, according to Husserl, is an original, synthetic act, and as such a very powerful tool of semantic research.[49]

Fine points in the debate over Husserl's reading of Kant are less relevant to this study than two of the broader historical contexts in which it took place. Imagination lay at the heart of the famous debate over the interpretation of Kant, held in Davos in 1929 and conducted between Ernst Cassirer and Heidegger.[50] This philosophical disputation directly touched on the status of neo-Kantianism so that in a sense it also related to the status of philosophical reason per se. The debate struck sensitive chords of Weimarian culture and came to be considered, culturally, as a milestone in Nazism's rise to power. Heidegger posited Kantian imagination at the center of his debate with Cassirer. In somewhat unrefined terms, it may be said that Heidegger claimed firstly that Kant had discovered the power of imagination as something lying under or behind rational knowledge, as a kind of abyss that no one before him had dared to behold directly, and secondly that the discrepancy between Kant's two editions was indicative of the entire philosophical tradition. In Heidegger's interpretation of Kant, then, the attempt to overthrow the neo-Kantian paradigm and show imagination to be the irrational basis underlying (or lying behind) knowledge is attempted circumspectly, albeit in a way that all those present immediately understood. Imagination, in this light, is a contested politicized subject and a category that unites diverse fields of discourse.[51]

Imagination also plays a key role in the tension between the authority of the term *race* and the fluidity of racial difference in reality. What happens when a man of Jewish origin "looks" exactly like a Norseman or a man of non-Jewish origin has a "Jewish" nose that identifies him as a Jew? What happens when accent, intonation, verbal expression, dress, or physical bearing—variables perceived as racial indicators of the period—can no longer identify Jews? What happens when reality fails to comply with racial perceptions to the extent that it undermines them? And what happens when identification based on inherent, genetically transferred, foolproof racial principles proves defective? To those committed to racial principles, these "inconvenient realities" do not alter the fixation of their principles but rather give rise to attempts to fine tune their distinctions and definitions and to overcome the obstacles in reality. Particularly important is the gradual replacement of physiognomic by internal, often latent, indicators not easily applied to identification and representation. Here imagination plays a key role.

Given the chaos of racial differences, the racial theorists sought a natural order that would apply to simple as well as complex cases and become authoritative. Imagination in racial literature, like in many modern philosophies, is the link between what exists in deficient form or is absent in relation to the "correct" or ideal form. Imagination enables movement from the actual to the ideal and vice versa: the actualization of something absent in reality or the imaginary removal of something actual from reality.[52] The writers who were committed to racial ideas may have never truly questioned their core beliefs, but they were not blind to the complexity of the surrounding reality. "Empirical" problems led them to search for ways to formulate the same elusive thing and establish it on a higher level of abstraction. Imagination enabled them to "rise" above reality, to speak about objects that did not exist, or that existed in the past, or would exist in the future, as they believed. Imagination enabled what was partially real to be completed and the "defective" to be restored to its ideal, correct form.

Imagination distinguishes between "simple" racial cases (in which the individual corresponds to the racial type) and "challenging" ones (in which the individual does not "neatly" manifest the type it "belongs to"). Imagination enables us to go from the general to the specific, and vice versa, to move from specific cases to general categories. It is an abstract formula, and in the following chapters we will see how these principles function concretely.

But apart from race as an attribute, another issue concerns imagination's role as a unifier or organizer of social life. Very little has been written about the role that imagination plays in social life, most probably because it is difficult to pinpoint and conceptualize imagination's role in the life of individuals or groups. One of the few attempts was made by the French-Greek philosopher Cornelius Castoriadis, who developed a theory of the role of imagination in social organization based on principles and a style of thinking very different from those of the racial writers—a theory that combined Marxist notions with Lacanian psychoanalysis. Be this as it may, Castoriadis and the racial writers shared the awareness that imagination plays a key role in social life. Castoriadis wrote that every society is based on fundamental questions as to who its members are, what constitutes them as a collective, and what they want or yearn for. These basic questions define the identity of the society, and the source of their answers is found not in reality but in the realm of the imagination.[53] Castoriadis stresses that these questions and answers can only be approached metaphorically. The questions are not explicitly raised and the "answers" are not received in the usual sense of the word. Life answers them de facto, and the answers are intrinsically contained in the acts of the collective.

Although Marxist and psychoanalytical conceptualization were foreign to the racial writers, their views on the social role of imagination are closely related to those of Castoriadis, albeit with one major and crucial difference: the racial writers viewed the role of imagination as closely connected to race. They believed—not always explicitly—in the constitutive role of imagination in social life. But race, for them, is the basis of the collective and also answers for the collective the question who and what they are, what makes the collective a collective, what unites (or divides) certain individuals, and what determines their wants and yearnings. One can claim that turning to imagination is an escape that testifies to a failure of defining, at the positivist (genetic, biological, or social) stratum, what members of one racial group have in common and what differentiates them from others. This view, however, overlooks that which the writers in the humanities and social sciences regard as deeper and more basic than physiognomic differences. Imagination enables one to rise above all the diverse forms of estrangement and to reach something of the sublime.

Can the unifying side of imagination be demonstrated by example? When an observer sees a solitary tree, and then another, and another, and another, how does he know that he is standing in a forest? This question diverges from a discussion of perception or cognition because it is not a question concerning quantification (the sensory absorption of a particular number of trees) or conceptual definition ("such and such" type of forest). An empirical "definition" for a forest is possible—a specific number of trees in a specifically measured area, their specific density, and so forth, or a botanical definition of the plants that are considered trees is possible (can a banana plantation be considered a "forest"?). Others will claim that the forest is a linguistic-cultural construct and that the transition from single trees to a forest is an acquired social construct. But the claim can also be made that the act of combining individual trees into a "forest" takes place in the observer's imagination as a synthetic, subjective act. Only there, at some point, do individual trees become a forest. The forest is indeed made up of individual trees, but this fact does not lessen by one iota the reality of the forest. The same applies to race. How does the observer know that a list of individual characteristics such as eye color, posture, gesticulations, intonation, way of looking at things, or spiritual form of a certain person—individual elements, according to the racial writers—belong to a person of a certain "race"? The answer is, through the synthetic act that takes place in the observer's imagination. Not every observer is blessed with this synthetic capacity, as it, too, is relative to race.

All of these elements contain the visual aspect and touch on the convergence point of visual and statistical logic. Francis Galton (on whom, see the next chapter) devised a statistical system of "correlation" that became the groundwork of modern statistics and was developed in the study of racial populations. Based on Galton's statistical notion of correlation, it was now no longer necessary to carry out exhaustive measurements of numerous physical features of group members. Based on statistical probabilities, an individual could far more easily be ascribed to a racial group on the basis of a limited number of statistically correlated measurements. Neither imagination nor statistical correlation require all the features; two are sufficient.

Just as the forest is a forest even if it has bald spots or individual trees of a different type that have infiltrated the great mass of trees, so too with race. The presence of single, foreign features in a particular individual do not detract from the legitimacy or reality of race. Racial imagination serves as a connecting thread among the individual elements, binding them into a unity. It is difficult to put racial imagination in its proper place: the act can occur only in the imagination of the individual, but it is not the individual's possession. Nor is it universal. It is common only to people of a particular race—it is a specific imagination, as shown by the analysis of photography in the following chapters.

At this point we can begin to understand how photography—with its realistic imprint—became so powerful a device in the hands of the racial writers. As an instrument of the imagination, it liberates the exclusively factual from its contingent status and grants facts the status of a potential ideal. Imagination can complete a photograph by the addition of what is absent from it or add nonreality to the photographed object. The essence of things, as the phenomenological tradition points out, is found in their potential more than in their actuality.[54] The profound dualism in photography's status in this cultural corpus can already be felt. On the one hand, photography's power stems from it being perceived as a "simple," transparent, realistic means of representation. On the other hand, given the connection to imagination, the important functions of photography can be understood as being nonrealistic and having less to do with representation or visualization of racial differences and the classification of individuals as members of a particular race (although all of these statements are true). In conclusion, the following chapters will show that photography was designed to serve an "imaginary" clarification: the rediscovery and enhancement of racial imagination. In a very different context, scholars have concluded that pressure of thought transforms conceptualization just as "the very act of imagining distorts the visual field."[55] The fact that racial imagination is purely imaginary in no way detracts from its social and political implications.

RACIAL PHOTOGRAPHY: FROM POLITICS BACK TO SCIENCE

Our excavation of six distinct cross sections has unearthed the historical and conceptual framework upon which will be constructed the analysis of the following chapters. Before embarking on these case studies, however, and by way of a conclusion to this introductory chapter, I wish to make a few points concerning both racial photography as a historical phenomenon and the appropriate mode of studying that phenomenon.

Let me begin by attempting to define racial photography as understood in this book. Pretty much immediately after its invention in the mid-nineteenth century, photography was incorporated into anthropology, and just as anthropology was deeply bound up with race, so too was anthropological photography. But not every anthropological photograph was racial in kind or degree.

The focus of this book is not anthropological photography as such, which has been the subject of previous research, but rather racial photography in a more limited and specific sense.[56] As should by now be clear, my principal interest is in fact the epistemological history of racial photography's methods; that is, my focus is on scholars who studied "race" and who employed photography methodically in these studies. This book, therefore, focuses on sections of scientific photography that were viewed by their users as grounded on race both within anthropology and within other branches of knowledge that were inextricably linked to various fields in the humanities, such as art history and linguistics, and from which, as the following chapters show, racial writers drew concepts, terms, categories and practices. As a result of these interconnections, any attempt to reconstruct the history of racial photography from an epistemological perspective is not only necessarily an interdisciplinary undertaking but also one characterized by several internal tensions. These tensions necessitate, in turn, a peculiar form of investigation.

The decision to study visual perception and the use of photography in the racial corpus stems partially from my wish to approach racial writing and visual images from a new, "diffracted" angle. Such an angle will, I hope, provide a stronger, more contextualized and nuanced historical interpretation than hitherto, one that not only highlights the tensions and contradictions involved in this history but that also reveals the wider, more pressing historical questions at stake. For various historical reasons, racial materials—first and foremost visual ones—have usually been identified in the literature as belonging to the history of propaganda rather than to the history of science. By shifting the epistemological status of this material, we are able to study the way in which its uses were related to the scientific ideals and practices and the types of argumentation, justification, illustration, exemplification, and demonstration of the period.

If approached from the perspective of the history of science rather than from a political perspective—as the following chapters illustrate abundantly—one finds that scientific racial photography underwent profound transformations between the 1880s and the Weimar and Nazi periods. These transformations are closely connected not only to the history of photography, but also to the history of ideas about race in this period. Only when the connections between the two are closely studied do the crucial changes in the use of photography— from illustration to observation and finally demonstration—become apparent. I have tried, therefore, to read the corpus of racist material from "within" in order to understand its effectiveness and connection to the ideas of the period. This approach has forced me to deal with conceptual and ethical questions, and it allowed repugnant texts to present themselves as stronger, more convincing, coherent, and closer to mainstream branches of science than I might have wished. Such an interpretive exercise, which is the essence of this book, bristles with the tension inherent in such an undertaking.

This tension pertains not only to the study of race but also to the realist understanding of photography with which, on the whole, writers of race were aligned. While the importance of visual considerations for understanding twentieth century history has been a sine qua non in recent decades, and a great amount of research and theoretical discussion have been devoted to photography's value in this context, most of the attention has focused on the tradition of critical writing and modernist views of photography and visual perception. Within science, however, another photographic tradition was much more influential in this period. That tradition was the realist one. On the whole, this photographic tradition has been studied specifically with regard to race, if at all, merely as an expression of politicized, simplistic, uncritical, pseudomystical positions whose complexity, profundity, and aesthetic strength fall far short of modernist and critical credentials. The racial writers with whom I deal considered photography to be an "objective medium" (although, as we will see, they had different definitions of what that meant) that faithfully depicts the object in front of the camera lens. All of the writers worked on the photographs—developing and printing them in unique ways, centering the subject by trimming the margins, editing the pictures at various levels of sophistication—but avoiding certain montage editing or other recognized artistic techniques that would have undermined the photographs' perceived scientific status. While my book is mainly concerned with the realist branch of visual culture, the confrontation between the realist branch on the one hand and the critical and modernist on the other underlies this study and has important implications.

The racial writers of the 1920s and 1930s were the cultural, and more often political, opponents of the modernist critics of photography. I believe it would be counterproductive to impose principles on the racial writers that were foreign to their projects and methods. In fact, by closely following the dialectic between notions of race and uses of photography, my historical study points to photography's unstable epistemological and ontological status and, consequently, to the appearance of questions that influenced the development, in the second half of the twentieth century, of those theories concerned with the ontological nature of the photograph. This is especially true regarding the issue of photography's deictic or indexical nature (Charles Sander Peirce, Roland Barthes, or Susan Sontag), which has special relevance for its use within science, where it appears as the epitome of "mechanical objectivity" or "trained judgment" (Daston and Galison).[57] The numerous cases reviewed in this book deconstruct these theories historically by showing that in some cases they are fully applicable, in others only partially applicable, and in still others they do not apply at all.

The racial paradigm can be discussed in several cultural contexts: art history, racial literature, anthropology, and even the photographic documentation of the murderous medical and scientific experimentation conducted on concentration camp victims.[58] The history of science, however, is the natural choice, because the authors with whom I deal viewed themselves, and were seen by their contemporaries, as operating within the scientific field. The book describes how racial perceptions, uses of photography, and notions of visual perception were indeed connected to the scientific fields of the period. However, this contextualization is only part of the picture. Other contexts, such as cinema, travel literature, atlases, picture books, and advertisements were also related to the general visual culture. In other words, with regard to understanding the culture of the Weimar and Nazi periods, the cinema's visual codes and everything connected to the cinema is equally as important and relevant as scientific photography. But reel film did not gain the same foothold as photography in the scientific and research fields mainly because it was impossible to obtain scientific control over it. Reel film was seen as a medium that in certain conditions might assist scientific observation but that could not serve for purposes of reconstruction, measurement, classification, or quantification and the conversion of data into statistical tables. Also, some of the basic metaphors of scientific films—the nexus between photography and hunting, for example—are different from those found in racial books.[59] Attention to scientific photography, then, far more than cinema, is essential for a history of the uses of photography in racial literature. Nevertheless, in addition to the scientific prestige derived from anthropometric and anthropological photography in the early twentieth century, the book shows how racial photography drew its inspiration from other, perhaps more important, sources.

The world of the racial literature (which includes racial photography) of the 1920s and 1930s is a modern world. The great discoveries of remote regions and primitive races were, for the most part, completed. The "white" man had penetrated the innermost depths of the continents and discovered alien cultures. The unknown and foreign were now locatable in an area that the white man had physically conquered.[60] This fact has implications not only for the way in which the foreigner and unknown were perceived but also for the way in which they were represented. Some of the conceptualizations of strangeness in this period seem to have been intentionally made in terms of proximity/distance rather than in terms of absolute alienness. Georg Simmel characterized the stranger as someone who is both close and distant; his pupil, Siegfried Kracauer, developed the idea of the "ethnography of the proximate"; Helmut Plessner developed the notion of the "alienating gaze" of the poet or scholar, which captures the long-familiar or self-evident; and under the influence of Russian formalism, Bertolt Brecht developed the effect of estrangement as *Verfremdungseffekt*.[61] Racial photography, as we shall see, is closely related to such concepts of the social presence of difference as the socially proximate, self-evident or hidden. Such concepts of racial difference are in fact closer to the heart of my analysis than are those contemporary exhibitions of deformed people and human "types" with recognizable ethnic markings.[62] This does not imply that such representations were any less stereotypical, but they do show that representations of the "Other" and the "I" are mutually linked in this corpus and require, in a sense, to be read in tandem. In fact, racial literature focuses much more on the "I" (the "Nordic," the "Aryan," the "German") than on the "Other" (the "African," the "Chinaman," the "Semite," and the "Jew"). The evidence that I present categorically refutes the view that it is always "Others" who are racialized and the white man who is "just" a human being.[63] Nevertheless, it should be kept in mind that ways of creating and representing the "I" also stem from the imagination and that they are as close to the stereotypes as are ways of creating and representing the "Other."[64]

An enormous increase in visual material occurred in the 1920s. One of the results was a flush of written texts.[65] Contemporary observers note that the written text was unable to attain the same effect after the surge in visual material—images and pictures can revitalize and enhance a text in a split second. This book traces the various relationships between the written text and visual images in one particular body of material.[66]

The development of the ideas traced in this book forms an inseparable part of the history of such key philosophical concepts as *Weltanschauung*, perspective, and style, the development of all of which are inherently linked to the history of relativism. Hayden White has noted that the "Nazis were anything but relativists."[67] In this book I show how far Hayden White erred and point especially to the extent to which important relativist moments are essential for a conceptual and historical understanding of Nazi racial literature.

The reconstruction of racial photography and its uses in its scientific and cultural context are directly related to the issue of its changing status as scientific evidence.[68] After Nazism these forms of photography lost their scientific legitimacy. But, from the end of the nineteenth century until the 1930s, the epistemological status of photography in the social sciences was fundamentally no different from its status in other disciplines, such as statistics. Photography was conceived as a powerful, legitimate medium for the study, observation, and illustration of supraindividual differences. As with statistics, intense controversies ensued regarding what is "normal," "typical," "individual," or "average." These controversies were especially intense because they touched on wider questions pertaining to modern society and the social sciences. Within this crowded paradigm, however, various conceptions of race and of photography were maintained, such as the polarized relationship between photography and the depiction of human differences. Photography was instrumental in constituting social epistemologies, forms of visual perception, and, especially in the German case, specific visual codes of "Jewish difference." But we face a dilemma when we attempt to situate this nexus within the larger historical picture.

It is an essential fact, at least with regard to the study undertaken in this book, that attempting to situate racial photography, in particular that of the Weimarian and Nazi periods, within wider historical coordinates, one is faced with a certain impasse. I address three closely interconnected and interfacing historical contexts. The first and the second, the core of this study, concern, respectively, the uses of photography and views of visual perception inherent to the use of photography in racial literature from the 1880s to the Weimar and Nazi periods and the link to science and culture in this period or the basic scientific ideas, ideals, and practices in this literature. The third, however, hovering continuously, as it must, above just about all aspects of the general discussion, touches on links between this discourse and the political field. The inconsistency lies in the fact that after 1945, racial literature was branded pseudoscientific because of its close association with Nazi culture and the principles of Nazi ideology—but the second and the first contexts point to the significant degree to which this literature was embedded in legitimate scientific and scholarly ideals and practices. The different contexts, therefore, cannot be fully consolidated. This tension is built into the historicization of racial photography attempted by this book.

*

This book has not been constructed linearly, though the chapters are arranged chronologically. The first chapter reconstructs key moments in the late nineteenth—early twentieth century history of racial photography as scientific evidence necessary for understanding of the later periods. The account that follows does not describe the growing radicalization of racial questions, though the roots of this radicalization appear in various chapters. 1933 was the critical year politically for every question on race, but the intellectual sphere is not the same as the political sphere, nor is its time line identical or parallel with the political one, although sometimes it is contiguous with it. In fact, most of the ideas discussed here were already around in the 1920s. But ideas are dependent on time and place, and under new conditions the same ideas become different ideas. Chronologically the book begins before the Weimar period, continues through Nazi Germany, and ends with the new Germany. The chapter structure, however, is different; each chapter addresses a slightly different set of historical and theoretical questions regarding photographic practices. Consequently, in each chapter the practical epistemological approach takes a different form. This introductory chapter has presented a historical, conceptual, sociological framework for the main questions treated in the book. The first chapter reconstructs three "threads" that converge in the photography of German culture in the Weimar and Nazi periods: the history of scientific-anthropometric photography, the history of "Mendelian photography," and the relativization of visual perception in the history of art. The second chapter focuses on the role of photography in the discourse on the Jews as a mixed race people from the inception of this idea by Felix von Luschan through Maurice Fishberg to Sigmund Feist. The third chapter is, in a sense, the key chapter. Here I concentrate on Hans F. K. Günther in order to argue that by using photography, Günther went far in establishing a racial-visual code of tremendous extension and influence. In this context, for example, I discuss his strategy of "estrangement." But my objective is to show that the code he created reflected wider historical and cultural anxieties and sensitivities and projected forward as well as backward on present and future visual materials, as well as changing the meaning of materials that appeared in the past. The racial phenomenologist Ludwig Ferdinand Clauß is the focus of the fourth chapter, which sheds light on several dilemmas in the nexus between racial photography and the history and philosophy of science. In that chapter I also try to show how deeply his photographic method is integral to the history of the humanities. The concluding chapter discusses racial photography in Palestine. Historically parallel to the third and fourth chapters, in this chapter I study Arthur Ruppin and Erich Brauer as well as pay special attention to the two-way flow between science and art.

CHAPTER 1

The Type and the Gaze: Racial Photography as Scientific Evidence, 1876–1918

Three genealogical threads converge in scientific racial photography of the 1920s and 1930s within the context of German society. These threads are intimately connected to transformations in the conceptualization of race from Carl Victor and Friedrich Wilhelm Dammann's *Races of Men* of 1876 to Hans F. K. Günther's *Rassenkunde des deutschen Volkes* of 1922.

In the first section of this chapter, I commence with a short note on Dammann's *Atlas* and then reconstruct three indispensable moments in the history of anthropometric photography: Alphonse Bertillon's anthropometric photography, Francis Galton's composite photography, and Rudolf Martin's standardization of anthropometric photography. A particular emphasis will be placed on questions of scientific control and transformations in the epistemological status of photography. Within this genealogy, photography is intimately connected to statistics, a connection that finds expression in the emergence of a specific visual code as well.

In the second part of the chapter, I shift the perspective to a parallel and lesser-known genealogy that was intricately tied to the emergence of Mendelian genetics. This genealogy, developed by Carl Heinrich Stratz, Redcliffe N. Salaman, and Eugen Fischer, could be termed *Mendelian photography*. Focused on the study of the gaze as a specific racial characteristic, this genealogy increasingly involved the question of the visible versus the invisible in the photograph. I argue that it is crucial for understanding racial writers of the Weimar and Nazi period.

In the final part of the chapter I turn to yet a different intellectual discourse. I examine the antimechanistic notions of "seeing" elaborated in German philosophy and history of art through the discussion of two books, one by sociologist and philosopher Georg Simmel and the other by historian of art Heinrich Wölfflin.

If anthropometric genealogy lent prestige to the use of photography as a major scientific device in the study of race, the photographic study of the gaze was appropriated and further developed by Weimar and Nazi writers toward racial observation. While photography was tied to anthropological notions of race, the idea that seeing was relative to race was not simply a pseudoscientific argument concerning neurological differences between races. Rather, this concept was based on ideas that were drawn from cultural relativism and in fact constituted a reinterpretation of ideas expressed by major intellectual figures such as Simmel and Wölfflin.

PART ONE: ANTHROPOMETRIC PHOTOGRAPHY

Alphonse Bertillon, Francis Galton, and Rudolf Martin could be described as the three most important figures in the history of anthropometric photography from 1880 to 1920.[1] Each was instrumental in establishing the use of photography as a measuring device in the study of race as well as the scientific prestige it came to enjoy within science and wider culture.

Photography was integrated into anthropological work from its very inception in the middle of the nineteenth century.[2] Carl Victor and Friedrich Wilhelm Dammann's famous *Races of Men*, arguably the most influential racial photographic book of the nineteenth century, can serve as a starting point in examining the transformations in the role of photography for the study of race and their shifting epistemological underpinnings between the 1880s and 1918. The "brothers Dammann's book" (in fact they were not brothers), which appeared in England in 1876, is conceived as a photographic atlas of the different races of man. In this album-sized book, each page comprises high-quality black-and-white photographic reproductions, which appear accompanied by titles and coupled with brief textual descriptions of central physical and mental traits of the respective type that appear in small print under the photographs. As customary in contemporary botanical atlases, individual humans were presented as specimens of their type and are seen to correspond with the central traits of the type. The atlas as a whole, conceived along an evolutionary scale, moves from what they perceived as primitive types to the highly developed ones: from the Polynesian to the Germanic.[3] Photographs merely illustrate for the general readership the gallery of types that make up the human species (fig. 1.1).

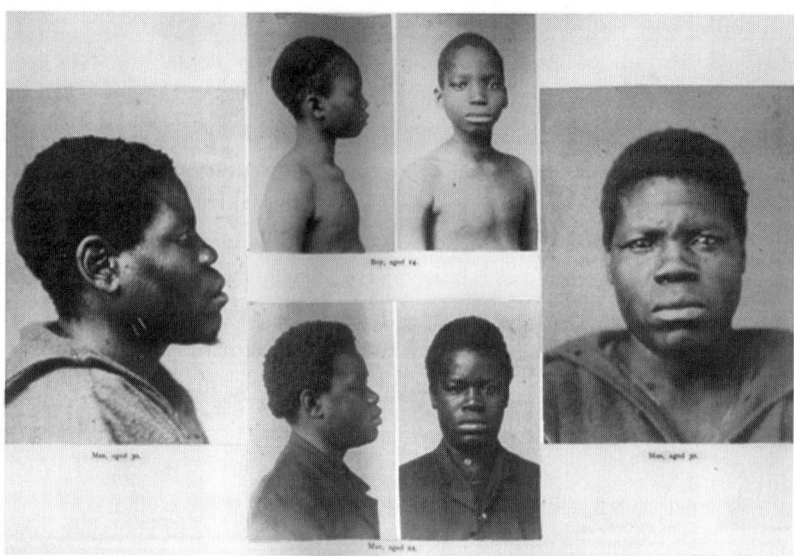

FIGURE 1.1. "Zanzibar Coast." Dammann's *Ethnological Photographic Gallery* consists of racial type photographs. Pages are composed of large photographs accompanied by short textual description of typical features ("the skull of the Negro is remarkably solid and thick, so that in fighting they often butt against each other like rams, without much damage to either combatant"). Carl Victor and Friedrich Wilhelm Dammann, *Ethnological Photographic Gallery of the Various Races of Men* (London: Trübner, 1876), plate 16.

But before I turn to their cases in more detail, I cannot avoid a short note on antisemitism in this respect. Both Bertillon and Galton were involved in antisemitic campaigns and expressed antisemitic views: as an expert on identification, Bertillon took an active part in the Dreyfus affair.[4] Galton, after completing a composite photograph of Jewish schoolboys, wrote on one occasion that he was struck by the diffidence of the Jewish gaze.[5] Antisemitism, therefore, while not central to their discussions of photography, was culturally and ideologically associated with this scientific genealogy.

Alphonse Bertillon: Measurement and Identification

Alphonse Bertillon was not an anthropologist or an academic, but as the founder of modern anthropometric photography, his contributions to the field are essential for a historical account of the development of anthropometric photography from the final decades of the nineteenth century through the first decades of the twentieth. Bertillon developed the anthropometric method toward different goals than Galton or Martin, emphasizing the significance of practical considerations in the development of scientific photography.

Alphonse Bertillon (1853–1914) was the son of Louis Adolphe Bertillon, a physician and statistician. After being expelled from the Imperial Lycée of Versailles, Bertillon held a number of jobs in England and France before being conscripted into the French army in 1875. In 1879, after his discharge, with no real higher education, his father arranged for his employment in a low-level clerical job at the prefecture of police in Paris. Bertillon, whose duty included copying onto small cards the descriptions of criminals apprehended each day, realized that what he was laboriously rerecording was practically useless for the purpose of identifying recidivists. With a general familiarity with anthropological statistics and anthropometric techniques from his father and brother, he devised a system of identification that relied on eleven different bodily measurements and the color of the eyes, hair, and skin. This system proved so successful in identifying recidivists that it was quickly adopted by all developed countries, and Bertillon himself became the head of the Department of Judicial Identity, created for the Paris prefecture of police in 1888.[6]

Toward the method of identification, Bertillon introduced photographic methods that revolutionized existing ones. Like Rudolf Martin later, as we will see, Bertillon was less interested in photography as a medium than in its service for criminological practices. He recognized that the main problems in identification of criminals lay not in making good photographic likenesses but in being able to classify or compare them. He was the first to develop, in 1883, a photographic archive: a card-based system that allowed easy retrieval of stored data based on strict uniformity. The card listed a range of physical characteristics, and from 1892 onward, fingerprints, to supplement exact one-seventh-scale photographs. He described his system in *Photography: With an Appendix on Anthropometrical Classification and Identification* (1890).

Bertillon's aims were practical and operational, the identification of individuals—a response to the demands of urban police work and the politics of fragmented class struggle during the Third Republic. While working toward different goals, the uses of photography by Bertillon, Galton, and Martin were in different ways grounded in the emergence and codification of social statistics in the 1830s and 1840s and relied heavily on the notion of the *l'homme moyen*. This term was coined by astronomer and statistician Adolphe Quetelet, who showed statistical regularities in rates of birth, death, and crime.[7] Galton and Martin both connected this notion to that of a "central type," while in contrast,

Bertillon did not address this issue, as his aim was practical: individual identification. Of particular importance for the use of photography by 1920s and 1930s racial writers was scientific control, the relationship between visibility and invisibility, and the relationship between photography and statistics.

Strict control was essential to Bertillon's anthropometric system. One of the major problems with earlier attempts at criminal identification concerned standardization of portraits. The precise method Bertillon invented to register offenders and classify the photographs to guarantee their retrieval is less important in our context than the fact, in the words of Allan Sekula, that Bertillon was engaged in a "two-sided taming of the contingency of the photograph."[8] In order to compare two photographs and to carry out and compare measurements with archived photographs, Bertillon realized, control was absolutely necessary. Control, however, was not unique to photography but rather to measurement as such. Bertillon insisted on a standard focal length, even and consistent lighting, and a fixed distance between the camera and the sitter. He introduced the profile view to cancel the contingency of expression, as the contour of the head remained consistent with time.[9]

In a recent doctoral dissertation, Josh Ellenbogen has argued that Bertillon's use of photography established an approach to identity that was "stripped of the idea of resemblance."[10] Bertillon's photographic method did not aim at images that would resemble the photographed but in fact were based on a certain form of alienation, on training the eye to know how to interpret. Bertillon systematically differentiated between the professionally educated eye and the layman's.[11] Bertillon's method was designed for the professionally trained eye, which "means nothing in relation to the world of quotidian appearance, is worthless to the inexpert eye."[12] As opposed to the racial writers of the 1920s and 1930s, however, Bertillon sought to establish the kind of "seeing" that he conceived as nonintuitive or innate but conventional and learned (fig. 1.2).[13]

For the untrained policemen, for friends or family of the prisoner, the frontal view of a face was more likely to be recognizable.[14] From the profile photograph anyone, including the prisoner himself, will have great difficulty at identification. The trained eye was dependent on particular conventions and their minute execution precisely because they were not based on memory or perception. For the nonspecialized eye, Bertillon claimed, such photographs were meaningless.[15] Bertillon's system wrests data from daily categories of experience (fig. 1.3).[16]

Bertillon constructed a strictly denotative signaletic system aimed at unambiguous translation of appearance into words.[17] Unlike Lombroso or Galton, as we will see, for Bertillon, the criminal body was not a site of meaning; it hid

Tafel 53.

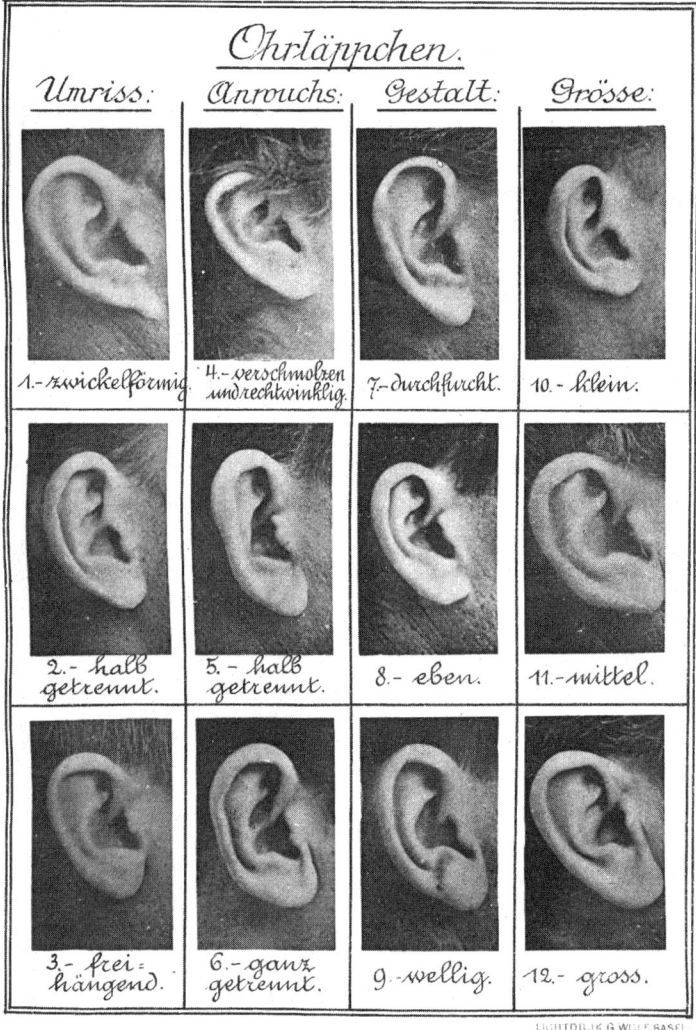

FIGURE 1.2. Ears. One out of numerous pages of photographs of ears. Alphons Bertillon, *Das anthropometrische Signalement*, trans. Ernst von Sury (Bern: Siebert, 1895), table 53.

no characterological secrets. Scars and other deformations of the body were particularly powerful clues not to any innate propensity for crime but to the individual's history, necessary for its identification.[18]

Bertillon's photographs were not oriented toward sense perception or conceived as its extension. If executed according to strictly followed conventions,

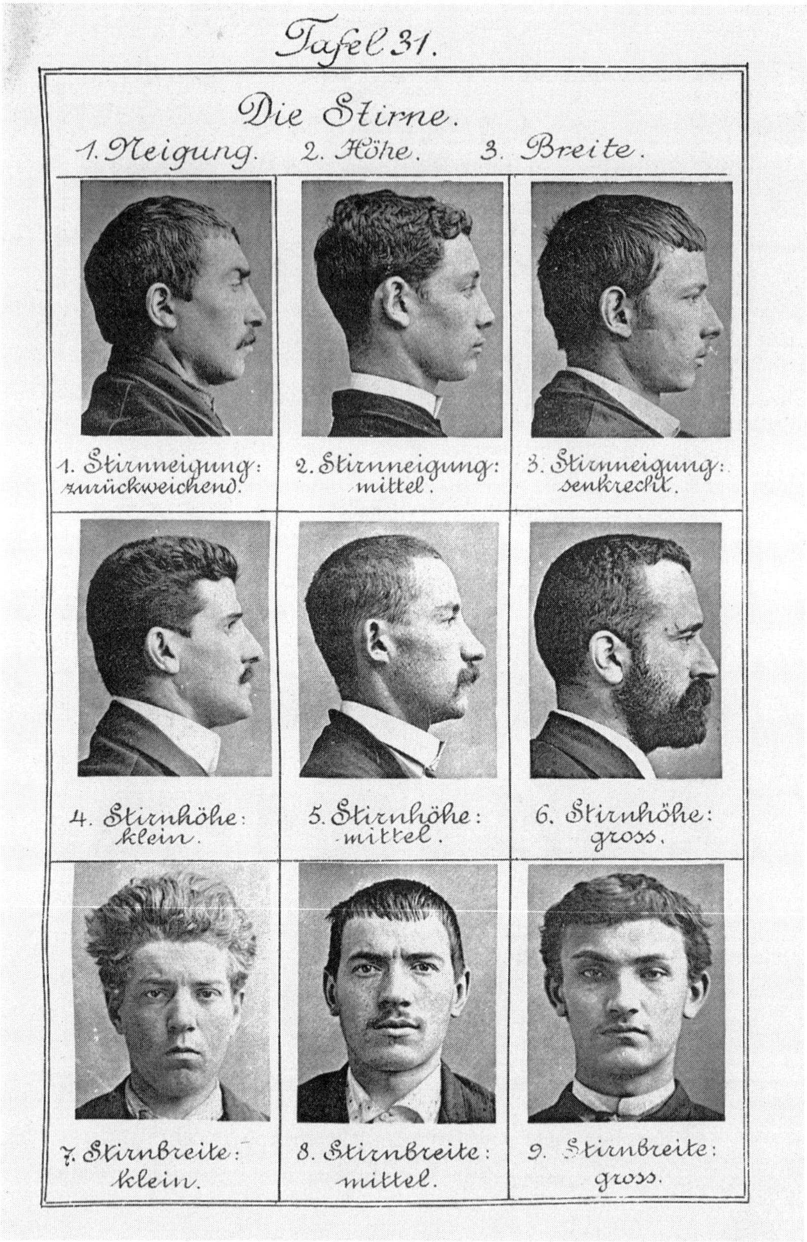

Tafel 31.

Die Stirne.

1. Neigung. 2. Höhe. 3. Breite.

1. Stirnneigung: zurückweichend. 2. Stirnneigung: mittel. 3. Stirnneigung: senkrecht.

4. Stirnhöhe: klein. 5. Stirnhöhe: mittel. 6. Stirnhöhe: gross.

7. Stirnbreite: klein. 8. Stirnbreite: mittel. 9. Stirnbreite: gross.

FIGURE 1.3. Foreheads. One page out of numerous focusing on foreheads. Alphons Bertillon, *Das anthropometrische Signalement*, trans. Ernst von Sury (Bern: Siebert, 1895), table 31.

photographs registered traces of the actual presence of the person decipherable only to the professionally trained eye. In this sense, Bertillon used imperceptible data (that the "lay eye" was unable to perceive) for the study of perceptible objects (the body of the criminal).[19] Photography characterizes individuals under uncharacteristic conditions, for a specific form of seeing.[20] In contrast, in the work of Günther and Clauß we encounter an alienating feature with regard to racial difference.

Before I turn to Bertillon's contemporary, Galton, I will briefly address the relationship between numbers and visual code in Bertillon's method. What is the relationship between photographs and statistical tables? Are they mutually exclusive, separate, or mutually convertible? Which is analytically superior? Bertillon developed a system in which photographs were intended to be fully convertible to numbers; photography, indeed, was a form of measurement. In fact, as no single measurement was sufficient for identification, the method depended on the principle of probability. In this respect, statistical tables were superior to the visualization photography offers. The relationship is mutual but asymmetric: scientific photographs can be expressed statistically, but not vice versa. Martin and Pöch, as we will see, give this asymmetrical relationship a visual expression.

Francis Galton: Photography as "Pictorial Statistics"

Francis Galton developed a method of "composite photography" that was not aimed at identification but at the presentation of "types."[21] Galton believed that physical features reflected specific inherited behavioral and racial traits. Composite photographs provided, in his view, a means of identifying and presenting the type's characteristics.[22] His photographic method, closely related to a specific notion of "type," cannot be understood without statistical principles and a specific interpretation of the "law of error."

Galton developed his photographic method in the final quarter of the nineteenth century in the context of the proliferation of new applications of photography enabled by the introduction of cheaper and easier methods of production.[23] Similarly to Bertillon, his use of photography was consistent with certain empiricist assumptions and methodological procedures of naturalism: the existence of pure facts beyond or before their identification, representation as an immediate function of reality, and observation as uncontaminated by subjectivism.[24]

Sociologically, Galton clearly belongs to the reactionary paradigm. His purist eugenic outlook was future oriented, based on an image of an uncontaminated distant past and haunted by a sense of the present as declining and in a state of exhaustion. Deeply influenced by his relative Darwin's discovery of evolution, Galton's hope for an improved future was checked by his early discovery that successive generations tended to regress back toward the mean. At the same time, Galton believed human "types" were rapidly disappearing.[25]

Galton's photographic method cannot be dissociated from his statistical contributions. Notions of probability, as Bulmer shows, occupied a central place in Galton's thought.[26] Normal distribution, known also as "the law of error" or "Gaussian distribution," was formulated already in 1810 by Laplace, at first to explain why errors of measurement, particularly in astronomy, were approximately normal.[27] This theorem was applied by Quetelet to human sciences, as previously noted, through the concept of the "average man." Quetelet's principle was that if the height of a building is measured twenty times, the value on each occasion may differ because of errors of measurement, but the building has a determinate height, and "the true mean" of the values of the different measurements is the best estimate of the building's true height. Quetelet observed that it was not always obvious whether a mean was a "true mean" that measured a real value or whether this was only an "arithmetic mean" with only descriptive significance. To decide between the two, he proposed normal distribution as the criterion.[28] In an influential review essay published in 1850, John Herschel codified a methodological distinction of immediate consequence for Galton. Herschel, fully convinced of Quetelet's application of the error law to real variation in nature, distinguished between two kinds of mean values: a true mean, which he termed *mean*, and a mere arithmetic mean, which he termed *average*. The scientific value of true means was in their representation of a type, and the deviations from these means were distributed in the characteristic form assumed by error. In his essay, Herschel made the following often-cited statement: "An average gives us no assurance that the future will be like the past. A mean may be reckoned on with the most implicit confidence. All the philosophical value of statistical results depends on a due appreciation of this distinction, and acceptance of its consequences."[29]

Galton's photographic method is inseparable from this methodological distinction. Galton appropriated from Quetelet his powerful statistical principle but interpreted it in a different manner.[30] Quetelet thought that only the mean value of an approximately normal biological variable is important. Deviations from the mean were meaningless errors of measurement, which should be eliminated from the analysis as far as possible.[31] Galton, in contrast, viewed differences as real, not as errors in measurement.[32] With regard to biological

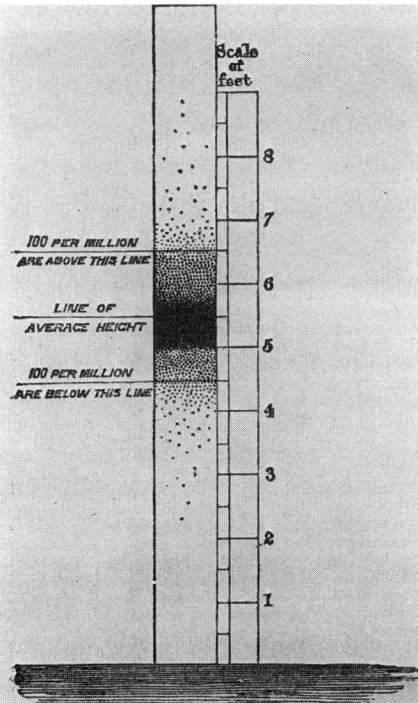

Figure 8.1. Galton's illustration of the "law of deviation from an average,"
showing the heights of a million hypothetical men. (From Galton, 1869, p. 28.)

FIGURE 1.4. Galton, visualization of hypothetical average height. Stephen M. Stigler, *The History of Statistics* (Cambridge, MA: Belknap, 1986), 269.

studies, Galton viewed as absurd the use of such expressions as "probable error," since variation in this domain was genuine, not a mere product of error of measurement.[33] Indeed, he labored to develop methods to represent this conception graphically. Means, in this interpretation, were not mere constructs; the mean of a normally distributed variable represented a biological "type," which was stable and persisted from generation to generation. "Type" was not a mere human construct but inherent to the things themselves.[34] Galton viewed statistical laws as empirical properties of the real world, not as context-dependent logical interpretations.[35] This is the conceptual context of Galton's photographic method (fig. 1.4).

Galton's photographic method cannot be understood without understanding his controversy with Charles Darwin. In the context of the emergence of evolutionary theory, Darwin proposed three different kinds of heritable variability: individual differences, discontinuous variation, and modifications due to physical conditions of life. Darwin accepted that all three played a role in evolution but argued that the first, individual differences, was by far the most important. Galton disagreed and insisted that the stability of type was related in part to the question of leaps in nature. Galton, who discovered reversion to type, deduced that there must be perpetual regression back to the original position when selection was relaxed. Small deviations from the type, contra Darwin, do not lead to evolutionary change because they are subject to regression to the central type.[36] Galton's understanding of race derived from these considerations.

Unlike later racial writers, Galton employed *race* as a synonym for *variety*, *genus*, or *species*, and in his later writings, *heredity*.[37] His understanding of race was closely related to definitions of type, particularly the stability of type and, to a large extent, on a form of "immediate knowledge."[38] In contradistinction to some of his German contemporaries, Galton did not oppose physical and psychical characteristics—both were inherited and together constituted "race" as opposed to "nurture."[39]

In *Inquiries into Human Faculty* (1883), Galton tied "race" directly to stability of "type": "It is the essential notion of a race that there should be some ideal typical form from which the individuals may deviate in all directions and toward which their descendants will continue to cluster."[40] In "Discontinuity in Evolution" (1894), Galton elaborated the definition of *race*:

> A race is taken to mean a large body of more or less similar and related individuals, who are separated from analogous bodies by the rarity of transitional forms, and not by any sharp boundary. . . . The type, or typical centre of a race, . . . is to be defined as an ideal form, whose qualities are those of the average of all the members of the race.[41]

Elements of this view are also found in the work of Günther, as we will see. But Galton's definition must be read as tackling the stability of the "typical centre of a race"—to which his photographic method was closely related.

According to Galton's own account, the composite method developed from a specific experience. In 1887 he obtained from the Home Office photographs of convicts that were classified into three groups according to the nature of the crime: murder and manslaughter, felony, and sexual offences. Familiarizing himself with the collection, he felt that "certain natural classes began to appear, some of which were exceedingly marked."[42] Galton set out to translate this impression into scientific terms. In time, he made composites of each of the three groups of criminals and later of lunatics, Westminster schoolboys, Jews, and phthisical patients.[43]

Conceptually, Galton's photographic method was an extension of his statistical notions. It should be understood with regard to contemporary science, which debated the best way to represent types whether through a single specimen or through a composition made from several individuals. Galton's method sided with the latter view.[44] He elaborated a controlled scientific procedure for the production of composite photographs through successive registration and exposure of portraits in front of a copy camera holding a single plate. Each member of the group was synthesized by his or her picture drawn on transparent paper. Successive images were given a fractional exposure based on the inverse of the total number of images in the sample—if a composite were to be made from ten originals, each would receive one-tenth of the required total exposure.

This method was based on the principle that deviation from any given type must be distributed in the same way as errors of observation. Providing that deviations from the type were accidental, flaws and eccentricities would average out. This technique could be used to discover distinctive human types such as the criminal or the Jew; it provided the viewer with a general picture, which resembled all constituents without being more like one than another.[45] Galton believed individual features faded away into underexposure and what remained was typical of the entire sample; "typical" based on the statistical notion of mean.[46]

Galton believed he had brought together the statistical and visual, translating the statistical error curve into pictorial form. Merging the optical and statistical within a single operation, Galton termed this method *pictorial statistics*.[47] The "types" his method revealed were true means in the statistical sense—pictures were equivalent to statistical tables.[48] Photography was particularly powerful in capturing such central types.[49]

In a sense, the composite method was a solution to the limitations of prior attempts at physiognomic typing, particularly the impossibility of measuring and comparing all the differences between men, as they were too many and too slight. Galton rejected the standard practice of atlas makers to photograph select individual specimens judged to be representative of the prevailing type because this tended to present exceptional and grotesque features, and hence the portraits supposed to be typical were frequently caricatures.[50] The composite method avoided the necessity of choosing a specimen, with its idiosyncrasies, to represent a type. Clauß, as we will see (following Husserl), went further than Galton in the epistemological criticism of empiricist sampling methods, emphasizing that a procedure such as Galton's ultimately fails to formulate how one knew in the first place that a certain individual belonged to a certain type rather than another.

Individuals "disappear" in the composite photograph; they disperse and regroup; photography allows one to fix, retrieve, and preserve memory of the origin.[51] Galton viewed the advantage of his method through the contemporary opposition between mechanical reproduction and subjective production. His method sided with mechanical objectivity, escaping the vagaries of the artist's hand. Yet the mechanical procedure inverted the view that only the trained hand of the artist could create true synthesis.[52] From the perspective of the later paradigms studied in this book, the centrality of the "imagination" here should be underlined.

The contrast between the imagination and sense perception and the specific role photography therein is a constant in this reactionary paradigm. In "Composite Portraits" Galton writes that his composites provide "a generalized picture; one that represents no man in particular, but portrays an imaginary figure possessing the average features of any given group of men."[53] As Sekula has claimed, Galton's composite photograph merged Quetelet's "mean man" with Kant's description of the psychological construction of the empirically based "normal idea" as images that "fall on one another."[54] Galton recognized the importance of the imagination in capturing the actual. In an earlier essay Galton examined analogies between mental images and optical processes. Based on the Weber-Fechner Law of psychophysics, according to which perceptual sensitivity decreased as the level of stimulus increased, Galton concluded that "the human mind is therefore a most imperfect apparatus for the elaboration of general ideas" when compared with the power and quantitative consistency of "pictorial statistics." "Pictorial statistics" had much in common with "abstract ideas," which he suggested be termed *cumulative ideas.*[55]

Galton believed that the composite photograph mechanically created the equivalent of a picture made by the artist with the highest imaginative power. It brought the type of the Jew or criminal before the eyes.[56] Such photographs, therefore, aim at a way to pass from individual to type.[57] In doing this, composites fused imagination with perception. Precisely because Clauß, as well as Günther, later merged perception and imagination in their racial photographs, the difference should be emphasized. Galton viewed neither the medium nor the perceiving subject as relative to race. The medium is neutral; the capacities of the perceiving subject are universal. From the perspective of racial writers of the 1920s, therefore, his method is not "racial" in the strictest sense of the term.

In 1882 and in 1885, Galton delivered papers dealing with composite photographs of phthisical subjects (tuberculosis) and of Jews, respectively. In the
first paper, Galton collaborated with medical doctor F. A. Mahomed, and in
the second with Joseph Jacobs. Mahomed and Galton concluded that their
results did not support the belief that any special type of face predominated
among phthisical patients (fig.1.5).[58] The composite of the Jewish type, on
the contrary, Galton and Joseph believed, successfully demonstrated the Jewish type. The answer to the question of whether by applying the composite
method to Jews, Galton was guided by antisemitism, remains unclear, but
there is no question that he thereby contributed to Jews' racialization.

In 1883, Galton was asked by Jacobs to make the composite of Jews. Jacobs
arrived in England at the age of 18 from Sydney, studied in Berlin, and completed his education in Cambridge. Jacobs, a prolific writer, whose publications include *Studies in Jewish Statistics* (1892), settled in 1906 in the United
States.[59] Jacobs approached Galton as part of his attempt to demonstrate the
existence of a relatively pure racial type of modern Jews, intact despite the
Diaspora. Jacobs recruited young (male) students from the Jews' Free School
and from the Jewish Working Men's Club in London. After the fact, Galton
and Jacobs agreed that a racial type had been produced but disagreed as to its
moral essence (fig. 1.6).

*

Galton believed that abstracting the typical from a group of individuals produced beauty, particularly so with regard to the Jewish type: "They were
children of poor parents, dirty little fellows individually, but wonderfully beautiful," Galton said in a later interview. "They are, I think, the best specimens
of composites, I have ever produced."[60] What did Galton mean by "wonderfully beautiful"? Galton believed composites were always more beautiful than
their components.[61] Expressing antisemitic views on the Jewish gaze, Galton
commented in a later interview that, "The feature that struck me most, as I
drove through the Jewish quarter, was the cold scanning gaze of man, woman,
and child. . . . I felt, rightly or wrongly, that every one of them was coolly ap-
praising me at market value, without the slightest interest of any other kind."[62]
American anthropologist Maurice Fishberg believed that the composite was
based on a small and arbitrary sample and that therefore on empirical grounds
it was scientifically meaningless. Unlike Fishberg, Jacobs agreed with Galton
concerning the fidelity of the composite, stating that it "gave the best available definition of the Jewish expression and the Jewish type." Furthermore,

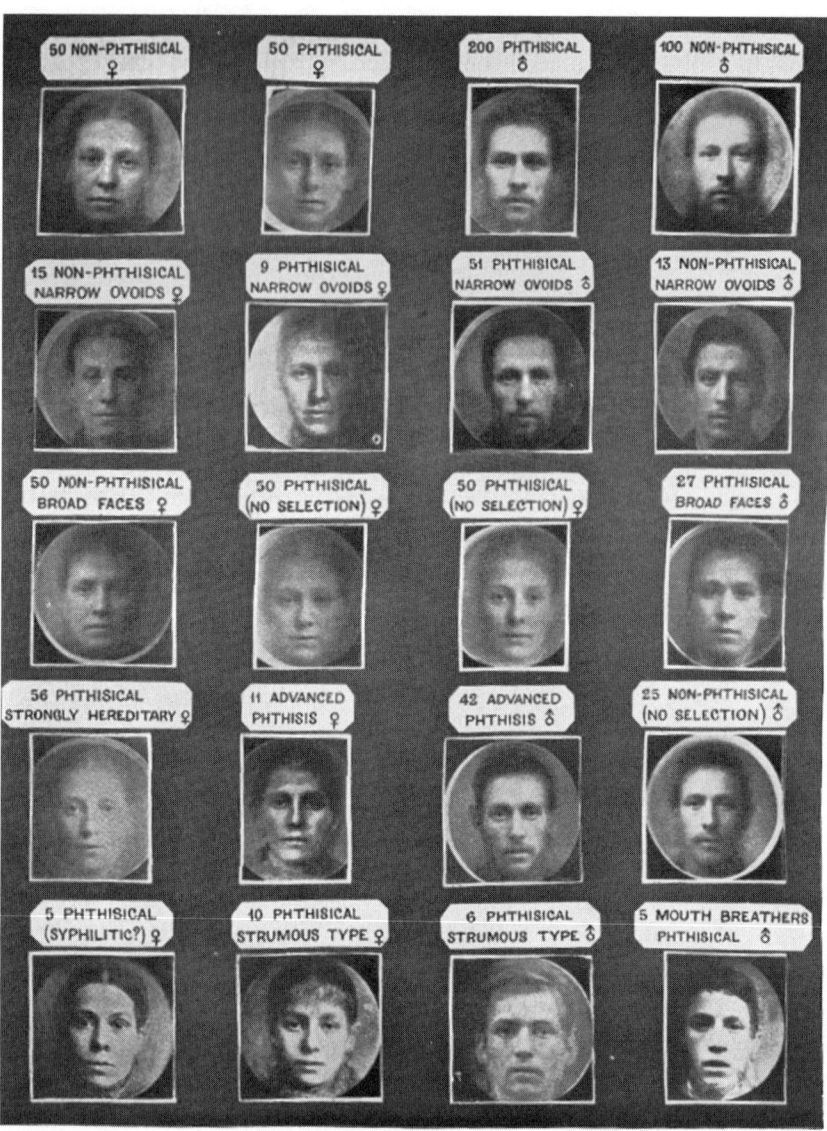

Composites of Phthisical and Non-phthisical Hospital Populations.

FIGURE 1.5. Composite of patients with tuberculosis (Galton). Karl Pearson, *The Life, Letters, and Labours of Francis Galton* (Cambridge: Cambridge University Press, 1924), vol. 2, plate 34, between pp. 290 and 291. Reprinted with the permission of Cambridge University Press.

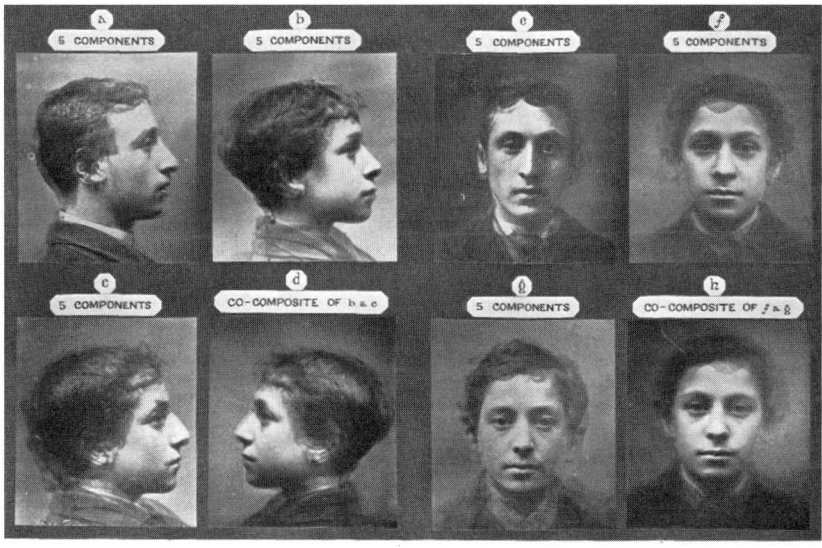

FIGURE 1.6. Composite of the Jewish type (Galton). Karl Pearson, *The Life, Letters, and Labours of Francis Galton* (Cambridge: Cambridge University Press, 1924), vol. 2, plate 35, between pp. 295 and 296. Reprinted with permission of Cambridge University Press.

he defined it in terms of demonstration, stating that " 'The best definition' said the old logicians, 'is pointing with a finger' (*demonstratio optima definition*)."[63] But Jacobs and Galton greatly differed in their judgement of the type. Jacobs realigned the composite of the Jewish type to his original intention: despite the Diaspora, Jews consisted of a stable type.[64] Galton and Jacobs agreed in one scientific register but profoundly disagreed on another.[65]

The success of the "Jewish type" convinced Galton that the future of composite photography lay largely in ethnological and genetic work. His composite method enjoyed wide prestige until about 1915. Its epistemic and technical sophistication were perhaps responsible for its only partial success. As Pearson later wrote, "it was perhaps a misfortune for composite photography that while it required really extraordinary care and patience, it was very easy to compound in an inferior manner."[66]

Rudolf Martin: The Standardization of Anthropometric Photography

Rudolf Martin's contributions to the development of anthropometric photography prove that in the history of science, the implementation of a technique is sometimes more important than the development of a theory. Rudolf Martin (1864–1925) was the focus of much less historical study than Galton or Bertillon, but the influence of his anthropometric photography has been undoubtedly greater than theirs, particularly in Switzerland, Germany, and Austria. Martin was the first professor of anthropology in Zürich, and in 1918, he succeeded Johannes Ranke at Munich University. He played an important role in the biologization of anthropology, in its scientific institutionalization, and particularly in the development, dissemination, and standardization of anthropometric photography. Addressing the conceptualization of photography and its epistemic status exposes the specific intrascientific dialectic of scientific photography from the 1900s to the 1920s.

Martin was not a theoretically inclined anthropologist. Rather, he gained influence by the introduction of concrete methods and their subsequent standardization as well as the development of instruments.[67] He was a devoted teacher,[68] and scores of his doctoral students later occupied professorships in German, Swiss, and Austrian universities. His *Lehrbuch der Anthropologie* was the single most important anthropological textbook in German between its first edition in 1914 to its last one, compiled by his student Karl Saller, in 1957.[69] Indeed, Rainer Knußman's 1980 anthropological textbook derives from this physical-anthropological and photographic tradition.[70]

Any attempt to place Martin in historical context raises a multifaceted set of historical and methodological questions. Martin conceived of himself as an impartial scientist. He was not a staunch opponent of racist science as was his contemporary Franz Boas, but as an individual he was deeply committed to humanist values and a liberal outlook of life.[71] One commentator wrote that the fate of Martin's method in Nazi Germany exemplifies the way a neutral method was ideologically abused.[72] But this interpretation, neatly separating a "neutral" method from its "ideological abuse," simplifies the way questions of method and ideological considerations met on grounds that underwent deep social, political, and ideological transformations. Martin conceived of scientific photography in a different way from Günther or Clauß. But in the 1930s, in Jena, nothing in Martin's method or project prevented Günther from using Martin's materials or teaching his method for the former's own goals.[73] Martin died in 1925, before the Nazis rose to power, but after Günther's influential publications had already appeared in print. In fact, Martin perfected his anthropometric photography in 1925 in a work published by Julius Lehmann's publishing house, which was also Günther's publisher and the leading publisher of antisemitic and racist literature in Weimar Germany. Their times and careers, therefore, do overlap.

Martin's legacy is reflected in the careers of some of his students, including Theodor Mollison, Josef Mengele's supervisor in conducting murderous experiments in Auschwitz-Birkenau, who took a keen interest in photography. An additional student was Karl Saller, who in 1935 was banned from teaching and publishing in Germany after he expressed reservations concerning elements of National Socialist race theory. The separation between "neutral" science and its "abuses," therefore, is fundamentally retrospective.

Martin conceived of "race" as integral to anthropology as a natural science discipline, not as its organizing principle, as Günther or Clauß did.[74] In the first part of the *Lehrbuch*, Martin defines the elementary anthropological concepts and terms such as *kind, variety, type,* and *feature complex* (*Merkmalkomplex*). He then turns to the hierarchy of primates, classifying human races based on the schemes of Ernst Haeckel and French anthropologist Joseph Deniker, with whom he had studied. While racial differences are prevalent already at birth, Martin states, compared with apes, human proportions are nonetheless entirely specific. Indicators of racial difference include hair, skin color, and characteristics of the ear or the nose as well as the skull, which Martin discusses at length. Through a comparison between humans and the Neanderthals, he ties the shape and size of skulls with brain capacity and intelligence. In an 1899 article Martin discussed the relationship between race and intelligence, arguing that spiritual and moral capabilities were no less inherited than physical ones. Every race shows a specific degree of intelligence, abilities, and tendencies, which should be understood within Darwin's law of natural selection and adaptation to the environment. Despite any mixture between peoples, Martin insisted, these tendencies remain in the genotype. As an example, Martin cites the case of the Jews.[75] The science of anthropology is urgent, Martin believed, given that human types are on the verge of extinction.[76] And it is not possible without recognition of the racial substratum.[77]

Martin's practical orientation fits well into recent history of science research, which has seen a shift from the study of scientific theories or ideas to the study of techniques and instruments—across fields and countries. Martin's practical orientation is evident in detailed discussions of instruments and references to specific makers and their addresses.[78] This practical orientation is coupled with a great emphasis on control. Whereas cinema was arguably the most important medium for understanding mass culture in the Weimar and Nazi period, within science, still photography was even more important. Considerations of control effectively impeded the widespread implementation of the moving image. Only in the 1957 edition of the *Lehrbuch*, compiled by

Saller, is the moving film discussed as a possible scientific apparatus, particularly suitable for instruction (*Lehrzwecke*) rather than scientific observation.[79]

Martin's photographic career ensued from his fieldwork experience. During his 1893 study in the Tierra del Fuego (Feurländerland) and his 1896 study of skulls of old Patagonians (*Alt Patagonia*), Martin noted deficiencies and inaccuracies in the then-current measuring techniques. From 1899 to 1904, as a result, he labored on improving measuring devices and methods for fieldworkers, thereby transforming contemporary anthropology.[80]

In his inaugural address at Zürich University, where he outlined his anthropological credo, Martin already lay out the logic behind the introduction of measuring devices, such as cameras, for anthropological work. Martin noted that in many publications, statistical generalizations are made on the basis of very few sampled cases. Worse still, measurements were based on imprecise methods, so even that small sample is inaccurate. But precisely when differences are minute or difficult to observe, he states, the human eye fails. There the introduction of precise instruments to replace the eye is necessary: "where the eye and language can no longer grasp, there measurement is of technical help."[81] This discrepancy between the eye and the instrument is at the core of Martin's efforts to invent, produce, and disseminate a host of measuring instruments. Unlike Bertillon, however, Martin claimed that instruments contribute to "training the senses through subtle observation of details."[82] The logic that underlies Martin's conception of photography is that through devices, observation gains control.[83]

Instruments, Martin repeatedly states, must be not only precise but also inexpensive and sufficiently simple to operate to be put into wide use. These considerations were closely related, in Martin's view, to scientific control: "What is the use of the measurement of thousands of skulls if each researcher chooses different points [*Messpunkte*] and uses different instruments?"[84] Martin developed numerous measurement instruments, which were used by prominent anthropologists: his hair-color table and his eye-color table were tested and used in the fieldwork by prominent anthropologists such as Eugen Fischer and Felix von Luschan (fig. 1.7).[85]

Martin's work in the field of photography closely follows the distinction between physical anthropology and psychical anthropology, dominant in German anthropology of the end of the nineteenth century. Physical anthropology (*Anthropologie*) dealt with the natural/physical aspects of the human body; it circumscribed the species of man (*Homo sapiens*) in its temporal and spatial extension.[86] Ethnology (*Ethnologie* or *Völkerkunde*) studied the *Völkerseele*, the

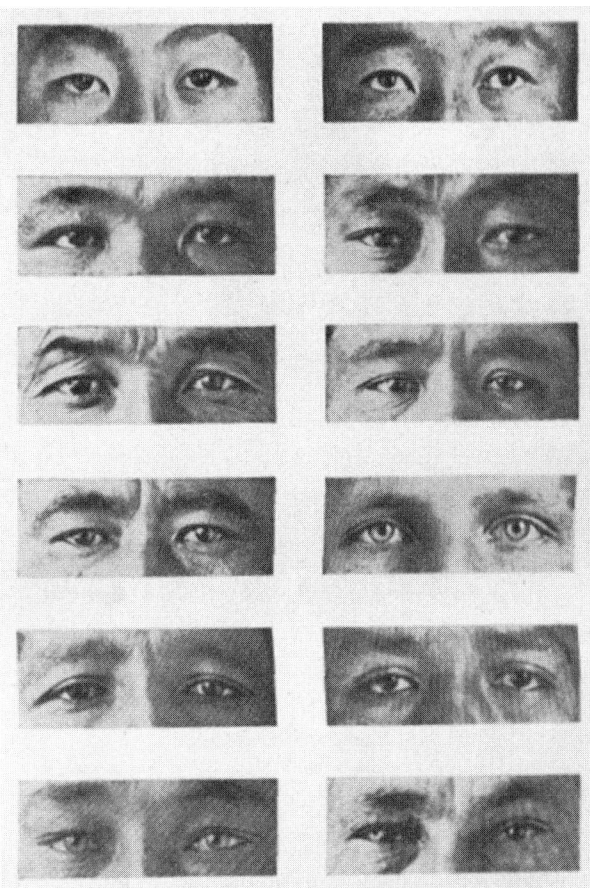

Fig. 2. *Verschiedengradige Ausprägung mongolischer Merkmale an den Augenlidern bei Völkerschaften des europäischen und asiatischen Rußlands.*

FIGURE 1.7. Eye tables for anthropological classification of types of eyes were employed by prominent anthropologists such as Felix von Luschan and Eugen Fischer. By separating the eyes from other facial features, a single trait functions as a synecdoche of the type. "Cover up every part of the composite A but the eyes, and yet I fancy any one familiar with Jews would say: 'Those are Jewish eyes.'" Josepha Jacobs on Francis Galton's composite of the Jewish boy, quoted in Sander Gilman, *The Jew's Body* (London: Routledge, 1991), 68. Rudolf Pöch, "Anthropologische Studien an Kriegsgefangenen," *Die Umschau* 26 (1915), 990.

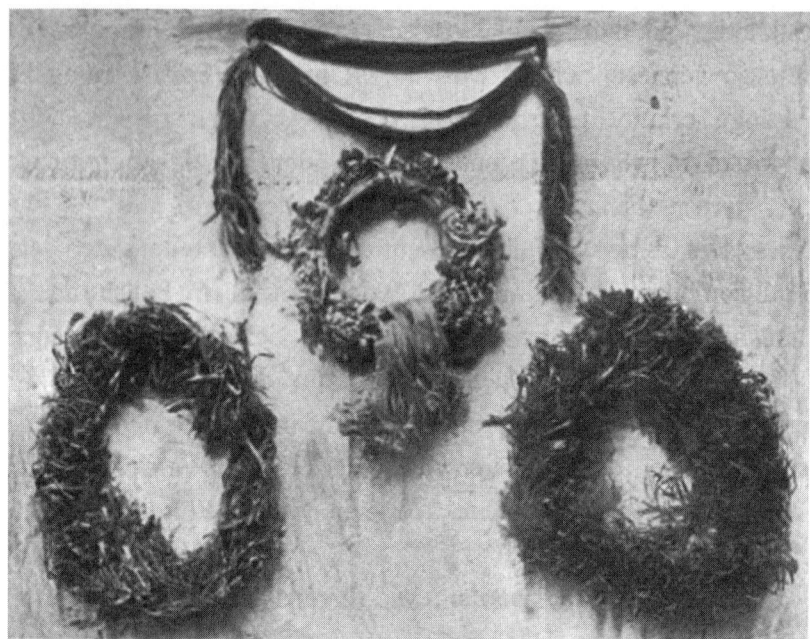

FIGURE 1.8. Photography for the reproduction of material culture. Rudolf Martin, *Die Inlandstämme der Malayischen Halbinsel: Wissenschaftliche Ergebnisse einer Reise durch die Vereinigten malayischen Staaten* (Jena: Gustav Fischer, 1905), 683.

Sitte and *Bräuche*, the cultural aspects of human existence, which developed in human relationships.[87] His photography followed this distinction closely.

In a 1905 book based on fieldwork in Malaysia, Martin employed photographs extensively for portraying the dress and ornamentation of the group under observation.[88] In this case photography was used mainly to reproduce objects taking two forms, either ornaments laid out on a neutral surface or worn by seminaked, mostly female group members. No particular attention is paid to issues of control (fig. 1.8).[89]

More important is his use of photography for physical anthropology. In 1910, as Martin was already working on the *Lehrbuch*, his student Mollison systematized the use of photography for human proportions measurement.[90] Mollison stressed that photographs could serve as anthropometric measurement only if executed with severe control (305). Mollison emphasized, however, that photography afforded no parallel projection; that is, by definition it distorts the reproduced object. Mollison described different forms of distortion caused by angle, distance, and size of object—points later repeated almost verbatim by Martin. Mollison emphasized that this distortion (*Fehler*) is greater than one is normally aware (fig. 1.9). His discussion does not disqualify photography but aims at its correction to ensure scientific validity.

Unlike Martin, Mollison enters into a detailed historical account of attempts to study human proportions in art throughout history, including Polyklet, Leonardo da Vinci, Albrecht Dürer, and Quetelet, a discussion linked to debates over the geometry versus perspective in anatomical drawings.[91] Perspective supplements metric data, providing invaluable morphological insights, according to perspectivists. According to geometers, however, perspective precluded measurement comparison.[92] Mollison attempted to bring together metric and perspective. Conceptually, the attempt was destined to fail. But visually, it was on the way to creating a new visual code of the "proportionate model" (*Proportionsfigur*), which he drew from Martin's 1905 book. Such proportionate models are necessary for deriving "types"; but, Mollison warns, on top of distortions caused by photographs, measurements made from

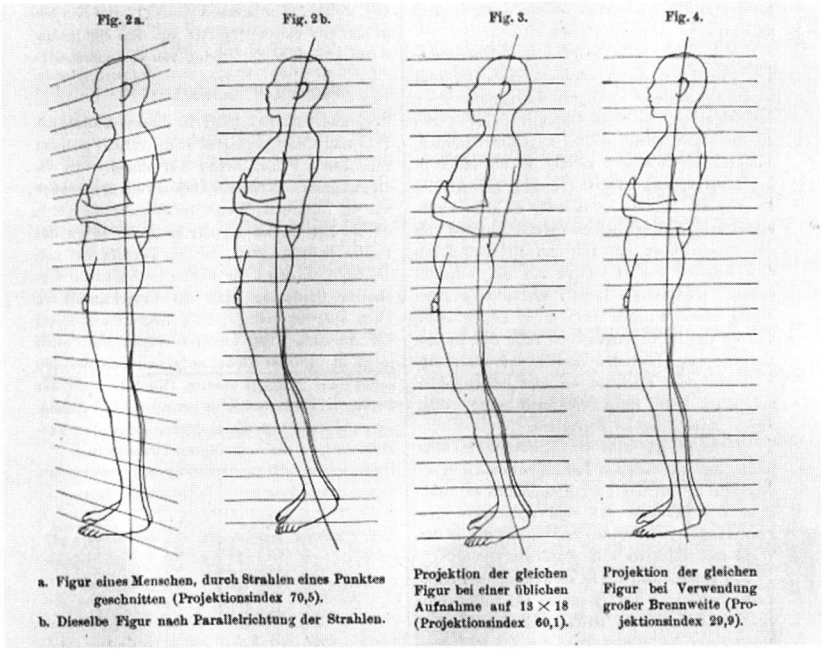

FIGURE 1.9. Mollison's analysis of photographic distortion: figure 2*a* illustrates an accurate representation of a human figure. The other three illustrations (2*b*, 3, and 4) illustrate degrees of distortion of different photographic methods or photographic apparatuses. For valid scientific measurements, photographic distortion must be corrected. Theodor Mollison, "Die Verwendung der Photographie für die Messung der Körperproportionen des Menschen," *Archiv für Anthropologie* 1910 (37): 307.

models are imprecise.[93] Nonetheless, despite distortions, photographs have the advantage over "living material": measurements can be executed under strict control.[94]

Martin's most important contribution in the field of photography is found in his *Lehrbuch der Anthropologie in systematischer Darstellung: mit besonderer Berücksichtigung der anthropologischen Methoden für studierende ärtze und Forschungsreisende*, intended, as indicated by the subtitle of the book, for students, physicians, and traveling scholars. Martin differentiates between methods for obtaining live (living human beings) or dead (e.g., bones) materials and methods of their reproduction (*Reproduktion*). Photography is a principal form of reproduction, together with drawing (*Zeichnung*), measurement and description, and statistics. Reproduction can be pictorial (*bildlich*) or plastic—in the first case the representation (*Darstellung*) is two-dimensional or flat; in the second it is corporal (*körperlich*).[95] Few drawings, according to Martin, can compete with photographic reproduction.

Martin's discussion, focusing to a great extent on control, is not intended for the general reader. His objective is to ensure the least possible distortion, specifically that caused by the angle from which the photograph is taken.[96] In the late 1860s, British anthropologists such as Thomas Henry Huxley and J. H. Lamprey developed systems to take exact scientific measurements directly from photographs, but these endeavors ultimately failed.[97] Martin, like earlier Berlin anthropologists, was not a naive realist with regard to photography. He did not believe that the camera reproduced an exact "parallel projection" of reality; what lies farther away from the lens (*Objektiv*) appeared smaller. The bigger the distance between object and lens, the smaller the distortion—the bigger the distance, however, the smaller the image.[98] To be useful for scientific ends, the image must maintain a minimal size, Martin stressed, allowed only by using a negative of a minimal size.[99]

Martin stressed that few photographs in anthropological literature were scientifically useful.[100] To secure control, anthropologists must provide the exact distance between the object and the negative. Martin provides a table with object size, distance from camera, angle, and negative plate size, then specifies the kinds and manufacturers of cameras, sizes of plates, kinds of lenses, aperture (F/6 or F/9 to F/12), and light conditions.[101] For particulars, he refers to Bertillon and Chervin.[102]

The best anthropological photographs are of living persons, Martin stresses. Individuals should be photographed standing, naked when possible. Photographs should be taken from front and sides, preferably from the back as well.[103] Head and breast are best photographed from front and sides; two or three photographs, from additional angles, are necessary. Martin offers specific instructions for execution, including angles, height, and sought gaze. His aim is to transport the accuracy of the scientific laboratory to the field, for which he invents a transportable "kit" of instruments. Photographs should include a number or name, age, and location of photograph; plates should be marked with chalk to avoid later confusion.[104] Within this context Martin critically discusses Röntgen photography (*Röntgenfotografie*) and Galton's composite photography (*Mittelbilder*)—questioning the assumption that the method uncovers the middle type (*Mitteltypus*).[105]

In visual terms, it is interesting to note, Martin's *Lehrbuch* insinuates important genres in the history of Western art. There is a certain circular irony here: whereas painters who were interested in studying the human body could only do so through certain religious subjects, Martin, who possessed a medical background (as well as most German physical anthropologists of his generation), embedded his knowledge of Western art history in the ethos of scientific objectivity (fig. 1.10).

Photography played a key role in anthropological studies of prisoners of war in German and Austrian camps during World War I. Several studies were based to a large extent on Martin's method.[106] One large study was initiated by Austrian anthropologist Rudolf Pöch, Wilhelm Doegen (an associate of Berlin psychologist Carl Stumpf), and prominent ethnologist and anthropologist Felix von Luschan. Andrew Evans has studied this project within the framework of the postcolonial history of photography, emphasizing the repressive power and the social control of this project. Evans examines the racialization of the enemy in the war context; the transformation of "race" from abstract notion to a "real and concrete" one as well as the construction of Central powers wartime identity.[107] Evans shows how the racialization of non-European soldiers was followed by the racialization of European enemies of the Central powers. He analyzes the political considerations involved in the peoples studied and the situation of Jews therein.[108] Significantly, Jews were not the only racialized objects in this project. Artist Hermann Struck, later famous for his drawings of eastern European Jewish types, worked together with von Luschan (chap. 2) in the observation of prisoners of war.[109] Racialization, therefore, was not always externally imposed.

The prisoners' project fired a controversy, mainly over technical questions, revealing the dialectic involved in the development of scientific photography. The controversy involved several scholars and branched out in different directions.[110] Pöch relied on Martin's method, but following his own field experience, he appealed for practical changes including sizes of plates and

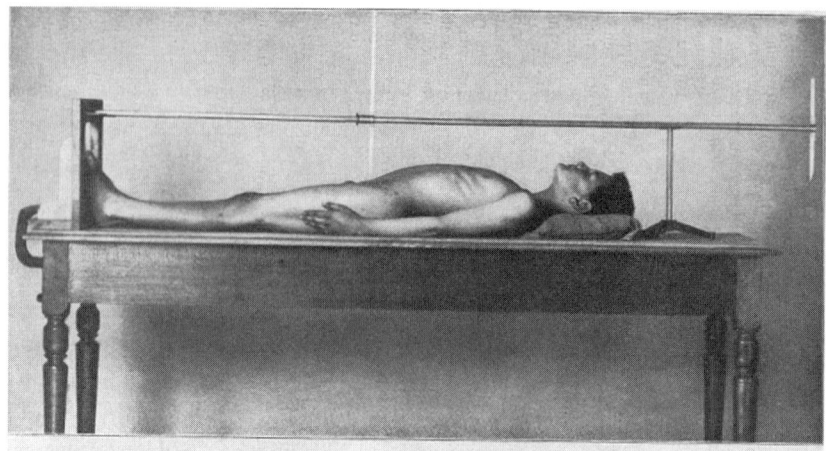

Fig. 30. Anthropometer zur Messung Liegender verwendet.

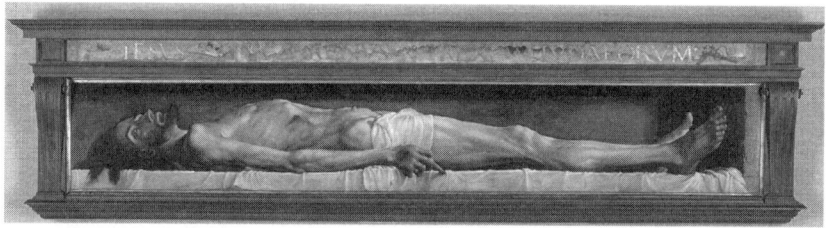

FIGURE 1.10. *a,* A photograph demonstrating measurement of a man lying down alludes to the genre of Jesus in the tomb. Rudolf Martin, *Lehrbuch der Anthropologie: In systematischer Darstellung* (Jena: Gustav Fischer, 1914), 114. *b,* Hans Holbein, *The Body of the Dead Christ in the Tomb,* 1521.

the addition of a third, side angle, which he believed should be obligatory.[111] Martin responded by defending his method, claiming that with smaller plates, details could not be observed. In response to the angle requirement, Martin, together with his son, developed a turntable (*Drehscheibe*) in order to standardize Pöch's proposal of an additional angle, which was incorporated in later publications. Controversy centered on practical considerations of technique— ethical considerations never entered the discussion.[112]

Martin was engaged in an additional controversy of a different kind. During the war, Martin responded to claims made in France and in Germany that science was specifically French or German. Later writers, as we will see, seriously considered the possibility that science was racially determined. A 1915 report in the journal *Umschau* claimed that Germans were better than their enemies in practice and in theory, and their science, too, was superior. French publications, in a similar vein, claimed French science to be superior on racial grounds. Martin viewed such claims as dangerous. In response, he stressed that science was by definition international and universal. No science is specifically national and, he claimed, one cannot distinguish between German, French, or English anthropology.[113]

Following the appearance of the *Lehrbuch*, Martin continued to develop his photographic method. Shortly before his death, he presented an elaborate version of his photographic method in the first volume of the *Handbook of Social Hygiene* (1925).[114] Martin repeats many details of the earlier publication but discusses photography not under reproduction of anthropological materials but as the "display [*Veranschaulichung*] of anthropological results." The aim of anthropometry, as conceived here, is to collect the characteristics of the human body in its totality. The procedure is inductive, from individual forms to analytic and synthetic analyses, advancing by way of calculation. In order to display results, Martin proposes three methods: photographic reproduction of the body from three angles, construction of proportionate figures, and statistical patterns of deviation.[115] Here Martin collapses the photographic and numerical, from which a "numerical skeleton-picture" emerges.[116]

Martin places photographs of two individuals next to each other. He then compares twenty-one of their physical features, arranged in three columns, followed by instructions for their drawing on measurements paper. Martin emphasizes that the smaller the paper (and the figure), the less noticeable the real differences. He further emphasizes that comparison should be made between individuals and their respective group average. In contradistinction to Galton, Martin warned that there are "no general valid norms"; that is, the average is not "normal." These "abstract" figures, made up of numbers, are "skeletal embodiments"; the impression of the photographed individual is replaced by a "numerical x-ray." Ultimately, these models allow for a comfortable representation of group features in a table for the study of their variability.[117] This method accomplishes an extraction of photography from the "noise" of visual idiosyncrasies (figs. 1.11, 1.12).

Martin differed from theoreticians of race such as Günther or Clauß in his conception of science because he believed that science need not necessarily have any practical application or even relevance.[118] Anthropometric photography was not, for Martin, an end in itself but a scientific tool or medium, similar in its epistemological status to other measurement instruments. Photographs do not even form evidence but merely reproduce anthropological data (fig. 1.13).

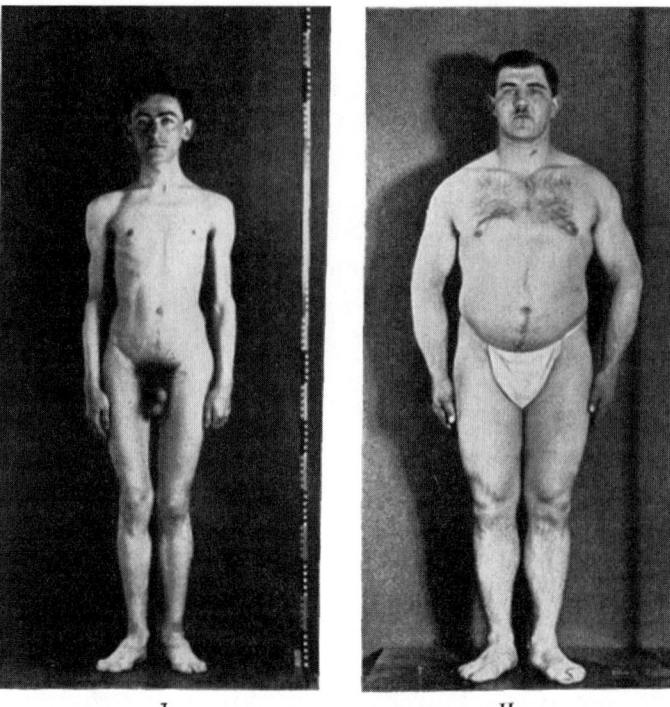

Abb. 17.   Junger Mann (I) und Schwerathlet (II) in Vorderansicht. $^1/_{18}$ n. Gr.

FIGURE 1.11. Valid scientific photographs eliminate individual features and capture measurable ones. Cultural markers must be removed. Standard gray and even background eliminate the environment. In terms of lighting, control was uneven, as only the right figure has an extended shadow. Both men are slightly stretched, facing the camera almost frontally. The subtitle states that the person on the right is an athlete. Rudolf Martin, "Anthropometrie," in *Handbuch der Sozialen Hygiene und Gesundheitsfürsorge*, ed. A. Gottstein, A. Schlossmann, and A. Teleky (Berlin: Lehmann, 1925), 298.

As a practically oriented anthropologist, Martin never discussed the question of sampling; namely, how does one know in the first place that an individual belongs to a certain class or type. Rather, he viewed that relationship to be self-evident. Martin did not seek typical specimens but measured individuals for variability of racial populations.[119]

Martin's *Lehrbuch* continued to be the principal anthropological textbook in the German language through its 1957 edition; the discussion of photography underwent no fundamental change. The illustrating photographs were replaced, following the same model, and the discussion of manufacturers was updated.[120] The situation with "race," however, is different. The 1957 edition, prepared by Karl Saller, presents a softened, almost critical concept of race.[121]

Unlike Günther or Clauß, Martin showed no particular visual sensitivity and was not guided by a concept of seeing. Granted, the dissemination of Martin's method within science reified "types" and "races" and endowed racial photography with scientific legitimacy and authority. His significance was in developing, standardizing, and disseminating particular photographic techniques toward anthropological ends. Martin was central in securing the scientific status of photography as a scientifically "blind," neutral instrument. Precisely these aspects enabled different ideologically committed methods of deployment of photography in the 1920s and 1930s despite important ideological and epistemic breaks.

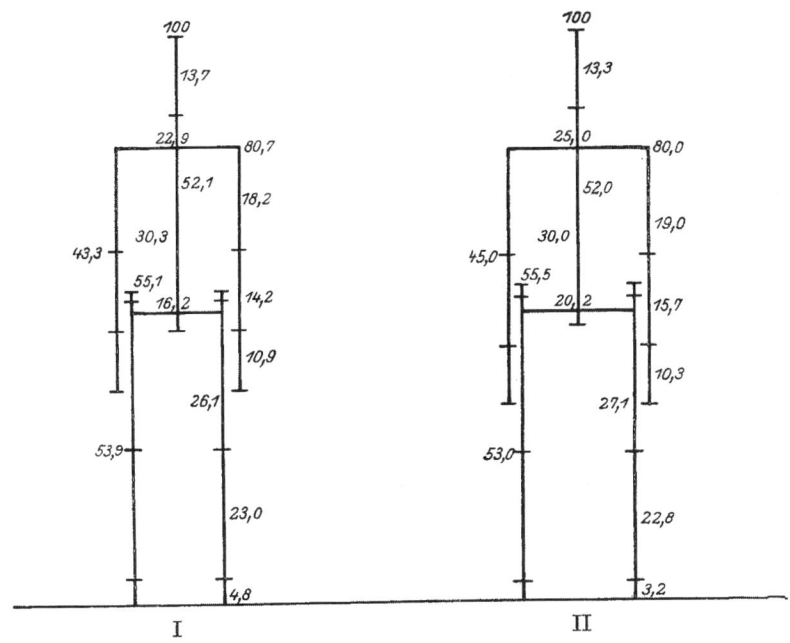

Abb. 18. Junger Mann (I) und Schwerathlet (II), Proportionsfiguren. Die Körpergröße ist bei beiden Figuren = 100 gesetzt.

FIGURE 1.12. The "numerical skeletons" or "numerical Röntgen images" abstract from photographs the essential information. Röntgen images affected the anthropological discourse in its quest for the typical under the individual visible surface. While abstracted from the surface of photographs, the diagrams suggest that the abstraction reflects their deeper typological structure. Rudolf Martin, "Anthropometrie," in *Handbuch der Sozialen Hygiene und Gesundheitsfürsorge*, ed. A. Gottstein, A. Schlossmann, and A. Teleky (Berlin: Lehmann, 1925): 300.

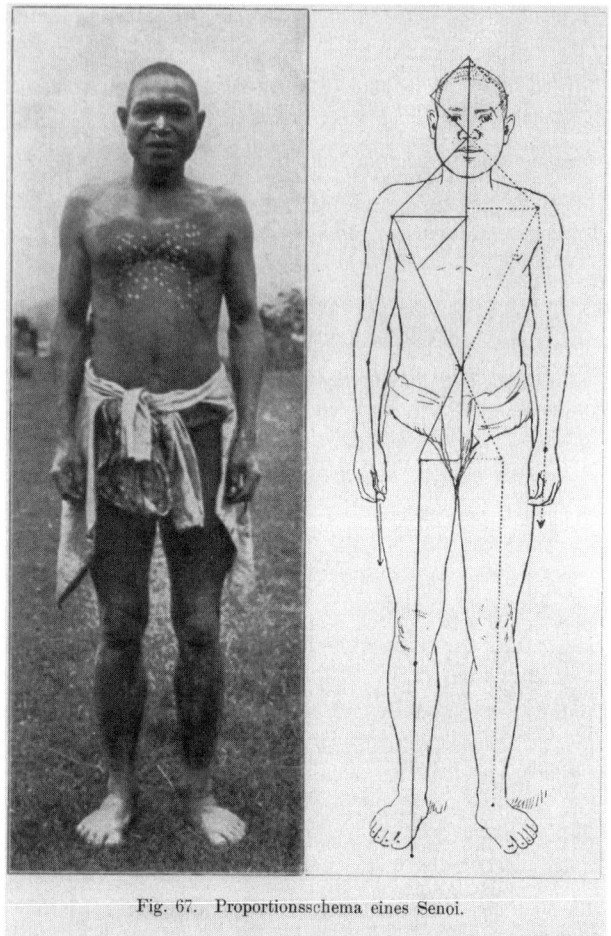

Fig. 67.   Proportionsschema eines Senoi.

FIGURE 1.13.   Abstraction of racial type from a photograph. Rudolf Martin, *Lehrbuch der Anthropologie: In systematischer Darstellung* (Jena: Gustav Fischer, 1914), 252–53.

Paraphrasing Peter Galison on the relationship between images and numbers in microphysics, the relationship between numbers and statistics from Bertillon's anthropometrical photography through Galton's pictorial statistics to Martin's anthropological reproduction could be described as "from number to image to number." All three started from and gave analytical priority to numbers over images. Yet they all incorporated images in their publications and elaborated on their methodological use. Even when images, and particularly photographs, were incorporated, numbers continued to be present in them, in fact, to such an extent that they were intricately interwoven into their visual patterns and fixed into their visual codes. True to a natural scientific ideal of science, numbers, in the end, remained analytically superior to the distinctly visual.

Allan Sekula concludes his seminal essay stating that with regard to Galton and Bertillon (with claims that could be extended to encompass Martin as well),

> despite their differences, both Bertillon and Galton were caught up in the attempt to preserve the value of an older, optical model of truth in a historical context in which abstract, statistical procedures seemed to offer the high road to social truth and social control. . . . With the general demise of an optical model of empiricism, Galton's hybridization of the camera and the statistical table approached extinction. Photography continued to serve the sciences, but in a less grandiose and exalted fashion, and consequently with more modest—and frequently more casual—truth claims, especially on the periphery of the social sciences.[122]

The cases studied in the following chapters challenge Sekula's interpretation. Günther and Clauß do not follow the optical empiricist model, and their scientific use of photography was anything but modest.

PART TWO: MENDELIAN PHOTOGRAPHY AND THE GAZE

The relationship between the emergence of Mendelian genetics after the turn of the twentieth century and the use of photography for the study of race has not hitherto been the focus of interest of historians of science. Viewing the gaze, in particular, as a genetic and racially determined trait, Carl Heinrich Stratz, Redcliffe Salaman, and Eugen Fischer were not interested in employing the camera as a measuring device and hence paid little attention to scientific control. Similarly, they were not interested in transforming photographic data into statistical tables or in the transformation of visual evidence into a quantified one. Rather, they developed a penetrating photographic method of observation. As their conceptualization of race was based on their understanding of the newly rediscovered Mendelian principles of genetic inheritance, these writers were particularly interested in questions of racial mixture, and all three viewed the camera as a particularly powerful device for its study. These writers viewed the photographic method as an objective procedure that afforded an accurate representation of reality. At the same time, however, their discourses can be seen as part of the general trend of developing ways to tie between visible traits and invisible essences.[123] But as the invisible essences were not easily visible on the photographic surface, all three believed that photographs necessitated interpretation and took into account in their writings the possibility that photographs could "lie," or at least that they did not necessarily tell the "truth." Thus, these writers were part of a larger attempt to impose a visual paradigm on phenomena they believed resisted the visual or at least did not succumb to it easily.[124] This was not, however, because the photographic medium was flawed, but because it could only represent the phenotype, which is visible to the observer. The genotype, the genetic code that is hereditarily transmitted, remained hidden from the eye. While photography is presented as a powerful and penetrating device in the service of observation, it is coupled with distrust or even suspicion, as it is possible that what you see, so to speak, is not what you get.

Compared with the Dammann brothers' *Atlas*, it could be said that Carl Heinrich Stratz, Redcliffe N. Salaman, and Eugen Fischer invert the role of photography. If the former used photography to depict racial types, the latter employed photography to deconstruct them; if the Dammanns employed photographs in order to illustrate racial types, Stratz, Salaman, and Fischer employed them for scientific observation. The latter did that based on the notion that individuals are not instantiations of types but rather are made up of distinct Mendelian traits. These three cases of "Mendelian photography" show the integration of photography into transformations in the theoretical understanding of race.

Carl Heinrich Stratz: Gynecology, Photography, and the Jewish Gaze

In the publications of Carl Heinrich Stratz photography, gender, race, colonialism, and Jewish difference all converge in a striking manner. His work, from a later perspective, constitutes a case study laden with ambiguities and inconsistencies, and it is an especially rich medium within which to address the question of whether this study of the Jewish gaze would have been possible without the photographic medium.

Carl Heinrich Stratz (1858–1924), the foremost authority on Wilhelmine body aesthetics and one of the main popularizers of racial theories of the period, was an Odessa-born gynecologist who studied in Heidelberg. Stratz became famous for several widely read books, in particular a book he wrote on the beauty of the feminine body.[125] Stratz's view of feminine beauty was closely related to notions of race.[126] According to Stratz, each human being had two lives, as an individual and as a member of a race. Women were less individualized than men, but as they possessed a "quicker imagination," they were said to have a larger effect on the life of the race. Stratz believed that the ideals of feminine beauty were relative to race, which he defined as the "hereditary common properties of a specific inherited physical and spiritual habitus."[127] Only white woman approached the superior Aryan ideal of beauty embodied in the art of Greek antiquity. Furthermore, Aryan feminine beauty was distinguished from other races' by its sexual disinterestedness—it aroused no mundane sexual interest.[128] In later editions of the book, Stratz introduced an increasing number of photographs (fig. 1.14).

Stratz's book *Women in Java: A Gynecological Study* (1897) combined, in Ann Stoler's words, "the sexual pleasures of scientific knowledge" and the "pornographic aesthetics of race."[129] Stoler's analysis of some of Stratz's dissonances and racial asymmetries makes plain that his later book on the Jews should be viewed in light of this earlier work. Through examining photographs of naked Javanese women, she stresses, Stratz searches for the hidden racial characteristics, even when the external physical appearance is European, as similarity only masks difference. But touching directly on his use of photographs, she writes,

> Commenting again on the lack of hair around the clitoris, he instructs his readership to the "particularly clear" view of this in a photograph he provides. But there is nothing clear in the figure at all. And this is just the point. The reader's gaze must be studied, because there is little to see in the profile picture. We must rely on Stratz's privileged view. Our gaze is pointed inward, to that which is not visible—but with Stratz's expert help—easily imagined.[130]

Stratz's 1903 photographic study of the Jewish gaze, therefore, cannot be dissociated from both the colonial and the gender aspects of his earlier work, in which he attempted to exemplify racial types through the use of photographs that did not follow the standardized anthropometric methodology. Rather, following a notion of the "image-forming gaze" (*das bildnerische Sehen*), Stratz accorded far greater credence to the intuitive gaze of the physician than to empirical measurements. But whereas Stoler concludes that the viewer imagines rather than sees, this division between imagination and observation is at the basis of the epistemological engagement of the current study. Many of Stratz's ideas on race and gender as well as his preference of the penetrating gaze (*Anschauung*) over the "sterile" anthropometric tradition can also be found in the later work of Günther or Clauß. But in political terms, Stratz was a committed liberal. According to Michael Hau, Stratz may have identified

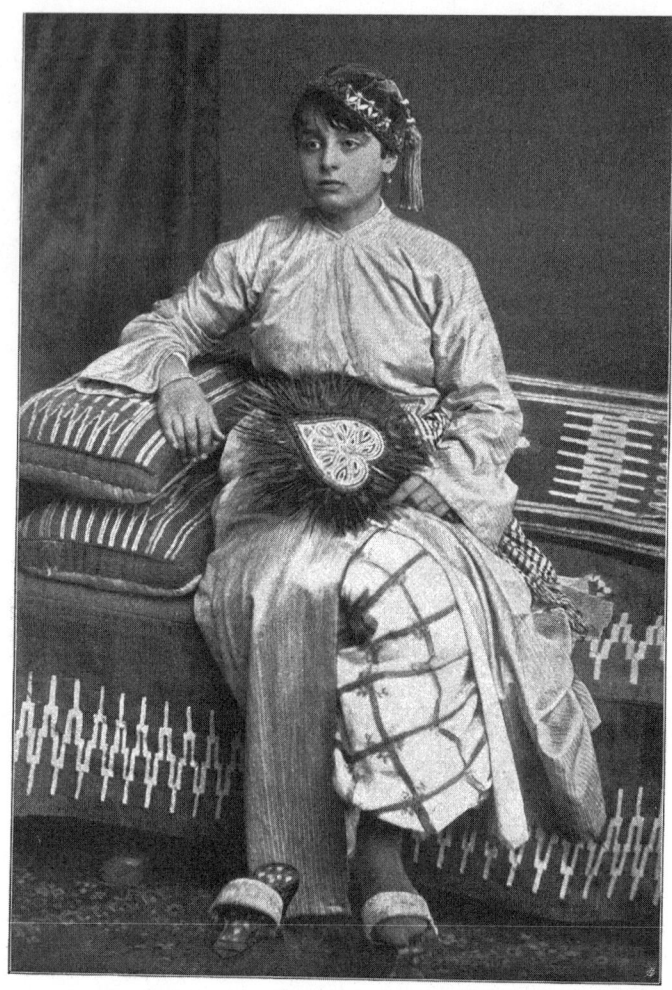

Fig. 9. Jüdin aus Palaestina.

stärkere individuelle Entwicklung und die geringere
körperliche Vollkommenheit bei erhöhter Ausbildung

FIGURE 1.14. Jewess from Palestine. Stratz focused his argument on the distinctive Jewish gaze. Rather than extricating the gaze and isolating it, however, this portrait depicts the whole person in her cultural environment. Heinrich Stratz, *Was sind Juden? Eine ethnographisch-anthropologische Studie* (Leipzig: G. Freytag 1903), 20.

with "everyday antisemitism," but he was so troubled by radical antisemitism (of the kind propagated by Ludwig Schemann) that the photographic study of the Jewish gaze was in fact aimed at its refutation.[131] Hau does not substantiate this interpretation with a footnote or quotation from Stratz, and Stratz's photographic study of the Jewish gaze, most certainly from a later historical perspective, reads as deeply ambivalent fig. 1.15.

In 1903, Stratz published a short book on Jews in which he argued that Jews' physical markers were not specifically Jewish but rather that Jews were marked by a specific gaze.[132] Galton, as we saw, believed there was a specific Jewish gaze, but his composites were aimed rather at generating the Jewish type as a whole. Stratz, however, employed photography specifically for the study of the Jewish gaze.[133] Based on the photographic study of the gaze, Stratz claimed he was able to identify "true" individuals of Jewish descent, although no longer part of the Jewish collectivity, whether in Spain, North Africa, or China (fig. 1.16).

Stratz used photographs in order to demonstrate this distinct gaze. Note the way imagination informed a circular photographic methodology. Stratz

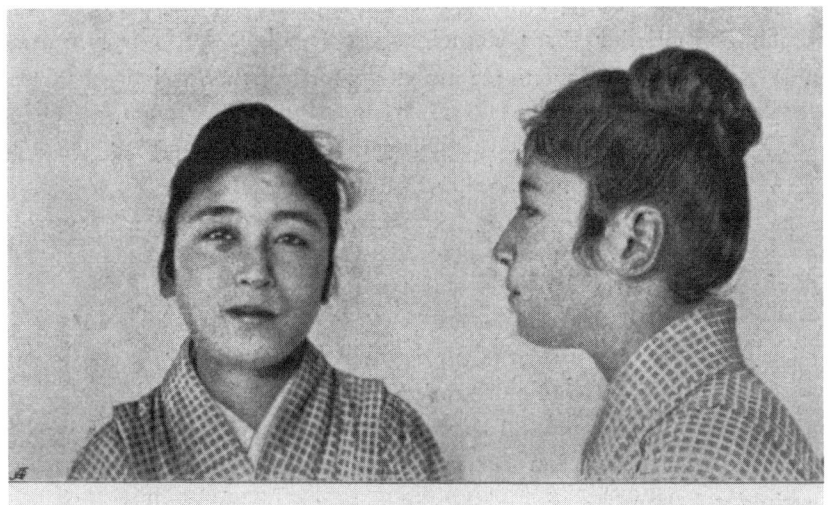

Fig. 10. Japanerin mit jüdischer Gesichtsbildung. (Phot. Baelz.)

FIGURE 1.15. Girl identified as Japanese. According to Stratz, this demonstrates the Jewish expression found among different races. While the photograph depicts the girl's face, her eyes clearly call the viewer's attention. Michael Hau, *The Cult of Health and Beauty in Germany: A Social History, 1890–1930* (Chicago: Chicago University Press, 2003) 88; Heinrich Stratz, *Was sind Juden? Eine ethnographisch-anthropologische Studie* (Leipzig: G. Freytag 1903), 22.

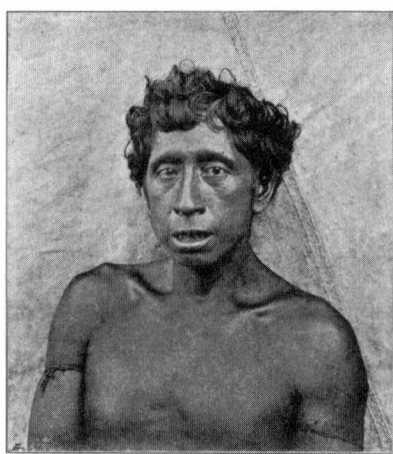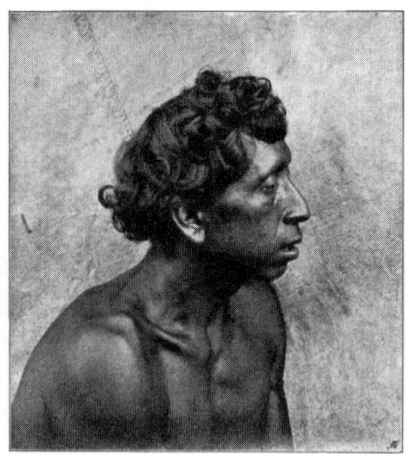

Fig. 11. Bakaïri mit jüdischer Gesichtsbildung. (Phot. Ehrenreich.)

FIGURE 1.16. Identified as Bakäiri (an Indian culture from the South American Amazon basin] with Jewish facial expression, Stratz took these photographs from Ehrenreich. They were later reproduced and further circulated by Maurice Fishberg and Hans F. K. Günther. Heinrich Stratz, *Was sind Juden? Eine ethnographisch-anthropologische Studie* (Leipzig: G. Freytag 1903), 23.

purchased his photographs during his travels and had to rely on the information provided by the sellers. In order to avoid being misinformed, Stratz confessed, he only purchased photographs that accorded with the impression he had gained during his travels. His use of photographs was aimed not at Jews' physiognomic traits but at a much more indescribable feature. Based on this research, Stratz concluded that the Jewish type was found among all human races of the earth—the gaze was not racial but the expression of intense individuality, the result of a long process of selection.[134] Without the technique of photography, it is not clear whether it would have been possible to argue for the existence of a specifically Jewish gaze let alone to demonstrate it. Hence, it is possible that the very idea of the racialized Jewish gaze (and most certainly its demonstration) was contingent upon the emergence of photographic technology.

Redcliffe N. Salaman, Eugen Fischer, and the Photographic Study of the Gaze as a Mendelian Characteristic

Unlike Stratz's use of photography to substantiate a specifically Jewish gaze, Redcliffe N. Salaman, who also attempted to use photography to ascertain the distinct Jewish gaze, grounded his study on Mendelian biological grounds. Salaman (1874–1955) was an English physician and biologist, a friend of the founder of Mendelian genetics William Bateson, and an early supporter of Zionism. Salaman's support of Zionism was closely linked to his commitment to his understanding of Mendel's principles of heredity.[135] Salaman pioneered the introduction of photography for the observation of biological variation. Before applying his photographic method to humans he viewed as racially mixed in 1911, he employed it for the observation of biological variation in potatoes in a study conducted in 1906.[136]

In a long article on the genetics of the potato, Salaman employed numerous photographic plates that were placed at the end of the article and referenced throughout his analysis. Salaman referred to these photographs as "representing," "showing," or "illustrating" the claims he made throughout the text.[137] In both the article on the potatoes as well as in the article on Anglo-Jews, Salaman refrained from discussing the reasons for introducing photography, the advantages and disadvantages of its introduction, its scientific status, or the significance of the visual medium for the biologist—indeed, his discussion of "the eye" is limited to the potato's eye (fig. 1.17).[138]

While the general strategy in these two articles is similar, including the employment of photography for the sake of the observation of biological varieties, I would like to point to several important differences. In the article on potatoes, Salaman discussed the difficulties in classification, noting "it must be remembered that it is a matter not only of considerable difficulty to classify the living plants according to shape and texture of their leaves, but that the personal element is paramount in such a classification. More particularly to such remarks apply to the consideration of texture and to the intermediate forms."[139] Salaman continued in qualifying the scientific validity of his observations, and while analyzing photographs of specific potato families, he noted their uniformity as well as the existence of exceptions, which he depicted in separate photographs.[140]

In his 1911 article, Salaman transported the photographic medium to the study of variation in humans. Without noting similar qualifications, he now attempted to argue for the validity of Mendel's discoveries on heredity for the study of humans.[141] Salaman was not the first to argue for the relevance of Mendelism for what he termed the "study of interracial mating"; he was also not the first to claim that the Jewish type remained racially pure throughout the centuries. But his attempt to employ photography specifically for observation of what he believed to be Mendelian racial traits is of particular interest. Salaman agreed with the claim that Jews possessed a specific facial expression, which,

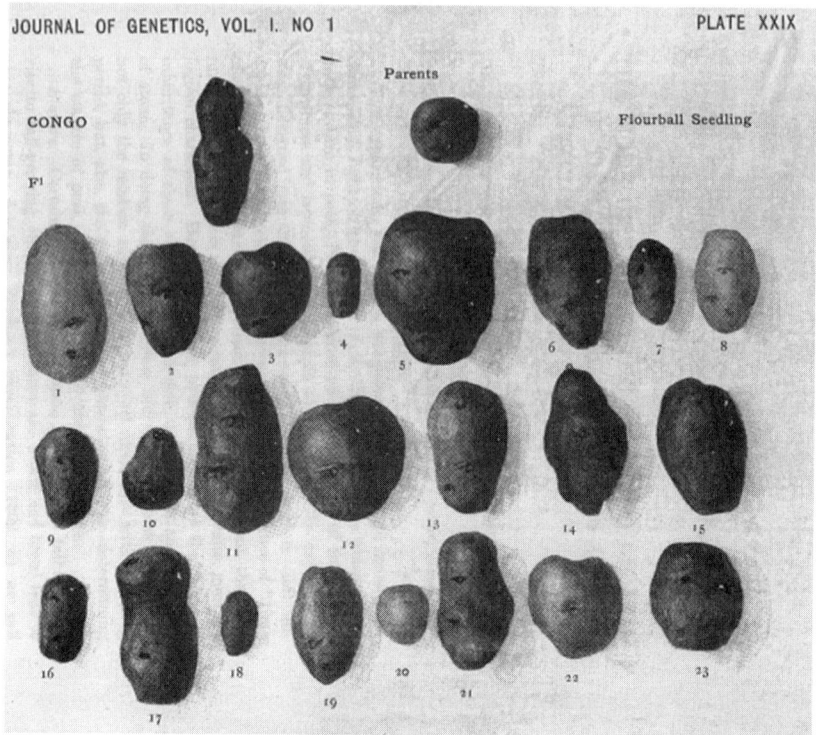

JOURNAL OF GENETICS, VOL. I. NO 1   PLATE XXIX

Parents

CONGO   Flourball Seedling

FIGURE 1.17. Photography for the observation of biological variety in potatoes. Redcliffe N. Salaman, "The Inheritance of Colour and other Characters in the Potato," *Journal of Genetics* 1, no. 1 (1910), plate 29.

as we saw, was a belief independently expressed by Galton and Stratz. Salaman did not refer to Stratz, Joseph Jacobs, or Galton, but his assertion—based on Assyrian sculptures, terra-cotta heads from Memphis, a photographic reproduction of an English sketch dated to 1277, and photographic reproductions of drawings by Rembrandt—that this expression was not a modern trait could be seen as a modification of Stratz's argument that it was the result of selection.[142] The viewer had no way to know that Salaman's choice was biased in the sense that other medieval images of Jews—such as those found in the Codex Manesse, created at the beginning of the fourteenth century—depicted Jews who could only be identified by the Jewish hat or by their beards, their physiognomy being indistinguishable from that of their Christian contemporaries. In fact, Salaman stressed that the facial expression was more stable than the skull shape of Jews, which underwent transformation.[143] Unlike Stratz, however, Salaman tried to break down the Jewish facial expression into hereditary elements.

Of particular interest in the context of this study is Salaman's attempt to study Mendelian lines in intermarriages in order to "obtain results comparable to those the genetic student has been obtaining in plants and animals."[144] Salaman produced and collected hundreds of photographs of children of mixed marriages between what he termed Anglo-Jewish and English Gentiles. He asked Jewish observers who were ignorant of both Mendelian theories of heredity and the purpose of the examination to identify those expressing the Jewish types. Salaman collected the replies, and based on this photographic experiment, he concluded that the Jewish facial expression was a Mendelian trait. Following Mendelian principles that hereditary units alternate but do not blend, Salaman found that these traits showed no blending and appeared to follow regular hereditary patterns. He demonstrated his findings to readers with photographic examples, out of hundreds available to him, that he provided in the appendix to the article. Salaman discussed "Jews who may be said to have a 'pseudo-Gentile' appearance"[145] and claimed that it was essentially different from its "Teutonic counterpart."[146] Based on Biblical passages, Salaman interpreted this feature as deriving from a Philistine element that was gradually absorbed into the Israelite nation but was never lost. Based on photographs, in what could be interpreted as a revision of Stratz, Salaman claimed that in Chinese Jews the Jewish facial type has been swamped by the Chinese (fig. 1.18).[147]

Eugen Fischer (1874–1967) was unquestionably a scientist of greater influence than Stratz or Salaman.[148] Years before his notorious Nazi career, Fischer's famous Rehobother Bastards study, which appeared in 1913 and was republished numerous times (up to 1961), focused on an African population of mixed descent.[149] This study was seen at the time as groundbreaking in shifting the scientific attention from the study of "racial types" to the study of "racial bastards." Unlike Stratz or Salaman, Fischer's use of photography is systematic and comprehensive; in the book itself he reproduces close to fifty photographs out of 300 that he had made.[150] Fischer focused his study on traits he believed were racial and that he conceived as Mendelian hereditary units (fig. 1.19).

In this book Fischer noted that the only serious attempt to date to study racially mixed populations had been carried out by Salaman, and that this study was scientifically flawed.[151] While praising "Salaman's beautiful material," Fischer criticized him for not focusing on racial traits and for his treatment of the

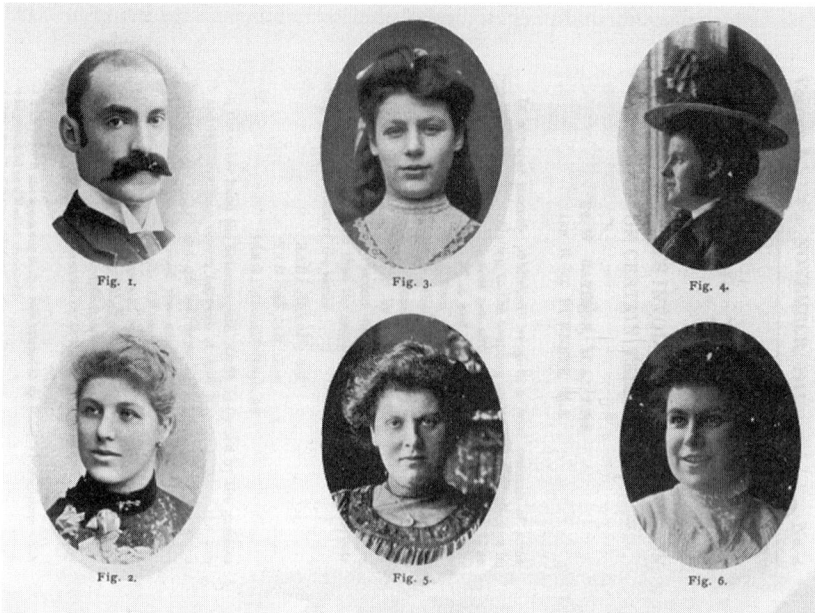

FIGURE 1.18. Photography for the observation of biological variety in humans. This page depicts mixed offspring of Anglo-Jewish and gentile English parents who, according to Salaman, possess the Jewish gaze. The viewer's attention is drawn to the eyes (shape, relative size, and dark pigment) and their soft but penetrating gaze. Redcliffe N. Salaman, "Heredity and the Jew," *Journal of Genetics* 1 (1911), plate 38.

facial expression as one hereditary unit.[152] In methodological terms, however, Fischer's use of photography is closer to Salaman's method than to Martin's anthropometric method. Fischer noted that for his purposes, photographs of the subjects' heads were sufficient: the undressing of observed objects was neither possible nor advantageous. Photographs, in his view, were not intended to replace or to extend measurements, which he carried out following Martin's method.[153]

Fischer's entire book closely examined a number of families of mixed descent.[154] His use of photography served to demonstrate his understanding of patterns of hereditary Mendelian characteristics within families. The photographs, therefore, focus on such families: what Fischer viewed as racially pure parents (Dutch or Hottentot, respectively) and their racially mixed children. Significantly, the photographs attempt to demonstrate the alternation of specific traits rather than whole types. As Fischer is interested in the genealogy of families, each photographed person is named. While it is not clear whether this was Fischer's intention, as the names of the mixed descendants are European,

Dutch, or German—Wilhelm, Friedrich, or Martin—the names, even more than the African features, add an effect of alienation. Fischer attempted to break down types into Mendelian traits through the portrayal of the mixed offspring next to their Hottentot or Dutch "pure" parents. The reader of the text and observer of the photographs is requested to identify different traits found in the offspring and to track them to only one of the parents (fig. 1.20).

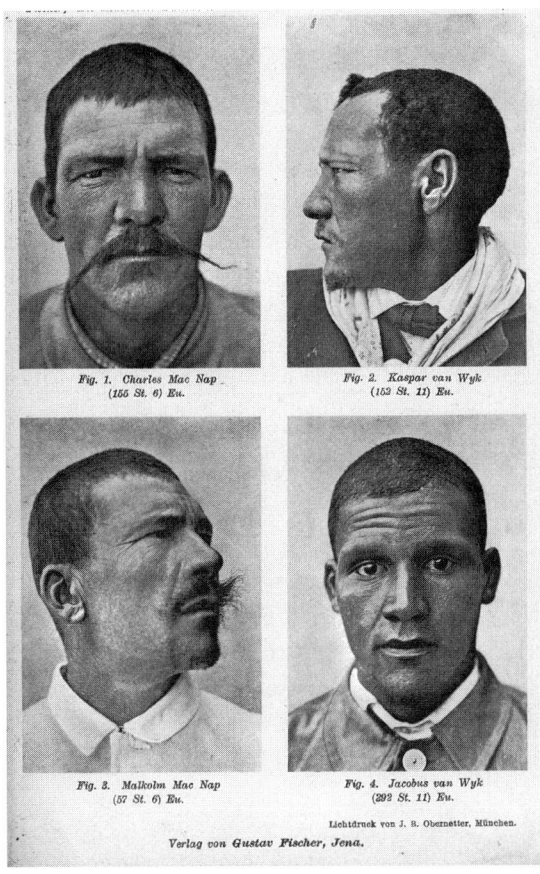

Fig. 1.   Charles Mac Nap
(155 St. 6) Eu.

Fig. 2.   Kaspar van Wyk
(153 St. 11) Eu.

Fig. 3.   Malkolm Mac Nap
(57 St. 6) Eu.

Fig. 4.   Jacobus van Wyk
(292 St. 11) Eu.

Lichtdruck von J. B. Obernetter, München.

Verlag von Gustav Fischer, Jena.

FIGURE 1.19. Photography for the observation of biological variety in humans. Individuals are named and their family trees provided separately. The photographs are aimed at determining and exemplifying Mendelian racial traits derived from one of the parents. In the subtitles Fischer determined to which of the two original parents the offspring is closer: *Eu.* (European), *Hott.* (Hottentot), *Mittl.* (middle), or *Unb.* (unidentifiable). This classification system continued into later racial photography. Eugen Fischer, *Die Rehobother Bastards und das Bastardierungs-problem beim Menschen: Anthropologische und ethnographische Studien am Rehobother Bastard-volk in Deutsch-Südwest-Afrika* (1912; Graz: Akademische Druck, 1961), plate 2.

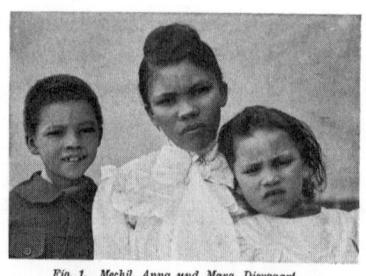
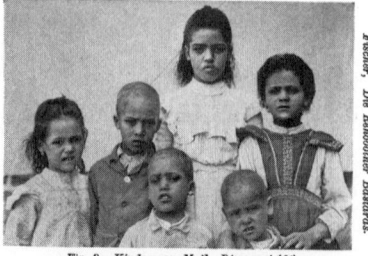
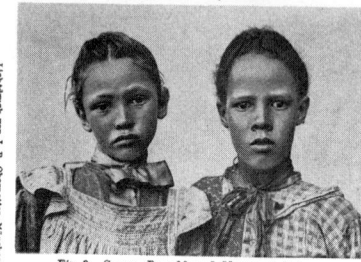
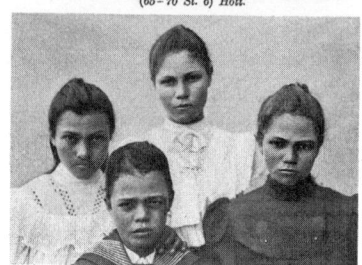

Fig. 1. Mechil, Anna und Marg. Diergaart
(105a. 105, 106 St. 6) Mittl.

Fig. 2. Kinder von Math. Diergaart (64)
(65–70 St. 6) Hott.

Fig. 3. Sanna, Dewald und Margarete Bok
(259 St. 17) Eu.   (257 St. 16) Hott.

Fig 4. Hendrina Rietmann, Otto, Emilie und Frida Johr
(33 St. 6) (47, 32, 46 St. 3) Eu.

FIGURE 1.20 . Plates of family photographs aimed at demonstrating the validity of Mendelian inheritance. The viewer is called to identify racial traits that circulate among the offspring of interracial relationships following Mendelian patterns. Eugen Fischer, *Die Rehobother Bastards und das Bastardierungsproblem beim Menschen: Anthropologische und ethnographische Studien am Rehobother Bastardvolk in Deutsch-Südwest-Afrika* (1912; Graz: Akademische Druck, 1961), plate 12.

Fischer focuses on individual characteristics in the Diergaart family and traces the evolving patterns of inheritance over the following two generations. The photographs of the children can thus only be studied in conjunction with that of their parents. In other words, the photographs are purely relational and are brought into conversation with the Diergaart's family tree history (provided as a separate sheet attached to the book's back cover). In fact, analytically, the photographs are subordinate to the family tree; no valid scientific knowledge about race or inheritance can be derived from independent photographs.

This use of photography epitomizes the Mendelian logic of traits that exist irrespective of a whole type; distinct traits that in persons who are racially mixed mix as "whole packages" but do not blend. Fischer's photographs are taken from the front or a half or full profile and yet are independent of measurements. It is clear that these photographs are free from strict considerations of control in terms of distance, angle, or lighting. Fischer was not aiming at transforming the photographs into statistical tables. In his short explanation of the tables of photographs, Fischer emphasized that the classification of individuals under the photographs introduced as "Eu.," "Hott.," or "Mittl."—which later became standard practice in Günther's work—afforded priority to the true (*wirkliche*) bloodlines over appearance (*Aussehen*). This is a critical moment in the history of racial photography: unlike modernist artists or cultural theoreticians who argued for the socially constructed nature of photography, Stratz, Salaman, and Fischer never actually discuss the epistemological status of photography as a medium. Their undermining of the "truth" of photography is based on their Mendelian reasoning, on the fact that recessive genes remain unseen and cannot be captured by the camera. But their practical social-epistemological effect, guided by very different assumptions or aims than their modernist counterparts, is that of undermining the unequivocal veracity of photographs. Photographs do not necessarily tell the truth; they can "deceive" and cannot be fully trusted.[155]

Following the publication of this study, Fischer increasingly turned to human genetics, becoming the founding director of the Wilhelm Kaiser Institute for Human Genetics in Berlin in 1927. Fischer employed photography on additional occasions.[156] Based on new photographs of the same individuals taken twenty years after the publication of his earlier work, Fischer revisited the mixed population study to observe the effects of time on racial crossing. At this stage, however, it is important to note that the racial laws of 1935 were based on the same premise: true genetic history was seen to hold priority over appearance (the phenotype). Other Weimar and Nazi racial writers, however, were not so ready to renounce the camera, instead developing alternative photographic strategies.

PART THREE: ON ART AND SEEING

I now shift the perspective considerably and attempt to show that ideas about seeing as specific to race—a key point in my account of the history of racial photography–stemmed from a cultural rather than a biological context. Importantly, however, as opposed to authors who took on these ideas in the 1920s and 1930s and advanced them with the use of photographs, the earlier proponents of such ideas were antagonistic to the photographic medium.

It is widely accepted that race was a principal category in the discourse of art history. Before I turn to the more specific idea of seeing as racially, historically, or culturally relative and the relationship between that idea and art, ideas developed during World War I, I exemplify with prominent Austrian historian of art Alois Riegl (1858–1905) how race was viewed as a critical factor throughout the trajectory of art history in a more general sense.

Riegl's famous concept of *Kunstwollen* (a complex and vague concept that is normally translated as "artistic will") is most relevant to our current discussion. I list those aspects of this concept that are essential for the history of racial photography. First, as American historian of art Christopher Wood notes, it could almost always be replaced with the word *style*, a term, as I show in chapters 3 and 4, closely tied to racial photography in the 1920s–1940s. Second, *Kunstwollen* meant viewing the work of art as informed by a structure or design that could be grasped only by an act of extraction that is at the same time an act of re-creation. Such an act necessitates a specific point of view. Third, *Kunstwollen* designates the aesthetic impulse within a culture, an impulse that is inseparable from a culture's "race" and "spirit."[157]

My treatment of Riegl certainly does not do him full historical justice, as I discuss articles he wrote between 1898 and 1900 but that were published only after his death as part of a tribute paid to him and that suggest a moment in his thinking that he did not pursue further. But the ambivalence that these articles express with regard to the role of race is indicative of the presence of certain strains of thought in art history that were later developed toward understandings of "seeing" as a constitutive and relative action, which are crucial for the trajectory of photography as scientific evidence.

Riegl's 1900 article on the Vapheio Cups offers remarkable examples of his understanding of the role of race throughout the history of art.[158] Riegl compares the cups with ancient Near Eastern art and claims that the Greeks, or "more cautiously expressed—their Indo-Germanic predecessors in pre-Homeric times" (rather than the Egyptians), invented artistic composition.[159] He interprets this development within the wider framework according to which the Indo-Germanic peoples participate in nature while the Near Eastern peoples tend to its practical exploitation.[160] Furthermore, Riegl determines that the Indo-Germanic cultures "largely dominated all subsequent human development . . . in politics and culture."[161] Riegl employs "race" in a rather vague fashion compared with later writers, but for us it is essential that in his interpretative framework the development of art is inseparable from that of race.

Within this wider context in which the development of art is viewed as closely related to race, I wish to move to a more specific idea according to which seeing is a complex and composite human activity that is not universal and while inborn is nonetheless relative to culture or race and intimately related to the development of art. In contradistinction to Bertillon, Galton, and Martin, Georg Simmel and Heinrich Wölfflin elaborated conceptions of "seeing" that were tied to art in a fundamental way.[162] In opposition to the anthropometric photographic genealogy or the Mendelian photography discussed above, Simmel and Wölfflin did not address measurement or the question of scientific control, and similarly, they did not focus on the transformation of images to numbers for statistical purposes. Their conceptions were appropriated by racial writers in the 1920s and are essential to the interpretation of the latter's use of photography.[163] To demonstrate discussions of seeing with regard to art, I examine elements of Georg Simmel's *Rembrandt* (1916) and Heinrich Wölfflin's *Principles in Art History* (1915) that address the relationship between art and seeing. In their use of photographs, racial writers of the 1920s and 1930s brought together elements of at least two discourses: they drew on the scientific prestige of racial photography while introducing principles from the history of art discourse. Racial writers appropriated the notion of *Stil* (style) from the latter discourse and connected notions of seeing from this discourse with scientific photography. Thus they created an elaborate theory of the "racial gaze."

Georg Simmel's Rembrandt: Art and the Subjective Activity of "Seeing"

Simmel's *Rembrandt: An Essay in the Philosophy of Art* (1916) was published a year after the appearance of Wölfflin's *Principles in Art History*. Published just two years before Simmel's death, it culminates a lifelong interest in art.[164] Written during World War I, this essay can be seen, in certain respects, as part of the "German Ideology."[165] Simmel's essay, written at the height of his vitalist phase, offers a strong antisensualist and antimechanistic interpretation of "seeing" from which emerges a powerful notion of a "gaze." From this rich book, I address two issues: the relationship between art, photography, and seeing, and the body-soul relationship.[166]

Rembrandt is a polemical essay. Simmel inverts standard relationships involved in seeing, such as between perception and deduction: "much of that which we believe we 'see' directly is in fact not seen at all, but rather, as one says, is 'deduced.'"[167] The essential part of what we see, according to Simmel, is accomplished through the imagination, not perception. This claim could be read as a radicalization of a point Simmel had made earlier in his programmatic essay "How Is Society Possible?" (1908). There Simmel developed a Kantian interpretation of society that stressed the constitutive role of subjectivity, where a visual image conveys the fragmentary structure of knowledge of others: "*just as we compensate for a blind spot in our field of vision so that we are no longer aware of it*, so a fragmentary structure is transformed by another's view into the completeness of individuality."[168] In *Rembrandt*, imagination does not merely complement sense perception; it precedes it. The subject underlies the constitution of the visual field.

In *Rembrandt*, Simmel couples this antisensualist view with the idea that artistic pictures are more "real" than their human referents. In this sense, Simmel inverts the standard relationship between the referent and the representation. The impression of life that emanates from Rembrandt's portraits, Simmel claims, stems from their *material nature*, not from the "referring back to its real existence in the sitter."[169] Simmel underscores this point by contrasting the art picture with the photograph.

Photographs offer only "surface verisimilitude," which is "able to lead beyond itself to the real image of its original," but, Simmel claims, there the sphere of art is "abandoned in favor of reality."[170] In the work of art, in contrast, the "content, stimulus, meaning" must be internal to the work of art itself: photographs are mere "bridges" to reality; artworks lead to a sphere of their own.[171]

Simmel ends his essay with an excursus in which he confronts realism directly: "What do we actually see in a picture that 'represents' something?," he asks.[172] Through the discussion of Rembrandt's paintings of a fur collar and a painted landscape, Simmel offers an antimimetic answer. The gist of his argument is that art does not imitate or "represent" nature or external reality. In what seems at first tautological, he claims that in a drawing we see "that which stands on the paper" because the work of art and "reality" are equally self-sufficient.[173]

"Reality" and "art," however, are not separate spheres, independent of each other: "it is not art that imitates nature, but nature that imitates art. That is to say, *in each epoch people see nature in the way their artists have taught them to*."[174] This constitutive aspect of art for seeing contains a deeply relativistic moment, because if art is not universal, but rather has a specific history, so does seeing (fig. 1.21).

In utterances not found in his sociological writings in *Rembrandt*, Simmel connects this relativistic moment in seeing with race. The contrast of "southern" and "northern" art traditions is standard in the history of art to this day, but Simmel goes beyond the racial moment sometimes implied in this differentiation. He suggests that the sonnet is a specifically Latin verse form that "closes off the path toward infinity of the northern peoples." The northern European, in contrast, seeks "to stretch into eternity."[175] Simmel also compares the Nordic and Latin forms of space, claiming that the Latin demands

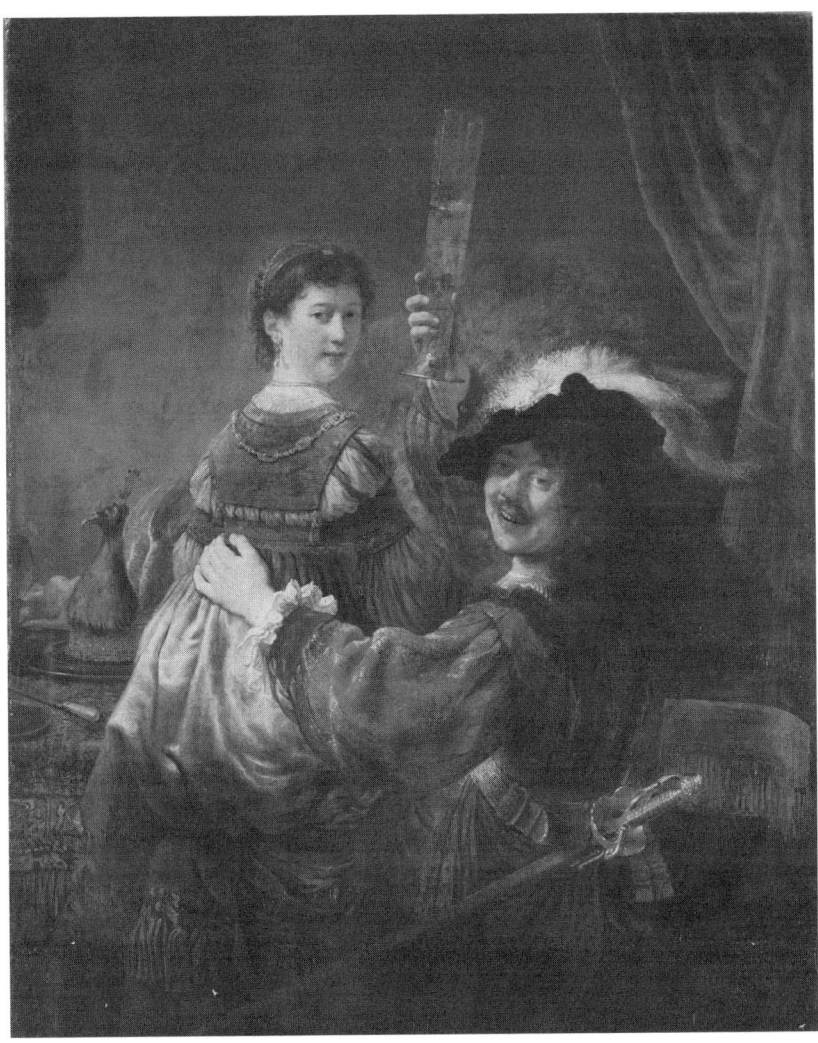

FIGURE 1.21. Rembrandt, *The Prodigal Son in the Tavern, Rembrandt and Saskia*. Unlike the cinematic moving image, the painted figure does not move, "so it can only be that the imagination of the viewer is aroused to complete the movement." Yet the movement depicted by Rembrandt is immanent to the gesture and is not simply the imaginary addition of a "before" or "after." In this sense, according to Simmel, "the work of art offers much more 'truth' than does the photographic snapshot." Humans, according to Simmel, are able to "see" life "in its transcendence" because seeing is a living-process, not a sequence of enclosed moments. Georg Simmel, *Rembrandt: An Essay in the Philosophy of Art*, ed. and trans. Helmut Staubman and Alan Scott (New York: Routledge, 2005), 38.

"the clear perspicuity of space," whereas the northern European "always appears to open itself anew."[176] The orientation of the Gothic soul is toward "transcendence that has left behind its life." Writing in the midst of the war, Simmel qualifies that "Gothic" refers first and foremost to the French.[177] Simmel also refers to "racially inferior Jews."[178]

Anticipating more elaborate notions of the "racial gaze" developed in the 1920s, Simmel portrays race as essential to the constitution of specific forms of gaze. He asserts—in a position elaborated later by Clauß—that the Mediterranean peoples "adjust their behavior to the presence of a viewer."[179] The German, in contrast, acts for "the sake of the expression itself."[180] Rembrandt's figures are essentially German, according to Simmel, as they never exist for the gaze. Simmel deduces "two conceptions of life" from the gaze.[181] These observations are endowed with weight because they come as part of a phenomenology of the gaze, the "aspatial [*raumlos*] gaze" of "immanent transcendence," a by-product of viewing a certain object, a certain "always-looking-farther," where interior and exterior come into contact.[182] Racial undertones are present in these observations, and only a slight shift in vocabulary or intentions, which occurred shortly after Simmel's death, was needed to transpose these observations into a specific theory of race.

The second strain of thought I wish to examine concerns Simmel's interpretation of the body-soul relationship. Similarly to Edmund Husserl and Henri Bergson (but also Ludwig Klages), Simmel attempts to revive an antimechanistic notion of humans. This notion rejects both religious views of the soul as transcendent as well as antihumanistic views that reject the notion of the "soul." Simmel develops an interpretation of the living body (*Körper*)-soul (*Seele*) relationship in a series of essays on the topic of the portrait.[183] In *Rembrandt* Simmel develops the idea that phenomena "appear": the portrait "presents" appearance on the surface, not beyond, beneath, or behind it. Expression is created through the interaction between the elements of the face.[184]

Each of the elements of the face observed is separate. But the smallest movement of one element alters their interaction and as a result transforms the expression as a whole. What "unifies" these elements, in Simmel's view, is the "soul." This is an ambivalent formulation, because "soul" here pertains to both observed and observing subject. "Being carried by a soul" Simmel views as inherent to the observation of a work of art.[185]

In the final essay in this series, "The Problem of the Portrait" (1918), Simmel emphasizes the difference between mechanical seeing, or what he terms "the mere optical," and "the visible" (*das Sichtbare*). The visible is seen as being closely related to the soul and expressed in the living body. The soul-like (*seelisches*) sets off (*loslassen*) the portrait's visibility. Consequently, "the movement of the body is a tool to step into the soul" only in art.[186] The meaning of the portrait is *on* the surface.[187]

Simmel was interested in the development of a theory, but based on similar principles, only a few years later, Günther and Clauß put similar ideas into practice through the careful analyses of photographs rather than paintings. But their appropriation of such ideas on the role of the "soul" in seeing and the development of the relativistic moment therein, at least conceptually, is not necessarily a narrative of radicalization. In terms of scope, at least, Simmel was more radical. Günther and Clauß developed the inherently relativistic racial gaze based on the notion of the "racial soul" that appropriated "nature" and turned it into "landscape." Hence they applied this notion only spatially, while for Simmel, the scope of these considerations applied to time as well.[188]

Heinrich Wölfflin's *Principles in Art History*: "Style" and the Relativity of Seeing

Simmel's *Rembrandt* is a philosophical essay, not a history of art book. In his monographs on Kant and Nietzsche he developed the idea that the typical is found in the idiosyncratic—Simmel was clearly after Rembrandt's unique "genius." Simmel contrasted style with individuality and claimed that with regard to the truly great works of art, we are indifferent to style.[189] The renowned historian of art Heinrich Wölfflin, on the contrary, was interested precisely in the meticulous development of technique over time rather than in the genius of individual artists.[190] Indeed, this is only one of many important differences between Simmel and Wölfflin. They did, nonetheless, share several ideas concerning the relationship between art and seeing. These ideas are relevant to our analysis, as they were appropriated by racial writers and are essential for their use of photography.

Wölfflin famously opens his book with an antirealist and antimimetic anecdote: four painters set to paint the same landscape without deviating from nature, and they end up with four entirely different pictures. Styles of expression, Wölfflin argues throughout the book, have little to do with immediate observations of nature.[191] Indeed, one of the book's central arguments shares Simmel's antimimetic view of art. Wölfflin claims that even in representative art, developments have much more to do with the history of art techniques than with observation of nature.[192]

Wölfflin structures the book on five pairs of independent concepts: from linear to painterly (from stress on the limits of things to a tendency to see them as limitless),[193] from plane to recession, from closed to open form, multiplicity to unity, and clearness and unclearness.[194] The concepts in each respective pair stand, therefore, in a multifaceted temporal and hierarchical relationship.

> The painterly mode is the later, and cannot be conceived without the earlier, but it is not absolutely superior. . . . They [painterly and linear style] are two conceptions of the world, differently oriented in taste and in their interest in the world, and yet each capable of giving a perfect picture of visible things.[195]

The concepts are, according to Wölfflin, analytical as well as historical-descriptive pairs, and the evolution from the one to the other is historically progressive.

Historians have noted that Wölfflin's book cannot be dissociated from the context of the war.[196] According to Martin Warnke, Wölfflin's book was critical with regard to the war rather than affirmative—a fact reflected in Wölfflin's restriction to the boundaries of formalism, in resisting nonartistic factors or political slogans, and in severing "all ties binding forms to historical life."[197] Ideas concerning the significance of race, however, pervade Wölfflin's book.[198] As if to follow Warnke's interpretation, Wölfflin conceived of "race" as apolitical. Wölfflin also never omitted references to race in subsequent editions of the book.[199] Wölfflin's attitude toward race is a contested subject. Rather than concentrating on Wölfflin's opinions, however, I want to focus on the reception of his notions of style both in the past and in the present.

Wölfflin does not employ race as an explanatory category, but race is nonetheless interlaced into his discussion of the book's central themes. Of particular significance is the connection between race, "style," and "national imagination." In the foreword to the seventh edition, for instance, Wölfflin notes that "visual attitudes" vary in different races.[200] Wölfflin views race as important in the classification of styles because "there is no question that the nations diverge from the very outset."[201] "There are peculiarities of national imagination which remain constant throughout all change. Italy always possessed a stronger instinct for the plane than the Germanic north, where it runs in the very blood to plough up the depths."[202] The difference between Germanic and Latin imagination is constant and innate.[203]

The history of artistic representation and the history of visual perception are not the same. Wölfflin, however, ties the peculiarities of the imagination to the relativistic aspect of seeing. "Vision itself has its history," states Wölfflin, "and the revelation of these visual strata must be regarded as the primary task of art history" (*Principles of Art History*, 11). The historicity and relativity of vision are central to the trajectory of art, according to Wölfflin. Similarly to Simmel, he claims that "people not only see differently, they see *different things*" (158; italics in original). Furthermore, he claims, with the progression from the classic to the painterly, "the idea of reality has changed as much as the idea of beauty" (229–30).

Differences in vision between peoples transcend art. They condition "the whole world picture of a people" (257). Wölfflin believes that "peoples and generations diverge" (233). While this might be the net effect of his writing, his expressed aim is not to undermine the humanistic notion of "Man": "However different national characters may be, the general human element which binds is stronger than all that separates" (257). Wölfflin's racial outlook resembles Simmel's, and for both, race is not at the center of his discussion but seems to be taken for granted.

Throughout the book, Wölfflin, like Simmel, expresses aversion to photography and an even stronger aversion to photographic reproductions of artworks in books. Photographs, he claims, fail to represent artworks in terms of perception and standpoint; they "miss the essential character" (56) of the work of art they reproduce (41). The situation is worsened in the case of book reproductions, which he terms "little zinc plates of our books," which are "reproductions of reproductions" (41). Art history concepts and ideas seeped into the work of racial writers in the Weimar and Nazi period: the role of the imagination in seeing, the notion of "style," as well as the use of visual materials for illustrative and demonstrative purposes. Writers, however, did not appropriate Wölfflin's or Simmel's aversion to photography.

Historians have noted the presence of race is Wölfflin's book. But it is questionable whether Günther, Clauß, or other racial writers drew on his ideas on race. They did, as we will see, appropriate his notion of "style." There is a fundamental tension in Wölfflin's narrative between style and progress, however, that is similar to that found in the racial discourse.[204] Wölfflin claims that styles are "ultimate forms"; hence, it is impossible to speak of "progress from the poorer to the richer form."[205] But he also portrays progress in the passage from one style to another: "the development from the linear to the painterly, comprehending all the rest," he states in one place, "means *the progress* from a tactile apprehension of things in space to a type of contemplation which has learned to surrender itself to the mere visual impression."[206] Translated into the racial discourse, the tension is between viewing races as ultimate forms or as organized on a single evolutionary scale. In the absence of a metaracial vantage point, according to the former approach, races are forms that are incomparable by definition. According to the latter, races can be organized hierarchically according to their relative value.

*

Three threads converge in Weimar and Nazi racial writers' employment of photography. From the works of Bertillon, Galton, and in particular Martin, Weimar and Nazi period writers appropriated not only the basic assumption that the camera was an objective device but also the basic elements of its underlying visual codes.

Stratz, Salaman, and Fischer were not interested in the camera as a measuring device but instead developed photography as a tool for racial observation, particularly with regard to racial mixture or the hidden or invisible characteristics of race. This development was closely tied not only to a colonialist context but to the reconceptualization of race along Mendelian lines that occurred in the first two decades of the twentieth century. While all three viewed photography as a realist medium, driven by the Mendelian differentiation between genotype and phenotype, their method questioned the veracity of photographic representation because it captured only surface phenomena.

Comparison of Simmel's *Rembrandt* and Wölfflin's *Principles in Art History* shows that through their respective discussions of art, they both developed antimechanical notions of seeing as subjective, historical, and inherently relative. In both cases, there is a clear presence of racial overtones.

Racial Photographs from Icons to Schemes: The "Case" of Central and Eastern European Jews, 1880–1927

At the heart of this chapter lies an intellectual experiment: what happens to our understanding of racial photography if, rather than looking at methods for the use of photographic techniques in the study of "race" (the subject of the previous chapter) we take as our starting point one "case" and examine closely and pragmatically the way photographs were used in actuality? As this study is particularly focused on the German and the Jewish contexts, I was led fairly quickly to choose as the case in question that of the idea of the Jews as a mixed race people, an idea developed in a string of interrelated publications between the 1880s and the 1920s and in support of which photography was employed massively. As we shift our focus from the methodological definitions of photography as scientific evidence to the actual use of photography in scientific and scholarly study and writing on race (in this chapter and the chapters that follow), it becomes even more important to pay close attention to the respective places of these various uses of photography in economies or processes of demonstration.

By *economy of demonstration* I mean the complex process by which scientists or scholars argue for a certain position and provide evidence to support a theory, a hypothesis, an idea, or an argument. Economies of demonstration involve assumptions, questions, methods, techniques, forms of argumentation, materials, and media that their authors believe can serve as reference, illustration, exemplification, or proof. Looking at the relative places of the medium of photography or of particular photographs in respective economies of demonstration, we will see that the same photograph, which could be described in one context as exemplifying a "racial type," is shown, analytically, to function in respective economies of demonstration in all of the following ways: as support for an argument made independently and earlier (which means that the photograph cannot be viewed as essential to the demonstration process), as an illustration of that argument, or as a reference to the demonstration of the argument. Photographs can also be shown to be essential to the demonstration of the argument itself. Paying particular attention to the position of photographs within economies of demonstration not only reveals the different statuses and functions of photography and photographs within scientific economies as well as the dynamic shifts within economies but also, ultimately, serves to bring to the surface the dramatic differences between different scientific economies even when they sometimes make use of the same photographs.

Studying the course of photographs in the scientific economy of Felix von Luschan, who first developed this idea in the 1890s, through Maurice Fishberg in the 1910s and Sigmund Feist in the 1920s, both of whom greatly expanded the use of photographs for its demonstration, in this chapter I destabilize the conventional interpretations of race as primarily pseudoscientific and of the uses of photography as inherently propagandistic.[1] More specifically, by following changes in the use made of photographs intended to carry the same idea, we come to see how the standard unit of racial photography changed in this period: from single photographs functioning as "icons" to serialized photographs functioning as "matrices."

In the conclusion to this chapter I offer some reflections concerning the wider significance of the emergence of such racial photographic matrices and the ways it brought about, in practice, the redefinition of *race* and *type*. The main body of the chapter, however, consists of a chronologically ordered analysis of the interconnections and dependencies between versions of science, racial beliefs, and photographs for advancing the idea of the Jews as a mixed race people in the discourse of our three authors. Following the use of the photographic medium leads to the mutual exchange between science and art through interaction between Felix von Luschan and the Orthodox and Zionist artist Hermann Struck during World War I. The narrative is interrupted by an excursus in which, drawing on my experiences in the archive of the Natural History Museum of Vienna, I attempt to elucidate a crucial aspect of the role of the imagination in the history of race and photography.

THE PHOTOGRAPH AS PERFORMATIVE INDEX:
FELIX VON LUSCHAN AND THE HITTITE
ORIGIN OF THE JEWS

Austrian born anthropologist, archeologist, and ethnographer Felix von
Luschan reported that he took the photograph in figure 2.1 in Turkey in 1883.
He identified the individual in the photograph as a Christian Armenian Turk
manifesting the "Armenoid type."[2] Then, in 1892, von Luschan published a
two-piece article in which he sought to demonstrate that Jews were not Semites but a racial mixture, the predominant part of which was Hittite in its origin
or, in physical-anthropological terms, of the Armenoid type. In this article,
though, he made no use of photographs. But in 1905 von Luschan was invited
to write the first article for the first issue of the first volume of the *Zeitschrift
für Demographie und Statistik der Juden* (Journal for the Demography and
Statistics of the Jews). In this article he reiterated in a shorter form his 1882
argument but now introduced the photograph in an appendix as an illustration of the type. The photograph then circulated widely in von Luschan's own
publications as well as subsequently in books by Maurice Fishberg, Sigmund
Feist, and Hans F. K. Günther. In doing so it came to be seen as the epitome of
the Jewish type. How could a photograph of a Christian Armenian from Turkey become the epitome of the Jewish type in Vienna and Berlin? Why was it
only used in the later article, and what was its scientific status?

Von Luschan, whose formal education was medical, moved from Vienna
to Berlin to take up a position at the Museum of *Völkerkunde* and was later ap-
pointed professor at the University of Berlin.[3] The idea of the Jews as racially
mixed, which was introduced in the Austro-Hungarian context in order to
undermine antisemitic definitions of Jews as Semites, became their dominant
scientific characterization between the 1890s and the 1930s.[4] If, as has been
argued, a medium asserts itself when it alters forms of visible perception,[5] von
Luschan's photograph of the Armenian Turk altered the perception of Jews.
But we will only be able to see how it did so once the relevant culturally and
socially patterned codes of resemblance, the assumptions about the veracity
of photographic documents and beliefs about race, and the form of statistical censuses in Austria-Hungary and in Germany are brought to the surface.
These steps are necessary for framing the analysis of this photograph in von
Luschan's own scientific economy.[6] To date, however, little scholarly attention has been paid to von Luschan's photography, and what little there is has
focused on the colonial context, which involves assumptions, considerations,

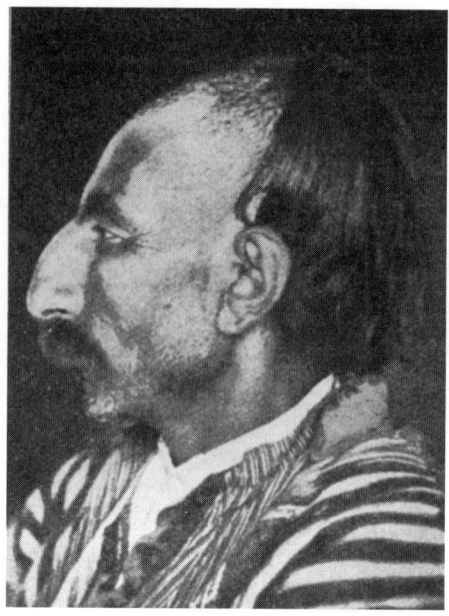

FIGURE 2.1. Photograph of a Christian Armenian Turk from the Ottoman Empire illustrating Jewish type. Von Luschan appended the photograph to his second article on the anthropology of Jews. This photograph became the epitome of the Jewish type. Felix von Luschan, "Zur physischen Anthropologie der Juden," *Zeitschrift für Demographie und Statistik der Juden* 1, no. 1 (1905), 21.

and sensitivities that are interrelated but different from those pertaining to Jews.

In one of the only treatments of von Luschan's photographic practices, Andrew Zimmermann has shown how in areas of German colonial influence, von Luschan's photographs participated in the imposition of the distinction between *Kultur* and *Naturvölker*. For example, Zimmermann describes an African man, one Bismarck Bell, who refused to be photographed in "native" dress and insisted rather on a collared white shirt and black suit.[7] Bell thereby infuriated von Luschan, who in private described him and similar individuals as *Hosennigger* ("trouser-niggers").[8] Von Luschan's fury was directed at the African colonial subject who undermined the "natural" (racial) boundaries between colonists and colonized, "whites" and "blacks."

When it came to Jews, however, von Luschan was evidently guided by different principles. According to the conventions of nineteenth-century European culture, photographing modern Europeans for purposes of physical anthropology was unacceptable. By excluding European Jews from being photographed, von Luschan was thus including them within the category of white Europeans, part of modern *Kulturvölker*. The photograph of the Armenian Turk, then, represents Jews, as it were, once removed, mediated through the notion of "type." This solution, from our perspective, involves a considerable amount of ambivalence because in the very act of photographic mediation, von Luschan both insisted on treating Jews as part of modern *Kulturvölker* while simultaneously linking them to a non-European population. This compound can neither be entirely disconnected from the colonial framework, on the one hand, nor easily collapsed into it, on the other. Nor is this the only tension involved.[9]

Three distinct tensions undercut von Luschan's work on the Jews. First, there is the tension in his model between races, languages, and peoples. While the relationship between the three was anything but simple, in the current context it is important that modern peoples were viewed as heterogeneous and races were by definition homogenous, while neither was identical with language groups. Von Luschan's intervention in discussions of the Jews amounted to the claim that the Jews were a people, and as such were made up of several racial components. The second tension flows from the application of two logics to processes of mixing, one anatomical and another cultural.[10] After the turn of the twentieth century, anatomical students increasingly followed Mendelian logic and adopted a cultural model based on the assumption that the stronger language, grammar, religion, mythology, and script imposed itself on the weaker one.[11] Finally, there is a tension between what von Luschan took as a descriptive (as opposed to deterministic) notion of race and his occasional racist remarks ("we are all more or less disposed to dislike and despise a mixture of Europeans with the greater part of foreign races").[12] These tensions underlie von Luschan's discussion of Jews and the introduction of the photograph and serve as a guiding thread in the following analysis.

Von Luschan on Jews

Given that von Luschan only introduced the photograph in figure 2.1 to a later article on the Jews, the most important question in the current context pertains to its respective place in his economy of scientific demonstration. Von Luschan's first article on the Jews appeared in 1892 and placed them squarely within the category of *Kulturvölker* and thus marking them as a distinct people. In defining the question of the identity of the Jews, von Luschan emphasized the importance of empirical, anatomical evidence,[13] but he also pointed to various different grounds for rejecting the classification of the Jews as Semites. He began by defining the subject in terms of religion: Jews are defined by their confession of faith (*mosaischer Konfession*).[14] Von Luschan then moved to a more general discussion concerning the relationship between language and race, criticizing the assumption that all the peoples speaking Indo-Germanic languages are of the same race. The purpose of this detour was to substantiate the Semitic as a linguistic (cultural) rather than biological category. Peoples who speak Semitic languages may and as a matter of fact do differ in their racial composition.[15] In fact, only the Bedouins are racially pure Semites. Von Luschan contrasted the long but sharp and narrow noses of the Bedouin with the "lay conception of the Jewish nose" as big, heavy, and flabby at its base. Based on the (anti-Darwinian) assumption of the fixity of types and on the basis of evidence he drew from photographs of monuments, he determined that most contemporary Jews were not Semitic.[16]

In this 1892 article, then, von Luschan defined Jews according to their religion but discussed them in terms of the Semitic language group. While he argued that race and language should not be confused, at the same time he assumed (rather than showed) a connection between contemporary European Jews and the Semitic language group. And he ended up discussing their anatomical features. By way of negation, therefore, von Luschan in fact connected religion, language, and anatomical features.[17] He ended the first part of this first article with the dramatic question, If they are not Semites, what is the origin of their short skull, their arched noses, and their blond hair?

In the second part of the article von Luschan sought an answer to this question by way of major archeological discoveries in Sendschirli in Anatolia. In earlier publications, von Luschan had already photographically illustrated ancient Hittite reliefs and reproduced them without considering how such stylized ancient reliefs could be used for the discussion of modern peoples. The source of the brachycephalic skulls of modern day Jews he now identified as the ancient Hittites, who had been largely absent in European conceptions of history before the end of the nineteenth century.

> A single look at the old reliefs from Sendschirli is enough to convince us that the people represented there belong to our armenoid race [*unserer armenoiden Rasse*]. Here we have the most beautiful anatomical evidence of the semiticization of a pre-Semitic people.[18]

The short skulls of contemporary Jews thus originated with the Hittites. Jews were comprised of a mixture of Aryan Amorites, Semites, and—predominantly—the Armenoid race.[19]

In the concluding remarks of his 1892 article, von Luschan emphasized the achievements of the Semites who founded great civilizations, praised the Hittite element in modern Jewry, and concluded by linking the scientific discussion with present-day concerns.

> The present day educated European recognizes in his Jewish fellow [*Mitbürger*] not only the living evidence [*Zeugen*] and inheritance of an ancient and admirable culture, but he observes and values him as his most loyal compatriot and colleague in the struggle for the highest good of this world, in the struggle for progress and for spiritual freedom.[20]

We can only speculate why in 1892 he did not use a photograph that according to his own account he had taken almost ten years earlier. We are able to deduce, however, that it was not necessary for his scientific demonstration. The second article (1905) reiterated the argument that in the middle of the second millennium BC a wave of Semites had invaded the Near East. This wave had been strong enough to impose the Semitic language, religion, and script but had not wholly transformed the racial components of the population. This explained, according to von Luschan, the presence of both long and short skulls in this area.[21]

The nature of the photograph and its relation to that of the type can be further clarified through von Luschan's statement on statistics of averages. Von Luschan mocked attempts to produce statistical averages of the various kinds of skulls found in the area as an attempt to compare the average temperature of two places, the one with a high summer heat and freezing cold winter, the other an Oceanian island with constant mellow spring temperature, day and night, winter and summer: the average weather may be the same, the actual weather quite different.[22] Unlike the average of their cranial index or a number of photographs of individual types, statistical averages were useless in capturing racial types.

The insertion of the photograph, then, introduced no new argument, but, by way of analogy, it did deepen the suggestion of the Jews' racial inferiority. This can be brought to light further if we turn to von Luschan's analogous discussion of the Hamitic theory, in which he made more extensive use of photographs.[23]

According to the Hamitic theory, a light-skinned population from North Africa had at some point penetrated "Black Africa" and, while their physical features had gradually disappeared, the superior language and religion of the newcomers had prevailed.[24] In the Hittite case a similar penetration of a superior language and religion occurred as with the Semites, according to von

Luschan. Structurally, then, the Hittites (and their alleged descendants, the Jews) are equivalent to the racially inferior black Africans in this equation. The immediate nature of photograph probably conveyed this more directly than a statement concerning cranial indexes.

Apart from the inclusion of the photograph into the later article, it is also important that the implicit meanings of the articles in Austria-Hungary (where the first was published) and Germany (where the second was) differed. In the multinational and multiethnic Austro-Hungarian Empire with its multiple minorities, the carefully crafted censuses classified citizens according to their religion and language but not nationality or race, and race remained secondary. Germany differed, however, not only because of the lack of a similar multiethnic and lingual reality and ethos but because of a stronger tendency to reduce religion or language to race as a more fundamental or deeper indicator of identity. The addition of the photograph to the 1905 German article only deepened the reduction of Jews' identity to race.

Moving from the general context of the photograph to that of its relative place in von Luschan's scientific economy, we face the question, what do others think with regard to his allusions to Jews? While photographs are often analyzed in terms of representation of the real, in this case a stronger framework of analysis has been developed in a different context by Margaret Olin.[25] Olin argues that photographs have the power to form referential relationships, what she terms *performative index*. Applied to the photograph of the Christian Armenian Turk, it could be argued that the photograph *formed* a referential relationship between the central and eastern European Jewish type and the Armenian one. In fact, von Luschan introduced the photograph on the basis of what he believed were cranial *indexes* of members of the type. His use of the photograph was thus closely related to the emerging logic of "the index," an idea that was developed in anatomy and anthropology, in linguistics (and photography, e.g., Peirce), and in economics.[26] In effect the photograph formed a relationship between two populations that had in reality nothing in particular to do with each other.

If this photograph only served as an illustration of an argument, an argument the demonstration of which depended on cranial indexes, elsewhere von Luschan took his lead from the photograph's ability to generate visual similarities. In a long methodological article from 1918, von Luschan warned against the danger of mistaking physical similarity for genetic relationship (acknowledging that the same individual could sometimes be classified differently).[27] In this same article, however, and by way of his discussion of another photograph, that of an individual named Abraham Platje, von Luschan in fact did exactly what he warned against. He found Platje's resemblance to the British-Jewish politician and onetime prime minister Benjamin Disraeli so striking that he inferred a Hamitic presence in contemporary Jews.[28] Indeed, von Luschan continued to see Jewish presence outside Europe in various parts of the globe. Based on reports of the existence of the Jewish gaze in New Guinea, he determined the existence of Armenoid influence there, possibly through the expansion of Islam.[29] If Stratz believed the "Jewish gaze" to be the outcome of certain social conditions, von Luschan interpreted it biologically.

In what seems rather similar to the photographic practice of the Dammann Brother's 1876 atlas (chap. 1), von Luschan believed one photograph captured the essential attributes of the type. With its dissemination the photograph of the Christian Armenian Turk came to be recognized as "the Jewish type," and the photograph gradually picked up iconic features: one photograph easily recognizable outside of its original context of production pointing to what is taken as transparent and fixed meaning.

Von Luschan employed this photograph as an illustration of a type he understood to be anatomical. Based on certain cultural-political assumptions, he refrained from taking and reproducing scientific photographs of European Jews. In truth, however, his study of Jews was mediated by photographs in two ways and in two directions. Based on his belief in anatomical evidence, a photograph of a distant population served as evidence for the Jewish type. Based on assumptions about European Jews' appearance, on the basis of photographic resemblance, distant populations were judged to manifest elements of Jewish biology. If his use of the particular photograph in question relied on existing and stable uses of photographic illustration, in reality von Luschan's practices involved assumptions specific to photography, which set racial photography on a course that was specific to the medium.

The Von Luschan–Struck Cooperation

Von Luschan's last publication featuring the Turkish Armenian photograph was a book entitled *Kriegsgefangene* (Prisoners of war), which he published with Herman Struck (1876–1944) in 1917. Their cooperation, then, brings us more closely not only to the question of the politicization of racial science during World War I but also, through the mutual exchange between photography and drawing, to the question of the specificity of the photographic medium in the emergence of the idea of the Jews as a mixed race people as well as, more generally, in the history of photography as scientific evidence in the study of race.

The photograph was reproduced as part of a two-page spread of photographs on pages 98–99, with Jews not mentioned in the text. Given its history, however, the photograph alluded implicitly to Jews. The subtitle noted that the nose is the same as found on representations of Hittites on Egyptian relics. On the facing page von Luschan placed a photograph he had taken of an individual identified as a "Kurd." The angle, distance, and lighting of the two photographs accentuate the similarity in the shape of the skull, the forehead, the nose, the cheeks, and the chin. This layout further corroborated the claim of the non-European origin of Jews.

Jews are discussed directly elsewhere in the book through Struck's drawings of Jewish prisoners of war. Here, as Margaret Olin has shown through a comparison with a parallel wartime project, Struck and von Luschan were delicately and yet deliberately subverting certain prevalent stereotypes of Jewish difference.[30] Struck's Jewish prisoners were drawn in their military uniform. An individual named Isaac Chotoran, for instance, was portrayed with a straight, almost Nordic nose and masculine military gaze. Jews were drawn straight on or from slightly above so that the viewer was forced to meet their gaze.[31] These were no stereotypes of poor, helpless, "ghetto Jews." The structure through which Jews were alluded to, directly and indirectly, involving photographs and drawings, visual and linguistic components, could not, in any case, be reduced to the subject of prisoners of war.

Indeed, a reader of the book is somewhat surprised by the incongruity between the title of the book, the text, and the images. Von Luschan's text is an exposition of fundamental anthropological questions; many photographs are not of prisoners of war but rather are drawn from existing repertoire, depicting populations quite uninvolved in the war. Indeed, von Luschan's statement that a visit to the POW camps is even better than a tour around the world implies that he viewed the war situation pragmatically and that the book in fact had nothing inherently to do with the war.[32] The initiative to create the book, however, certainly did.

The idea to "create a collection of the different types of prisoners of war in our hands" was Struck's.[33] An orthodox Jew and a prominent artist, a committed Zionist as well as a German patriot,[34] Struck contacted von Luschan for his assistance with regard to the anthropological aspects of the project.[35] Struck's tension-riddled identity, his military position, and his artistic work during the war are thus crucial elements of the collaboration of the two men.

With the outbreak of war in 1914, Struck volunteered for military service.[36] Being 38 at the time, his request was first declined, and he was only conscripted in July 1915. In March of that year he was assigned to serve in the *Oberbefehlshaber ost* (eastern high commission HQ), where he worked as a censor and translator. At his own request, from November 1916 to July 1917 he served on the Russian front, where he earned the Iron Cross Second Degree. From July 1917 until his discharge in January 1919, Struck was the military supervisor on Jewish affairs on the Eastern front. It was here that he encountered eastern European Jewry in its environment for the first time.[37] In this period Struck continued to draw, focusing mainly on the local Jews he encountered. He published his works in Jewish newspapers and in a widely circulated book that appeared in 1920 as *The Face of East European Jewry*, accompanied with a text by Arnold Zweig. In terms of artistic technique, style, subject matter, and the general context, there are important parallels between this publication and von Luschan's *Kriegsgefangene*.

Struck extensively drew both eastern European Jews and prisoners of war. His drawings of eastern European Jews were marked by deep ambivalence. This derived, at least in part, from the Buberian perspective of a German Jew observing eastern European Jews as the living embodiment of authentic Jewishness.[38] But this was deepened by the war situation. Struck served in an area in which the relationship between Jews, Lithuanians, and Poles was directly affected by the policies of the occupying German authorities that he served. The German authorities supported the Lithuanians to a certain extent against the Poles and wished to keep the Jews as neutral as possible, while their actual actions caused tremendous suffering to a large Jewish population.

Struck drew the Jews he encountered as *Ostjuden*. Attempting to capture and document their inner essence, he naturalized their identity in visual terms. He was maybe not driven by a notion of race, but he was certainly not opposed to it.[39] While deeply sympathizing with these Jews, he represented them as an obsolete anachronism unfit for modern society (not unlike some of the types of August Sander in his *Face of Our Time*[40]). The context of Struck's cooperation with von Luschan, then, was not only two million POWs detained in Germany but also a war situation that sharpened and corroborated the visibility of the Jewish minority in every possible sense.[41]

Struck also toured POW camps for almost a year, and the joint book, which appeared with the permission of the ministry of war in Berlin in 1916 (apart from the deluxe edition, the war situation was reflected in the poor paper quality), was aimed at the general public as well as anthropologists. It was composed of "type" photographs, etches made by Struck, and a lengthy introduction by von Luschan.[42]

In his textual exposition von Luschan defined Struck's drawings as art-works that possess scientific value and did not treat them as fundamentally different from photographs. Von Luschan also referred to Struck's drawings directly on several occasions in the body of the text.[43] On some occasions facing pages feature frontal and profile photographs; on most occasions the name of the photographer is provided; on numerous occasions the photographed persons are named.[44] On the bottom part of the page von Luschan denotes the geographical origin of the photographed subject and describes somatic features. The specific location of photographs or drawings is often arbitrary with regard to the textual exposition.

Von Luschan did not introduce photographs of Jews even into this book (only drawings). The photograph of the Armenian Turk was reproduced, without referring to anything Jewish. Could, by the time of this study, the Armenian Turkish Christian have been retransformed into a non-Jew, or would readers of the *Kriegsgefangene* book now be conversant with the earlier claim about the relation between the Jewish and the Armenian type? This structure, then, closely intertwined cultural, political, and scientific considerations as to whom to study and in what way interspersed with questions concerning the medium of representation. Attempting to unpack this structure, I end this part of the chapter addressing more directly, first, the politicization of racial science during World War I, and second, the specificity of the photographic medium.

In *Anthropology at War* (2010), Andrew Evans not only shows how German POW studies racialized political enemies but also points to concrete political considerations that the selection of objects for photography followed.[45] Evans analyzes photography as a technology that "captured" the physical character-istics of the subject alongside measurement and other quantifiable data. Ev-ans also emphasizes the ways that mug shots of foreign and colonial soldiers, particularly dark-skinned ones, represented POWs as menacing and in need of control and confinement, allowing Germans and their allies to see the face of the foe and identify him as someone physically different from themselves.[46] This definition of what qualified as "German," implicitly defined by race, de-fined also who was excluded from that class. But Evans's analysis assumes an internal contradiction between politicization and scientific objectivity, and my contextualization shows that in certain respects their project stood at the forefront of a certain version of scientific objectivity.

On the basis of evidence that von Luschan gave Struck instructions in the depiction of certain individuals, Evans concluded that von Luschan and Struck's "type" was an artificial, subjective construct.[47] Evans also criticizes the selection of individuals as nonempirical, because by "their choices, the anthropologists in effect created the categories of 'typical' for each group."[48] But Struck and von Luschan believed that following their changes, the type under observation was better encapsulated and portrayed.[49] Evans is here presuming a specific form of objectivity from which they diverted. But von Luschan and Struck display elements of what Daston and Galison term *trained observation*, a form of objectivity that recognized the subjective but, as such, essential role of the scientist's trained judgment in the identification, classification, and representation of types.[50] Situating their negotiation in a comparative framework, their type can be shown to approach both Weber's "type" as a methodological idealization or contemporary astronomers and botanists who recognized their constitutive role in the establishment of types. The difference between von Luschan and Struck and physicists or astronomers has to do with the nature of the politicization of their object of study, humans. In terms of the form of scientific objectivity, their willingness to concede to the scientist's subjectivity could be taken to reflect wider tendencies in the history of objectivity.

It is also possible to unpack this structure through the specificity of photography as a medium of representation.[51] True, even the differentiation between drawing and photography is not entirely stable as, from one side Struck made some of his drawings from photographs rather than from direct observation and, from the other side, James Henry Breasted abstracted a drawing from the photograph of the Armenian Turk.[52] But differentiating between the two is crucial for pinpointing the specificity of photography as scientific evidence.

In *Production of Presence*, Hans Ulrich Gumbrecht defines the medium in terms of materialities of communication, that is, "all those phenomena and conditions that contribute to the production of meaning, without being meaningful themselves." Gumbrecht raises the possibility that different media do affect the meaning of what they carry.[53]

From the perspective of meaning, von Luschan's photography and Struck's drawing are grouped together as they observed, studied, and represented "types" based on shared assumptions and a visual language of "types," which encompassed both photography and drawing. If, from a methodological perspective, von Luschan's photographs could be contrasted with Martin's anthropometric photography or Fischer's photographic observation, then, nevertheless, their shared medium of expression grouped them together and defined their status as a form of evidence. If a new medium asserts itself when it alters forms of visible perception, with regard to the perception of European Jews, von Luschan's photograph of the Christian Armenian Turk did precisely that.

EXCURSUS: THE PHOTOGRAPH IN THE
ARCHIVE AND IN THE IMAGINATION

Attempting to trace some of von Luschan's photographs in archives reveals an aspect of the relationship between the archive and the imagination in the history of racial photography.[54] In many book publications the source of the photographs is stated under the photograph or in the acknowledgment section of the book, but attempting to locate the originals often proves to be hard. Reconstructing part of that process sheds light on the fluctuating definition of racial photographs and points to some of the ways the archive affected the imagination.

The photographs that were reprinted in books in this period were garnered from various archives, private collections, or other book publications. The references to photographs, however, unlike references to books or articles, normally did not state the exact name of the archival collection or its number. While photographs were referenced or quoted, the question of the "original," quite literally, has aspects specific to it.

Von Luschan began his academic career in Vienna, and given his central role in the Viennese Anthropological Society and the close relationship between this society and the Museum of Natural Science in this city, I initially assumed that the originals of his early expeditions would be found there. I quickly learned, however, that he kept the originals in his private possession. Furthermore, his relationship with the Museum of *Völkerkunde* in Berlin became strained in the latter years of his life, and so rather than donate the photographs to the museum, after his death his widow sold parts of his *Nachlass* to collectors in the United States.[55] The originals are now scattered and can no longer easily be located.

While not the originals themselves, I nevertheless attempted to trace some of von Luschan's reproductions in the archive of the Museum of Natural History in Vienna. The museum possesses thousands of photographs. The small glass reproductions, in small brown bags on which a paper reproduction of the glass is glued on the outside, are organized in small wooden boxes that are numbered and placed in large wooden cupboards. The photographs are carefully listed in four large folio handwritten catalogs. They are listed according to the geographical location in which they were taken as well as by the names of the peoples who were photographed. The photographer is not normally stated, and it is only on the basis of some knowledge of what anthropological or ethnographic expeditions operated where and when that one can deduce which collections are related to which photographer or anthropologist. The list in the first part of the catalog, however, refers one to a more detailed list where the photographs are numbered and the name of the photographer is often given. With the guidance of anthropologist and historian of anthropology Dr. Margit Berner, I was able to identify several collections as belonging to von Luschan.

But the old folio catalog is no longer valid. To find out whether the photographs really do exist in the archive one has to turn to a second, later, smaller-sized, and thinner handwritten catalog that was compiled in the 1970s. From this we learn that several of von Luschan's collections of reproductions have been stricken from the shorter catalog and are no longer in the possession of the museum. While the exact reasons are not given, the short catalog opens with several typed letters dating from the early 1970s that state that parts of the photographic collections, now no longer classified as anthropological photographs, have been handed over to other institutions in Vienna, such as the Museum für Völkerkunde, the Österreichisches Museum für Volkskunde, the National Gallery of Portraits, or have been destroyed.[56] These letters point to an important if silent moment in the history of racial photography as scientific evidence—a moment in which practical decisions redefined scientifically valid photographs, and in so doing contracted their definition.

In the German and the Austrian context of the early 1970s (a time when physical anthropology was still led by students of anthropologists of the Nazi era), these redefinitions also served to "clean" the archive of materials increasingly recognized as problematic (indeed, the general list of materials that were removed from the archive include photographs of prisoners of war from World War II). The scientific redefinition and the cleaning of the archive cannot be separated.

One collection, closely related to von Luschan's reinterpretation of the Jews' racial characteristics and made up of roughly twenty reproductions from his expedition to the Near East, is present. The best way I can describe the experience of attempting to match photographs from this collection with re-productions in his books is that of a memory card game. These photographs of individuals—here dark-skinned bearded men with thick hair, there dark-skinned bearded bald men—create a thick network of similarities and differences that make matching photographs found in the book with those of the collection a much harder task than one would intuitively imagine (fig. 2.2). By selection of individuals that clustered around certain physical features, by employing photography that was free of formal control but tended to standardized use of strong lighting (eliminating much of the photographic detail that characterizes studio portraits), close distance (tending to distort and enlarge

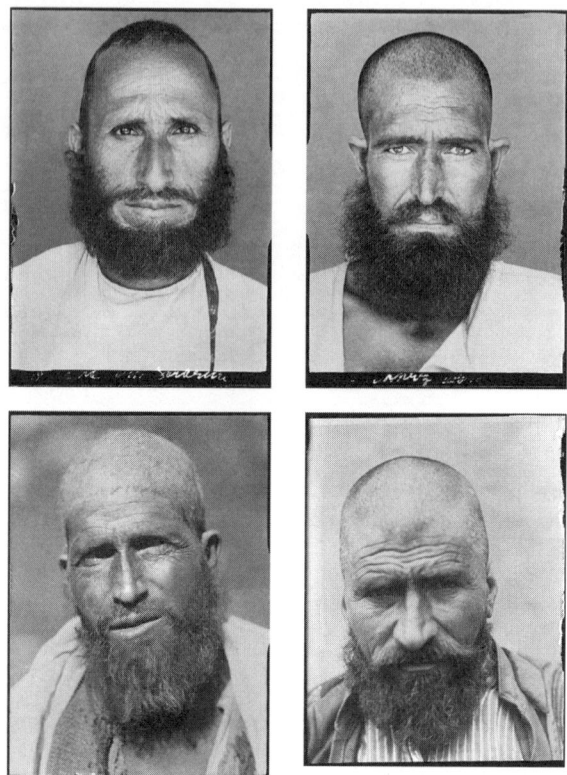

FIGURE 2.2. Felix von Luschan, four type photographs from one collection: 1903 Türke, 1904 Darde (Kaukasus), 1912 (Kurde), 1914 (Kurde). Felix von Luschan, the collection of Die Anthropologische Abteilung, Naturhistorisches Museum Wien.

the nose), and low angle, his photographic patterns strongly corroborated the reality of "types." In chapter 4 I address in more detail the crisscrossing of differences and similarities through the genealogy of Wittgenstein's notion of "family resemblance"; here I wish simply to point to the invisible and imaginary but powerful structure of the archive as standing behind reproductions found in books or exhibitions.

My difficult experiences matching photographs from the book and the archive collection seems to corroborate a thoughtful argument put forward by German historian Andreas Mayer concerning the sociological functioning of the archive in Europe in the first decades of the twentieth century. Mayer argues that when a photograph from an archive was used in an exhibition or a book, the implication for the reader or viewer was that there are, in the archive, many other similar photographs that could replace the one exhibited. The photograph is a "fact" that demonstrates an idea (about race) that could have easily been demonstrated by a different but similar "fact."[57] The knowledge that the archive possesses numerous similar photographs is subconsciously present in the imagination and is part of the observer's experience. Von Luschan's collection of Near Eastern types demonstrates powerfully the assumption of the audience that photographs of types are typical because any single image could be replaced by numerous others.

The work of the imagination was not free, however, but conditioned by political and social assumptions. Why were central European Jews perceived as resembling Egyptian mummies, depictions on Hittite monuments, or Armenian Christians in Turkey, and not, say, Sri Lankans or Native Americans? The reasons relate to the cultural imagination of Europeans, which while unstable and dynamic nevertheless possessed certain historical, geographical, and religious anchors. The process of association between two references was a culturally constitutive process. In 1904 anthropologist, traveler, and anatomist Gustav Fritsch concluded an article on the "Oldest Representations of Egyptian Types of People" (by which he included reproductions of Egyptian monuments, statues, and mummies) by suggesting that a resemblance existed between some of the types on the monuments and contemporary Jews.[58] This link was later reiterated many times, especially with regard to the El Faiyûm portraits, by Günther, Eugen Fischer, and others.[59] Photography played a major role in establishing connections on the basis of alleged similarity; but similarities were only found where, based on cultural and geographical assumptions, they were sought. In the meeting point between wide cultural belief in the deictic nature of the photograph and assumptions about the abundance of photographic facts in the archive, then, the belief in types was solidified.

MAURICE FISHBERG AND THE SERIALIZATION OF PHOTOGRAPHIC EVIDENCE

Maurice Fishberg extended von Luschan's argument concerning the Jews racially mixed character. Unlike von Luschan, however, he introduced, for the first time in this context, photographs of Jews to corroborate this idea. Moreover, by introducing multiple pages of serialized photographs, Fishberg played a significant role in redefining the unit of the racial photograph.

While Fishberg's areas of writing were much narrower than von Luschan, his use of photographs in the Jewish context was massive. Fishberg (1872–1934) was born in Podolsk, Russia, immigrated to the United States in 1889, and in New York received in 1897 a degree in medicine.[60] Before I turn to his principal publication involving photographs, I review two journal articles in order to underline a change of emphasis in Fishberg's writings that occurred between 1902 and 1911.

Fishberg's 1902 article on the cephalic index of the Jews extends von Luschan's arguments discussed above: Fishberg shows that near all the contemporaneous peoples known to be of Semitic origin are dolichocephalic (a relatively long skull) while European Jews are mesocephalic (an intermediately shaped skull) or brachycephalic (a relatively short skull).[61] The particulars of Fishberg's article are less important than several features of his physical-anthropological assumptions and in particular the centrality in his thinking about Galton's study of correlations.[62] Exclaiming that "the problem becomes more complicated the more thoroughly we attempt to analyze it," Fishberg ends his article calling for the study of correlations that may help in the identification of races.[63]

The second article (1903) moves on to study correlations of pigmentation.[64] While the main focus of the article is physical-anthropological measurements, he integrates into his analysis linguistic and literary interpretations of biblical passages and other religious documents.[65] His assumptions about types are important for his later use of photographs.

> Typical representatives of a race show a constant interrelation between the color of their hair and that of their eyes; for example, in the blond northern races their light hair is usually accompanied by blue eyes, while in the brunette races the dark hair is usually accompanied by brown eyes. Individuals who do not exhibit such interrelation, having dark eyes with fair hair, or the reverse, are considered as "mixed types."[66]

Based on the assumption of pure types, evidence of blond hair among Jews leads Fishberg to infer the "infusion of Teutonic blood."[67]

In the 1911 book *The Jews: A Study of Race and Environment*, Fishberg made extensive use of photographs.[68] The 578 page book, which declined to oppose race and environment or to reduce the identity of Jews entirely to their historical and geographical environment, is accompanied by 141 photographs distributed throughout. The German version, *Rassenmerkmale der Juden* (Racial characteristics of Jews), only 272 pages, is structured differently, although its main arguments are the same. Differences in quality, shape, and size of the photographs as well as in the paper used for them (the German version's paper being shinier, bringing out more details), results in the photographs in the

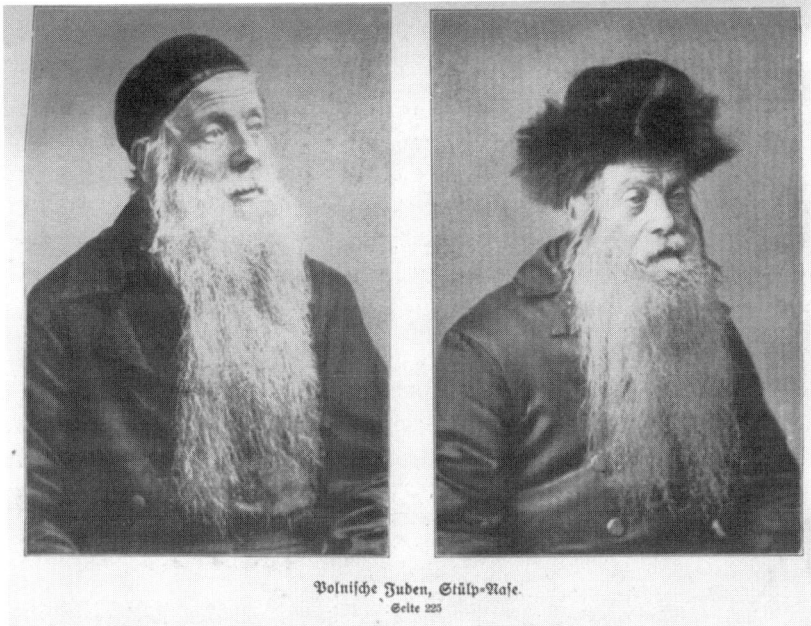

Polnische Juden, Stülp-Nase.
Seite 225

FIGURE 2.3. Fishberg's incorporated photographs as an integral part of his demonstration of Jewish diversity. Maurice Fishberg, *Rassenmerkmale der Juden: Einführung in ihre Anthropologie* (Munich: E. Reinhardt, 1913), 16.

German edition having a far more photographically realistic impression on the reader-viewer.

The main purpose of photographs is to contest the view that Jews are a pure race, propagated, Fishberg states, by antisemites and Zionists. While the idea of comparing skin color in black-and-white photographs might seem absurd to contemporary readers, the 42 pages of photographs in the German edition depicting Jews from virtually all corners of the earth are geared to demonstrate the Jews' racial variation. Underlying captions provide geographical origin or the source of the photograph, the depicted type, and sometimes direct the viewer to observe particulars such as the differences between noses among Jews.

While the photographs comprise several genres, including ethnographic photographs, portraits, private photographs, and racial type photographs, Fishberg uses them all to advance his argument. It is fair, therefore, to assume that he employed photographs he believed served his argument.

Observe the photographs of two elderly orthodox Jewish men with long white beards and dark elegant suits and head coverings (fig. 2.3). While the

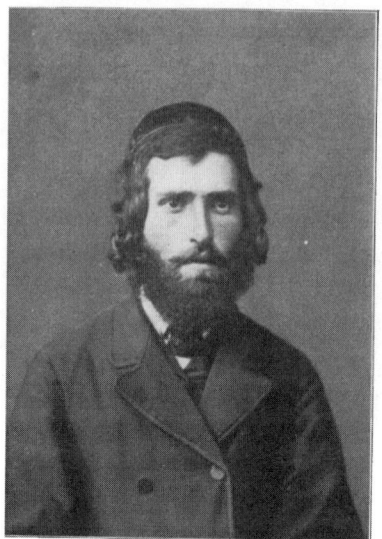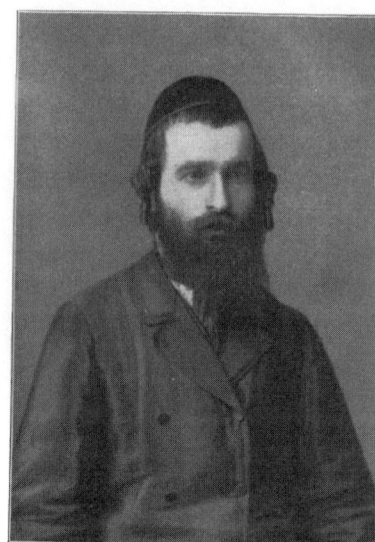

Polnifche Juden, Sephardi-Typus (I).
Seite 190

FIGURE 2.4. Photographic contingency could be endowed with racial meaning. Maurice Fishberg, *Rassenmerkmale der Juden: Einführung in ihre Anthropologie* (Munich: E. Reinhardt, 1913), 16.

source of the photographs is not specified, it can be discerned from their quality, large frame, the even gray background, the form of soft lighting falling on the face, the semiprofile positioning of the body (where one shoulder is closer to the camera), and the direction of the gaze away from the camera to a point outside the frame that these are studio photographs taken by a professional photographer for private ends. Fishberg simply characterizes these two subjects as "Polish Jews." In the language of historian Marc Bloch, these are "unintended sources," photographs produced in one context made to function as scientific evidence in another.

The photographs of two younger, elegantly dressed orthodox Jews, characterized simply as of "Sepharadic Type," exemplify how photographic contingency could be endowed with specifically racial meaning (fig. 2.4). The photograph on the left seems to have been made in a studio, but its photographic quality is weaker than that of the elderly individuals discussed above. The person is photographed in an almost frontal angle, closer to police or anthropological angles, his shoulders at almost equal distance from the camera lens and the eyes fixed in the direction of a slightly elevated point to the right of the camera. The photographer caught the widely open dark eyes gaping, objectless. Günther, with his greater (and more vicious) visual sensitivity than Fishberg, would later interpret this photograph as an epitome of the objectless "Jewish gaze."

Some of the photographs raise the question of whether features that are discussed in the text can actually be observed on the photographs or only projected. A case in point is the two facing monochromic photographs that Fishberg identifies as a "White Jew, Cochin, East India" and a "Black Jew, Cochin, Malabar" (fig. 2.5). Here even the keen observer finds it hard to observe a difference in the color of the skin.

Fishberg gave serious consideration to such racial categories as "pseudo-Jews" and "crypto-Jews."[69] Somewhat like Stratz, discussed above, he agreed that there may be such a phenomenon as a Jewish gaze, but he believed its source was social rather than biological. Indeed, he claimed that assimilating Jews in Scandinavia or England no longer bore the mark of the ghetto and could no longer be identified as Jews.[70]

A rare contemporary review in the *New York Times* (1911) paid special attention to Fishberg's use of photographs to demonstrate the "diversity of type

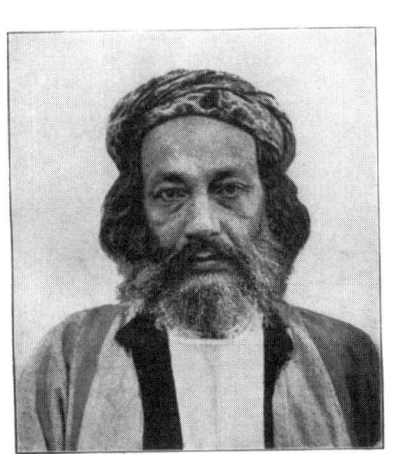
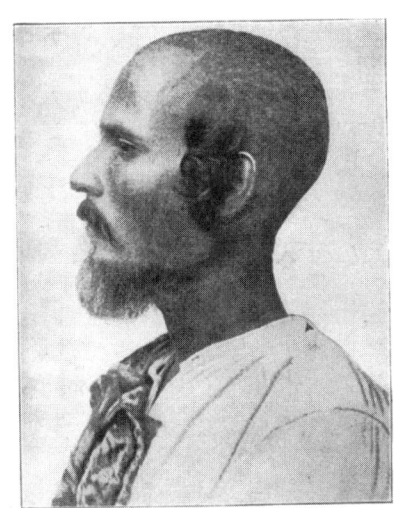

Weißer Jude. Cochin, Malabar.

Schwarzer Jude. Cochin, Malabar.
Nach Originalphotographien von Emil Schmidt.

Seite 206/7

FIGURE 2.5. Fishberg compared what he considered to be a "white Jew from Cochin, East India" and a "Black Jew, Cochin, Malabar." Maurice Fishberg, *Rassenmerkmale der Juden: Einführung in ihre Anthropologie* (Munich: E. Reinhardt, 1913), 160.

height and brachycephalic, conforming to the *race orientale* of Deniker. This would tend to indicate that the blonde-

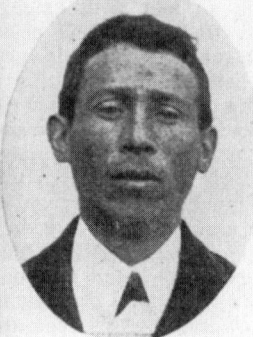

Fig. 64.   Fig. 65.
Figs. 64, 65.—POLISH JEW, MONGOLOID TYPE.
[*Photo lent by Elkind.*]

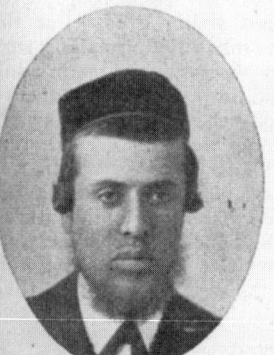
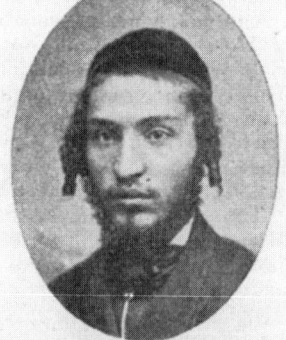

Fig. 66   Fig. 67.
Figs. 66, 67.—GALICIAN JEWS, NEGROID TYPES.

ness was acquired in Europe while living among these races (Figs. 72-73).

FIGURE 2.6. Fishberg describes several individuals he identified as Polish Jews as manifesting "Negroid" or "Mongoloid" types and that occasionally, Jews are mistaken for mulattos. Maurice Fishberg, *The Jews: A Study of Race and Environment* (New York: Walter Scott, 1911), 117. plates 64–67.

among the Israelites."[71] Rather than asking about the status of photography as such, however, the reviewer questioned "the proportion of Jewish faces which have no sign of Jewish physiognomy, before accepting these interesting reproductions as bearing on the problem of the parity of the Jewish race."[72] What was questioned, then, was not the legitimacy of photographs as scientific evidence as such but their selection.

Fishberg's refusal to reduce Jews' identity to society or geography as part of a wider framework that opposed racial purity was reflected in his choice of photographs. Fishberg describes several individuals he identified as Polish Jews as manifesting "Negroid" or "Mongoloid" types (fig. 2.6) and even claimed that occasionally Jews are mistaken for mulattos.[73] In the next chapter we will see how the selectivity he practiced in mounting photographs to demonstrate his argument eased their overturning by Hans F. K. Günther. In a framework that emphasized racial purity, the same photographs would demonstrate the Jews' mongrel racial identity.

SIGMUND FEIST: THE VISUAL AS PHOTOGRAPHIC

In 1925 Sigmund Feist (1865–1943), mainly remembered because of the orphanage for Jewish children that he directed in Berlin and his work in German linguistics, published a widely circulating book titled *Stammeskunde der Juden: Die jüdischen Stämme der Erde in alter und neuer Zeit: Historisch-anthropologische Skizzen* (A history of the Jewish stock: ancient and modern Jewish tribes of the world: historical-anthropological Sketches).[74] This book terminates the anti-antisemitic use of photographs to advance the idea of the Jews as a mixed race and forces to the surface asymmetries built into the nature of the racial photograph as well more general assumptions concerning visual evidence in the study of race.[75]

Feist, who earned his doctorate from the University of Strasbourg for a study of ancient Germanic etymology, moved to Berlin in 1900 to become the headmaster of a Jewish orphanage. He nevertheless continued his linguistic work, gaining the reputation of a serious critic of amateur excess, *völkisch* antiquarianism, and the view that "all that is worthwhile in world civilization is of Germanic origin."[76] Of particular interest here are his respective interventions in the linguistic Germanic and the racial Jewish discourses and the argument over linguistic versus visual types of evidence.

In 1910 Feist argued that the language of the old Germans, the *Germanen*, was a mixed language (*Mischsprache*), and that as a people the old Germans were racially mixed.[77] In 1927, two years after the appearance of his major

publication on the Jews, Feist posited a theory based on the existence of ambiguous and conflicting linguistic evidence that "Celts" and "Germans" from
the Rhine area were members of a third community of Celto-Germans.[78] Almost instantly, members of the German Philological Society rallied to have him
expelled for having supposedly impugned national pride, and in 1928 Feist indeed resigned.[79] After the Nazi party rose to power, Feist became for antisemites a representative of the destructive Jew within the field of Old Germanic
studies.[80] In 1939 Feist finally left Germany. He died in Denmark in 1943, just
before the rest of his family was evacuated to Sweden.

While "race" and "type" are central to Feist's 1925 book on the Jews, in no
place does he define them. Indeed, biological and, most notably, Mendelian
principles are absent from his discussion. The chapters move from discussion
of the Jews as a race in ancient times to a discussion of geographically ordered
Jewries to the discussion of modern Jews as a race, and the book's structure,
therefore, corroborates the argument concerning the heterogeneity of the Jews
as geographically spread and as anthropologically diverse.

In his preface Feist explains that in order to keep the book from becoming
too expensive, he had to limit the amount of images but that he had made every
attempt to authenticate these photographs and detail their exact source. Many
of the photographs he eventually employed were taken by a handful of individuals, such as Hermann Burchardt, Max Burchardt, Samuel Weissenberg,
and Felix Rosen, the author of a travel guide to Ethiopia. Hermann Burchardt
(1857–1909) was a German-Jewish explorer, a pioneer in the field of Jewish
ethnography who, from 1890, traveled to remote places in Australia, Asia, the
Middle East, and North Africa (he was murdered in Yemen in 1909), photographing and documenting Jewish and other traditional ways of life, customs,
and folklore.

Feist's text is followed by thirty-eight pages of photographic tables, each
made up of several black-and-white photographs. The general quality of the
reproduction is uneven, and several of the photographs are almost illegibly
blurred. The photographs include reproductions of ancient Egyptian and Hittite reliefs, a painting by the Dutch painter Dierk Bouts, and a map of skull-
shape distribution across Europe as chartered by William Ripley. The bulk
of the photographs, however, emphasize the difference between Jews living
in distant cultural and geographical environments—from Lithuania, Ethiopia,
Kurdistan, Caucasus—between light-skinned and dark-skinned Jews from India, and the similarities between Jews and their respective environments.

The photographs intersect several scientific genres. A great number belong to the ethnographic genre, that is, individuals or families photographed

in their natural and cultural environments, in their traditional dress, or while performing social activity in their respective social milieu or cultural or religious rites. Other photographs are closer to the "racial type" genre, focusing on the physical features of individuals as representatives of types from frontal or profile angles in which the subjects were asked to pose in profile but the profile angle is not full. Examples of these include the full-page profile of a man identified as a black Jew from Kochin reproduced from Fishberg's German edition and two young men identified as Jews from Damascus reproduced from the *Zeitschrift für Ethnologie*. The photographs demonstrate that Jews can hardly be conceived as a homogenous race.

One of the most interesting aspects of the book is Feist's assumptions with regard to visual materials, pointing to the degree to which indexical understandings of photography defined visual material as scientific evidence. After providing historical evidence for mixture between non-Jews and Jews throughout history, his basic thesis throughout the book, Feist asked whether this process had already in ancient times aligned Jews with the peoples among whom they lived. This question, Feist wrote, is not easy to answer because of the scarcity of visual material (*Bildmaterial*).[81] Feist's first assumption, therefore, was that the question was a visual one.

Franz Boas, to whom Feist turns explicitly in his conclusion, ruled out on methodological grounds the ability to know what previous types looked like. Feist here argues differently: because of the state of empirical evidence, the question centers on the appearance of Jews in the medieval period. As opposed to ancient Hittite and Assyrian monuments, Feist claims, medieval Christian and Muslim visual illustrations do not provide "truthful depictions of Jewish types" (*naturgetreue jüdische Typen*). He here mentions several medieval sources in which, he insists, depicted Jews cannot be identified through their physiognomic features but only through social markers attached to them.[82] While this, precisely, could corroborate his argument concerning Jewish mixture, Feist in fact chooses to rule out the realism of these images and declares that only with early modern painting, specifically with Rembrandt, Rubens, and van Dijk, did representations of Jews *regain* their ancient realism; only here did the realistic character of Jewish faces and Jewish forms (*jüdische Gestalten*) reappear in art.[83]

Feist operates with different assumptions when the nature of the evidence is legal rather than visual. He interprets medieval Muslim and Christian decrees concerning the Jewish hat or armband as evidence that otherwise Jews could not be physiognomically discerned as distinct from their environments.[84] Indeed, following the same line of reasoning he also asserts that the modern

Jewish type stems from the ghettoization of the Jews, which diminished the vitality of their race (*verkümmerte die Rasse*), rather than from the continuation of an unbroken ancient type.[85] These contradicting assumptions with regard to visual and legal records point to the degree to which the veracity of visual materials was now defined by indexical assumptions about photographs.

Written three years after Günther's first major book, Feist's understanding of Günther is of particular interest. Feist opens his conclusion by stating that the study of Jewish racial difference has been deeply politicized, and he then proceeds to identifying directly contrasting interpretations of the subject. His account of Günther's postscript on the Jews is perhaps not inaccurate, but it is rather soft in its tone.[86] Feist at least twice criticizes Günther's ideas of racial intuition and blood consciousness as "nonscientific," "mystical," and "fuzzy."[87] He does not identify, however, the degree of Günther's antisemitism. In a similar vein, Feist situates himself on the side of science when, in contrast to students of the Jewish gaze, he draws a clear differentiation between physical traits, which can be scientifically studied and represented, and the gaze, which cannot.[88]

The opposition between Feist and Günther may also be discerned by way of the interpretation of individual photographs. Feist ends his photographic appendix with portraits of individuals he describes as German-Jewish with Nordic characteristics, including his own family (fig. 2.7). Two rows of young men and women, smartly dressed in suits or military uniform, serve to illustrate the degree to which German Jews are not only culturally but biologically German.

I will return in the last chapter to similar liminal moments in which the boundaries between the scientific and the personal are crossed. Here, though, I'd like to look at some possibly unintended outcomes of Feist's layout. At the top left corner of the page (fig. 2.7), Feist places a photograph of an individual whom he identifies as a German Jew with a Negroid "touch" (*Einschlag*). According to semioticians, in languages written from left to right, when layouts make significant use of the horizontal axis, the elements placed on the left are presented as "given," the elements placed on the right as "new."[89] The given is Jews as a non-European other. Feist's choice of photographs and their layout produces a far more ambivalent picture of German Jewry than his written account.

While I have argued that the meaning of photographs is overdetermined, and therefore open to opposing interpretations and uses, it is difficult to free oneself from the impression that Feist's photographic representation of Ethiopian Jews emphasizes their primitiveness. Photographs show their poor

Tafel XXXVI.

Tafel XXXVIII.

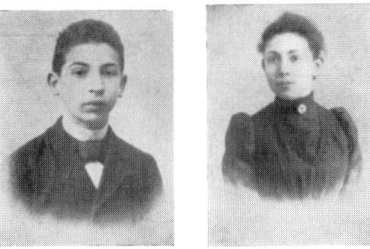
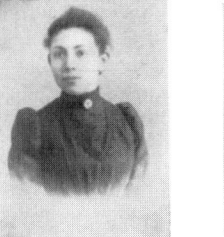
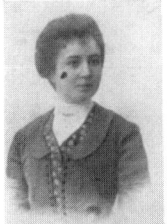
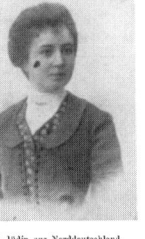

79—80. Junge deutsche Juden mit negroidem Einschlag.
Originalphotographien.

87. Jüdin aus Norddeutschland.   88. Jüdin aus Wien.
Originalphotographien.

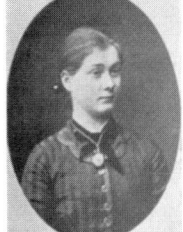
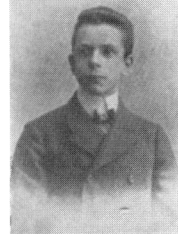
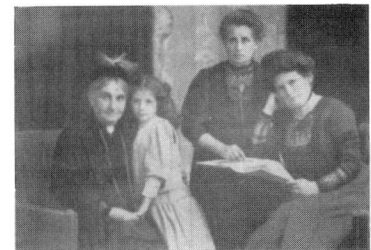

81—82. Junge deutsche Juden mit nordischem Einschlag.
Originalphotographien.

89. Vier Generationen (jeweils die älteste Tochter) einer Familie.
Originalphotographie.

FIGURE 2.7. *a*, Feist ends his photographic appendix with portraits of individuals he describes as German Jewish with Nordic characteristics. Sigmund Feist, *Stammeskunde der Juden: Die jüdischen Stämme der Erde in alter und neuer Zeit: Historisch-anthropologische Skizzen* (Leipzig: J. C. Hinrichs, 1925), images 79–82 in tables 36–38. *b*, Feist includes photographs of his own family members. Sigmund Feist, *Stammeskunde der Juden: Die jüdischen Stämme der Erde in alter und neuer Zeit: Historisch-anthropologische Skizzen* (Leipzig: J. C. Hinrichs, 1925), images 87–89 in tables 36–38.

material conditions, the synagogue being a simple hut with a straw roof, and, in an even starker contrast to German Jews, a group of Ethiopian Jews dressed in rags, the breast of one woman exposed, in agreement with the representation of nonwhite peoples and specifically the iconographic tradition of representation of mulattas.[90] Crowding in the center of the photograph, their lost gaze is fixed unemotionally at the camera, their eyes blinded by the strong sun behind the photographer (fig. 2.8).

Feist followed Boas's line of argument that Jews were deeply assimilated into their respective environments.[91] But he was not willing to rule out that contemporary Jews across the globe still betrayed certain shared recognizable features. He exemplified this view with a photograph of Herzl alongside two Jews from Caucasus (fig. 2.9), which allegedly shows their similarity.[92] *Showing* similarity relied on the use of photographs. With this particular photograph, Feist asked readers to project the similarity of the Jewish type into several distant contexts. Fishberg referred readers to the eye color and skin color of individuals in black-and-white photographs, where any such observation could only take place in the imagination. One is thus compelled to ask whether Fishberg and Feist meant something else than skin color when using the words *white* and *black*.[93] In his uses of photographs Feist both cultivated

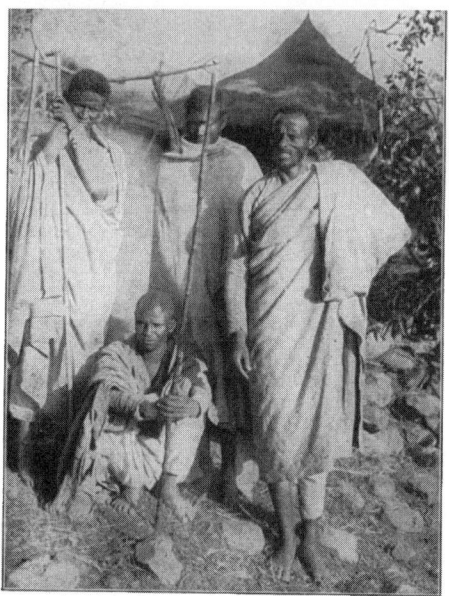

52. Faläschä (abessinische Juden) bei Gondar. Nach F. Rosen, Eine deutsche Gesandtschaft in Abessynien, S. 428.

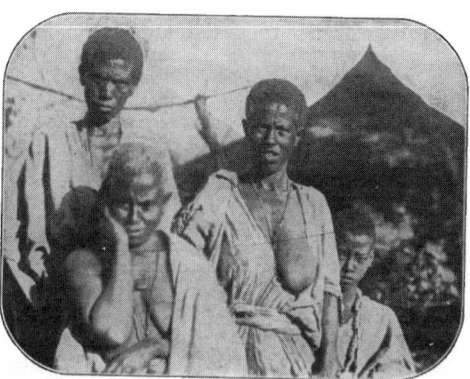

53. Faläschäfrauen aus Gondar. Ebenda.

FIGURE 2.8. Photographs of Ethiopian Jews emphasize their primitiveness. Sigmund Feist, *Stammeskunde der Juden: Die jüdischen Stämme der Erde in alter und neuer Zeit: Historisch-anthropologische Skizzen* (Leipzig: J. C. Hinrichs, 1925), image 52 in table 22.

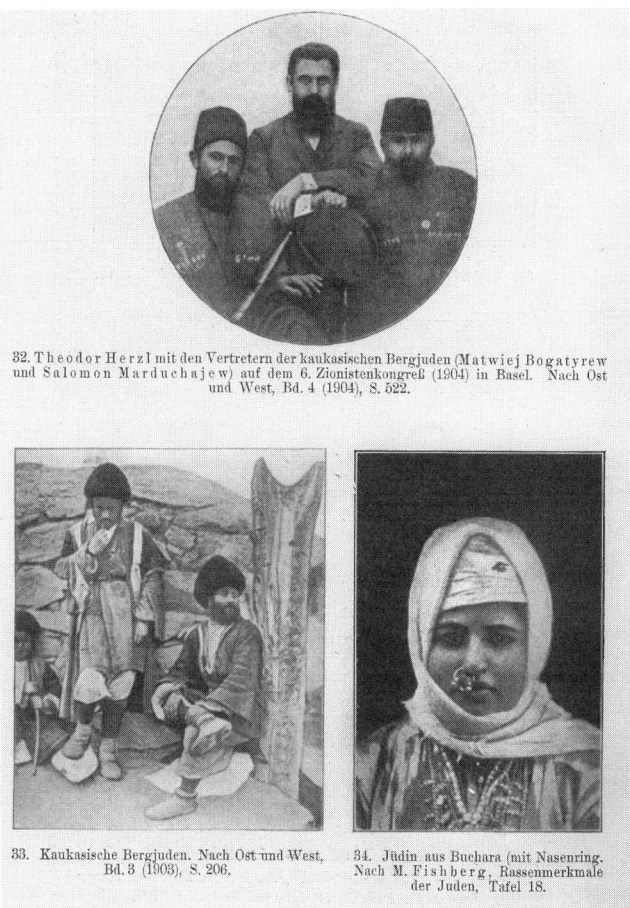

32. Theodor Herzl mit den Vertretern der kaukasischen Bergjuden (Matwiej Bogatyrew und Salomon Marduchajew) auf dem 6. Zionistenkongreß (1904) in Basel. Nach Ost und West, Bd. 4 (1904), S. 522.

33. Kaukasische Bergjuden. Nach Ost und West, Bd. 3 (1903), S. 206.

34. Jüdin aus Buchara (mit Nasenring. Nach M. Fishberg, Rassenmerkmale der Juden, Tafel 18.

FIGURE 2.9. Feist was not willing to rule out that contemporary Jews across the globe still betrayed certain shared recognizable features. Feist exemplified this view by placing a photograph of Herzl alongside two Jews from the Caucasus, allegedly showing their similarity. Sigmund Feist, *Stammeskunde der Juden: Die jüdischen Stämme der Erde in alter und neuer Zeit: Historisch-anthropologische Skizzen* (Leipzig: J. C. Hinrichs, 1925), image 32 in table 12.

his readers' visual observation and used photographs as the site of projections, thereby closely mixing the two.

FROM ICON TO MATRIX: HYBRIDIZING INSTRUMENTS

Through the case of the idea of the Jews as a mixed race people. we have witnessed major transformations in the status and in the nature of the racial

photograph between the 1880s and the 1920s. With the increase of photographic reproductions in the decade between 1910 and 1920, their status was
transformed. Photographs that earlier were taken to illustrate scientific arguments based on anatomical facts were increasingly integrated into the process
of argumentation. If in the beginning of this history photographs belonged to
the surface and the deeper structure was understood to be medical, physical-
anthropological, by its end the two were far more inexorably mixed. In this
same process the very nature of the racial photograph was redefined.

If the single photograph von Luschan introduced to characterize the Jewish
type—the Christian Armenian Turk—quickly needed no caption to be recognized, it could be defined as "iconic."[94] The decrease in reproduction costs of
photographs in publications following the introduction of new printing technologies such as the haplotype brought about a proliferation of photographic
reproductions between 1910 and 1920. Individual photographs were replaced
by series of photographs. And a new, specifically photographic racial genre
emerged. The racial-photographic matrix, I wish to suggest, functioned as a
schema.

We have no evidence that this change was planned or that authors were
aware that the change in practice redefined *race*, *type*, or *Jew* or had consequences for their visual perception. Serialization undermined the idea that
a race could be defined and photographically represented by a single photograph. Types were now dissected by photographic series, broken down into
traits or characteristics—*type* was in practice redefined. Multiple photographs
placed next to each other created relations between the individuals depicted
in the photographs. Even more accurately, they created relations between their
physiognomic characteristics. Rather than a photograph of *the* nose of the Jew
or *the* eyes of the Jew in the singular, for instance, viewers were exposed to
photographs of numerous Jewish noses and Jewish eyes. This shift in photographic practice implied that the type existed only in a much more fluid and
harder to detect plural, heterogeneous form.

On the spectrum of simplicity and complexity, iconic representation was
closer to the former and serialization to the latter. The photographic matrix
did not rely on strict control, was hardly informed by genres and visual codes
of Western art (which still informed the discourse of Mollison or Martin), and
made use of multiple photographic genres and repertoire irrespective of their
original social function.

It is difficult to think of a more ambivalent example for the power of mental
schemes that dynamically discriminate and exclude information, corroborat-
ing preexisting beliefs and stereotypes as well as generating new ones, than

ideas of "race." The serialized photographs fashioned a cognitive framework or concept that (re)organized and (re)interpreted information that was defined as racial. The result of the shift in photographic practice, it must be emphasized, was *not* to undermine the idea of race or type but to modify it and its social perception. An instrument embodies already acquired knowledge and helps produce the phenomenon or object of investigation.[95] Serialized photographs, in this sense, segregated and visualized what were believed to be distinct racial components, and in this sense functioned as a powerful hybridizing instrument.

The shift from the icon to the scheme was not unidirectional. In the 1920s, following the success of certain genres of racial literature, more popular and inexpensive editions of these same books were issued with fewer photographs. Major examples include Hans F. K. Günther's book on the German people in Germany and Gustav Kraitschek's similar book on Austria. In this situation photographs regained some of their iconic property.

These transformations in the nature of the racial photograph were not unrelated to major tendencies in science. One of the observations that Daston and Galison make in their book *Objectivity* pertains to the balance that writers attempted to strike between "realism of individual" and "realism of type," that is, to the representation of a kind through that of an individual that was intended to express the kind's common structure.[96] If individuals fully manifesting or instantiating the type were not easy to find, photographs that were intended to capture racial types were by definition marked by a certain internal contradiction. The reason for this is that photography possesses an inherently individualist aspect; it always reproduces the image of the specific individual that stood before the camera lens. Serialization further tipped the balance toward the realism of the individual at the expense of that of the type. Given that no two individuals are identical, the plurality of photographs created a network of similarities, together composing a thick description of type which also, at the same time and by the introduction of photographic practices, undermined "type" in the singular.

Photographic series created patterns of partially overlapping features without a necessary and sufficient core set of properties. The discourse addressed here shows elements of the stages of representation found in other fields. Von Luschan started from a notion of type, which he believed he depicted photographically (mechanical objectivity). There is less evidence in this discourse (only in its statistical branch) of relational invariants, but the serialized photographs indicate the transition from mechanical objectivity to trained judgment of families of objects.[97]

None of the writers addressed in this chapter reflected on photography's epistemological status, its advantages or disadvantages, and the practical negotiation of the epistemic status of photographs as scientific evidence was invariably tacit. While the meaning of photographs was overdetermined and could be mobilized to different ends, their overdetermination was not entirely free and indeed was, in an important sense, asymmetric. Opponents of racial determinism and of antisemitism gradually recognized the drawbacks involved in realistic uses of photographs, and from the middle of the 1920s at the latest, racial photographs in book publications were used solely by antisemitic writers, and the realist use of photography was aligned with proponents of racial determinism. But proponents of racial determinism also had to face the tension built into the use of photographs between the drive for scientific legitimacy and biased or motivated uses. No author made more sophisticated use of serialization than Hans F. K. Günther, to whom we turn in the next chapter. If in historical hindsight his photographic practices are deemed entirely political, ideological, and propagandistic, our first task is to recontextualize them in such a way that will reembed them in the wider cultural and scientific contexts to which they responded and from which they drew.

Serialization as Construction of Meaning: The Photographic Practice of Hans F. K. Günther in Context

Hans F. K. Günther, leader of the Nordic Circle and Himmler's mentor, is widely recognized as the most prominent theoretician of race in Nazi Germany. Extensively using photographs to combine scholarship and ideology, Günther created a multifaceted racial visual language embedded in a structure of racial asymmetries with clear political and social implications. It would probably be no exaggeration to state that Günther was seminal in providing Weimar and Nazi Germany with a visual code that was so sophisticated and powerful that it projected itself forward as well as backward, determining future uses of photography (such as those of Walter Scheidt or Peter Heinz Seraphim) and reinterpreting already existing ones (Rudolf Martin, Arthur Ruppin, or Eugen Fischer). Günther differed from Eugen Fischer or Ludwig Ferdinand Clauß in his ability to simplify or concretize existing ideas, rendering them accessible to a wide public. Through the use of photography, Günther concretized the racist and antisemitic intellectual tradition of Houston Stewart Chamberlain, who in his book, *Foundations of the Nineteenth Century* (1899) developed the idea of a specific form of Nordic or Aryan seeing. Eugen Fischer, as was elaborated in chapter 1, used photography in order to identify the distinct traits of individuals of mixed descent. His conception of scientific objectivity, however, was devoid of a notion of relative visual perception. Günther married Fischer's "Mendelian photography" with Chamberlain's concept of *Anschauung*, of specific innate forms of vision, thus merging scientific objectivity with racial relativism. Günther, in contrast to Fischer, was not content with demonstrating racial difference; he attempted to force readers to experience "race." After

1945, Günther's work was relegated to the status of vicious pseudoscience, but in the 1920s and 1930s his publications were read by many as serious scholarship. It is not possible and probably not advantageous to isolate the ideological and propagandistic aspects of his work. Rather, in this chapter I will analyze from a practical epistemological perspective Günther's use of photographs while pointing to ways in which they partook in and reflect wider scientific and cultural practices, ideas, and values.

Hans F. K. Günther (1891–1968), the son of a chamber musician from Freiburg (Baden), studied German and comparative linguistics, receiving his doctorate from the University of Freiburg in 1914. Günther fantasized about the heroic nature of Nordic men but was declared unfit to serve as a soldier in World War I. In 1919, after he passed the Gymnasium teaching examination, he was briefly employed as a substitute teacher before he was commissioned by the extreme right-wing publisher Julius Friedrich Lehmann to write a book on the racial science of the German people after Rudolf Martin had declined the offer. Lehmann later recalled that he was particularly impressed by Günther's intuitive perception of racial physiognomies of the farmers and hikers they had encountered on a joint hike in the Bavarian Alps.[1]

Günther's writing career stretched from the early Weimar period to the West German republic, but his principal study, *Rassenkunde des deutschen Volkes* (Racial science of the German people), appeared in 1922. Günther wrote this book motivated by a profound sense of a crisis of modernity and by his contempt for the Weimar democracy. Key to understanding his employment of photographs was Günther's staunch belief, from before the rise of the Nazi party to power, that the majority of living individuals were racial bastards.[2] The "reactionary logic"—which he shared with founders of the racial paradigm, such as Gobineau, Galton, or Chamberlain—can be inferred from his employment of photographs that contrast the distant and idealized past with the corrupt present.

Visual considerations were at the core of Günther's racial project from its very inception and were closely tied to his attempt to use photographs in order to create a visual text parallel to the written one, thus transforming readers into viewers. The visual components in Günther's publications are both closely interwoven into the written account as well as independently organized and structured, connected to the verbal text but not subordinated to it.[3] Günther's use of photographs was derived from the academic discipline of art history at least as much as from the field of anthropology. While comparing Günther's work with that of Rudolf Martin and additional contemporaries, I will argue that the former initiated and established a specifically photographic racial visual tradition. The text accompanying these semicoded naturalistic photographs as well as their specific organization within the book indicates how they should be viewed.[4]

The ingenuity of Günther's use of photographs, his racist ideas, and his unmistaken political commitment could easily lead one to interpret his publications as mere ideology or propaganda. True, Günther never questioned his racist premises. But to view Günther's work as ideology rather than scholarship is far too convenient. Günther stood on the border between scholarship and ideology. While the ideological moment is certainly important for analyzing his use of photographs, the emphasis of my analysis is on the intersection of ideology and science. Ideology and science were not necessarily in opposition. Moments of intersection, conceptual as well as rhetorical, question the comfortable separation between the most prominent theoretician of race in Nazi Germany and "untainted" scholarship, which characterized post-1945 German culture.

Günther was clearly aware that the forming of patterns by means of repetition played a key role in what he viewed as the "reeducation of the Nordic eye."[5] For Günther, pattern was more important than any single image in itself, and he masterfully created such photographically subtle and sophisticated patterns. A close study of Günther's series of portraits reveals that these patterns are often implicit, thus escaping easy encoding. Further complicating the interpretation was Günther's habit of inserting unexpected photographs that destabilize the reader's expectations. This practice of defamiliarization can be compared to that of his contemporary Berthold Brecht's *Verfremdungseffekt*. In contrast to photographs of types referred to in the first chapter and in contradistinction to Günther's own written accounts, patterns acknowledge no "center," in Rudolf Arnheim's terminology.[6] In this vein I also show certain similarities between Günther's practice and Wittgenstein's notion of "family resemblance."

Günther possessed not only a keen interest in but a remarkable sensitivity to questions of perception of racial difference. While he clearly wanted to sharpen his readers' visual perception, his use of photographs was ultimately aimed at the creation or formation of a specifically Nordic "racial imagination." I conclude this chapter by arguing that racial imagination in Günther's work attempts to achieve a certain form of "totalization" of aspects of race, including the perceptual, intellectual, and political. Photography, consequently, was at the core of Günther's project.

EMPIRICAL SOURCES AND PREDECESSORS

The question of how Günther selected the pictures he included in his publications from among those that were at his disposal is difficult to answer. In his 1922 study of the racial elements of the German people, he employed 409 images, the vast majority of which were photographs. Günther did not produce his own photographs, instead obtaining them from a network of scientific institutions, museums, and private collections all across Europe. Günther's estate was destroyed, and it will remain unknown exactly how many photographs Günther had in his possession or were available to him. But the sheer amount of photographs he used in his publications leaves no doubt that he spent a significant amount of time, energy, and resources on obtaining and employing photographs.

Günther employed photographs as well as other visual materials, such as reproductions of works of art, of sculptures, of maps, and of instruments, from numerous sources: private collections; scientific publications; ethnographic collections from Vienna, Sweden, or Finland; anthropological photographs of prisoners of war; newspaper clips of famous individuals; and reproductions of works of art or details of works of art. Günther was guided by a sophisticated form of realism. By inserting a painting of Heinrich Heine (who died more or less at the same time as photography was invented) into a series of photographs, for instance, Günther was implying that the "truth" of race preceded that of the invention of photography.[7] In his attempt to establish a new racial methodology that transcended separate fields of knowledge, Günther attempted to convey to his audience that the ideas he expressed through photographs, a modern medium, had a much longer history. The integration of contemporary photographs of individuals as if chosen at random together with portraits of famous historical characters implied that the notions or terminology may be new, but that earlier generations, who did not have photography or an explicit notion of race at their disposal, were no strangers to race.

In order to understand the status of photography in his work, it is crucial to observe the discrepancy between the centrality of photographs in Günther's racial paradigm and an almost complete absence of its discussion, empirical or methodological. As it is most likely that Günther was aware of the concurrent literature on photography, it can be assumed that he avoided its discussion in order to endow his photographs with a realistic transparent status.

When Günther did discuss photography, his discussion was limited and local. Günther was familiar with the works of Galton and Jacobs, Stratz, Ruppin or Fishberg, Salaman and Fischer, and Mollison and Martin. While his use of photographs drew on these works, he nonetheless distanced himself from their photographic practices and theorization. For example, Günther referred to Stratz in detail in his discussion of several physical racial markers of Jews while avoiding mentioning Stratz's discussion of the gaze or the use of photographs.[8]

In 1930 he removed the appendix on the Jews in Germany from the 1922 racial study of Germany and expanded the appendix into a separate book on the racial characteristics of the Jewish people. In this book he discussed Redcliffe Salaman in detail and acknowledged his permission to reproduce his photographs as well as referring to Salaman's article(chap. 1).[9] Günther defined Salaman as a Jewish race scholar (*jüdische Rassenforscher*), cited him as the first to have employed Mendel's laws of inheritance to Jews, mixed Jews (*Judenmischlingen*), and their offspring. Günther criticized Salaman's methodology, claiming that Salaman treated Jews and English as "races," whereas these two groups are racial mixtures.[10] Salaman's statistical findings therefore were seen to be insignificant. Yet Günther accepted Salaman's assertion that certain specific characteristics were hereditary "Jewish." Günther made parallels between Salaman's work and Fischer's bastards study but said nothing of their common use of photographs for the sake of observation,[11] only mentioning in passing that Salaman's use of photographs strengthened his conclusions. Günther therefore rejected Salaman's conclusions about the hereditary Jewish traits but stated that Salaman's study can only serve as evidence of Jewish mixed marriages.[12]

Günther rejected Galton's photographic method but nonetheless reproduced his Jewish type series.[13] Günther in fact began his discussion of Galton's composite photography with an analysis of the works of Joseph Jacobs, whom he termed a Jewish race scholar. Günther judged the attempt to create such composite "facial-average" photographs as having little value from the point of view of the study of race. The composite photographic method, stated Günther, represented outdated racial studies of the past, when it was believed that through measurements or arithmetic means an average (*Mittelwerten*) of the "race" could be reached. Günther then proceeded to a lengthy quote from the Jewish encyclopedia's entry on "Type," based on Jacobs' study.[14] Günther rejected Galton's form of racial photography as scientifically obsolete but nonetheless reproduced an image from the Jewish type series. He determined that the children depicted were southern Jewish rather than eastern Jews, which explained, in his view, why the result of the series was overall deficient.[15] Rather than racial averages, Günther's explicit intention was of affording an overall picture (*Überbild*) of the entire hereditary pool of the Jewish people.[16]

A similar pattern of partial acceptance combined with criticism is found with regard to Eugen Fischer, who featured prominently in Günther's 1930 book. Günther heralded Fischer, together with Gregor Mendel and Francis Galton, as a pioneer of human heredity and presented his own method as deriving from Fischer's. Günther appropriated from Fischer the study of phenotypic racial characteristics through photographs as well as his method of subtitling the photographs he presented, but nowhere did he discuss Fischer's photography. Thus we can see that despite the centrality of photography, Günther carefully avoided its discussion.

HERMENEUTICS OF SUSPICION

Compared with both earlier and contemporary uses of racial photography, images in Günther's book enjoyed a greater independence from the written account. In Ruppin (as we will see in chap. 5) and in Walther Scheidt's works, photographs primarily illustrated the written text and thus were analytically subordinate to it. In Clauß's style (as we will see in chap. 4), the photographs could not be understood without their detailed textual exposition. In contrast, in Günther's work there was a complex relationship between text and image. The written account was independent of the images; in contradistinction to Clauß, Günther generally refrained from directly referring to the images he presented. Conversely, the images were also more independent of the written account. Günther's publications, and in particular the *Kleine Rassenkunde*, the shortened and cheaper version of the 1922 book on the racial characteristics of the German people, enjoyed massive circulation, far greater than any of the other publications discussed throughout this book. Accordingly, it is likely that many glanced at the photographs without carefully reading the text. Thus, visual aspects that were independent of the written narrative are of particular interest to this analysis A characteristic specialty of Günther's employment of photographs can be revealed through what Paul Ricoeur has termed the "hermeneutics of suspicion."[17]

Examining the works of Nietzsche, Marx, and Freud, Ricoeur argued that these authors taught readers to regard with suspicion their conscious understandings and experience, because behind the surface lay causal forces that explained the conscious phenomena, laying bare their true meaning. Ricoeur's argument concerning the reduction of one order to a more fundamental underlying one is also applicable to Günther's interpretation of "race." Indeed, the writers studied in this book often viewed "race" as a concept parallel to and competing with Marx's "class." But Ricoeur's term applies particularly well to

Günther's employment of photographs. Through careful choice and layout of photographs, Günther transformed his audience into suspicious observers.

In chapter one I have shown that the genealogy running from Stratz through Salaman to Fischer had already transformed readers into suspicious viewers, as it called onto them to actively observe racial traits in photographs of racially mixed descendants based on the Mendelian distinction between the visible phenotype and the possibly invisible genotype. Yet these writers limited their use of photographs to depict well-circumscribed groups (individuals of mixed Anglo-Jewish and Gentile descent in Salaman's work or of Hottentot and Dutch individuals in the photographs presented by Fischer) and appeared at the end of the publication as an appendix. While aligning himself conceptually with this tradition, Günther integrated the photographs into the main body of the text and extended the scope of suspicion from carefully chosen and determined individuals of mixed descent to race in general. His use of photographs subtly yet effectively demonstrated what the written account could only attempt to explain. The joint effect of the integration of the photos within the text as well as the extension of their use beyond well-circumscribed groups had a somewhat paradoxical effect.

Whereas Günther's written account attempted to develop a comprehensive and disambiguating interpretation of race, photographs more often generated ambiguities. Günther's visual suspicion was heightened by his belief that 97 percent of the German population was racially mixed, that is, made up of the various European types, and thus practically every individual was subject to his suspicion. His photographic series could further be categorized by specific forms of suspicion. Günther often decontextualized his photographs, for example showing a photograph documenting intimate family life as evidence for the dangers of racial intermarriage. Günther's concept of suspicion was closely related to the Mendelian logic of a racial genotype hiding behind the phenotype; what is seen on the surface can potentially mislead the viewer. This form of suspicion was deepened by the tension between Günther's belief that racial essence was readily recognizable and his equally strong belief that in modern conditions, it was increasingly becoming camouflaged. Photographs sharpened observational skills while reproducing the belief that racial characteristics were often subtle and necessitated careful observation. Günther's practice of accompanying photographs with subtitles followed by question marks also expressed a form of suspicion by questioning the very presence of a certain racial characteristic. Günther viewed the observation of "race" not as a simple and harmonic process but rather as being dynamic, rife with frictions and possible inconsistencies, as racial difference often had to be sought behind that which was seen. The joint effect of Günther's visual practices, much more than the written account, generated a hermeneutics of racial suspicion.

This leads us to a central question concerning the specific role of photographs in the generation of suspicion. Günther employed photographs *as if* they were a transparent medium but at no point discussed the photographic medium as such. So what generated this suspicion? It might be possible to follow Günther's design and view the source of suspicion in physiognomic differences irrespective of the photographic medium. Yet such an interpretation perpetuates Günther's overhauling photographs directly to physiognomic racial differences. Bruno Latour's actor-network-theory methodology proves itself particularly powerful at this point. Günther presented photographs as transparent, but in truth they are material objects, an "actant" in Latour's terminology.[18] Suspicion pertained to racial difference but was generated by means of photographs. The irony is that a realist and naive understanding of the photograph as immediate and transparent was necessary for the generation of that form of suspicion. At this crucial moment the historian cannot ignore the possibility that this specific form of suspicion could not have been generated without the medium of photography.

*

Günther employed photographs as instrument of unmasking (fig. 3.1). Günther brought together photographs from different sources ranging from anthropological photographs of prisoners of war to family albums. Beneath the thin mask of national culture or religion (expressed in clothing, glasses, beards, suits, hats), Günther intended to indicate that one could easily identify the elementary racial components. The oversized protruding ears, dark gazing eyes, thick lips, curved heavy nose, high forehead, and long, rounded skull, remained constant despite age, dress, and nationality.

The layout, appropriated from art history books, emphasized similarities and differences. On the left-hand side appeared five images, four individual portraits and one family portrait with mother, baby, and child; on the right-hand side, six individual portraits. The arrangement evaded standardization in terms of distance from the camera, lighting, background, direction of glance, age, gender, clothing, or religious affiliation. The angles employed included full frontal (top left of the right page) through three-quarter profile in the bottom left of the right page, to full profile (top right of the right page).

The images were styled after different photographic practices: some conformed to standard portrait poses (face facing the camera, shoulders slightly

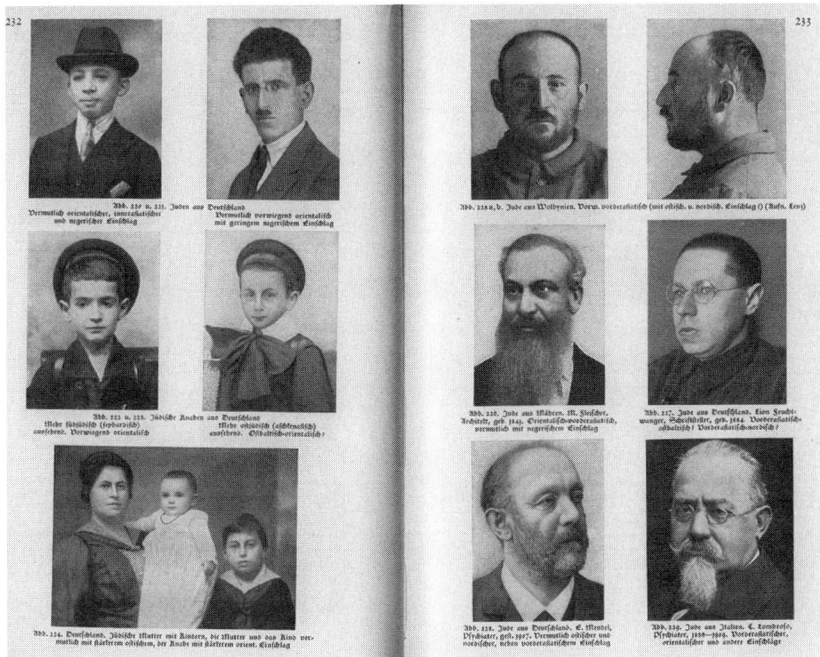

FIGURE 3.1. Serialization for unmasking and rhetorical suspense. Hans F. K. Günther, *Rassenkunde des jüdischen Volkes* (Munich: Lehmann, 1930), 232–33.

tilted, only one ear being visible) and were probably executed in a photographic studio, such as the image of Cesare Lombroso at the bottom right corner of the right page, or a semiprofessional pose (top right of the left page and bottom right of the right page) where both ears can be seen (an angle viewed as a unflattering). Some were likely taken from family albums (bottom of left page); the top of the right page was probably taken from a prisoner of war camp. Günther purposely chose photographs of individuals whose clothes range from uniform (top right page) to suit (top left page and bottom right page). No two individuals glance in the same direction, looking straight at the camera.

The rationale behind the layout of this series in figure 3.2 is difficult to discern. Appearing on page 236, Günther constructed this series that is made up of three sets of frontal and profile photographs from an unidentified source. On page 237, six images of individual young, mature, or old women appear from different angles. I am particularly struck by the inclusion of the photograph of the child, identified as a Jew from Sweden at the bottom of page 236, that appears within a series of images of women. Why did Günther include

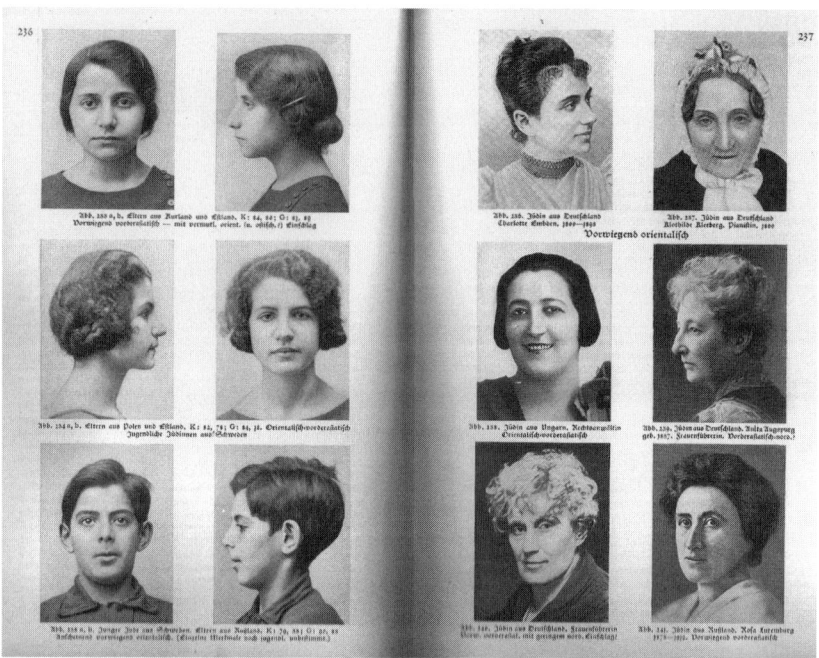

FIGURE 3.2. Layout principles are not easily discernible. Hans F. K. Günther, *Rassenkunde des jüdischen Volkes* (Munich: Lehmann, 1930), 236–37.

this photograph, and why did he insert it precisely here? Let us look at the woman on the right-hand page, middle row, left-hand image (picture no. 238), identified as a Jewess from Hungary: her mouth is open and stretched. She is portrayed with a smile that betrays her knowledge of being photographed and acknowledgment of the photographer's perspective. It is likely that this photograph was taken at a family event. The difference in lighting between the sides of the face and the smoothness of her white skin suggest that this was a painting made on the basis of a photograph. Photographic films of a later period brought out more detailed skin texture, and to achieve smoothness, films then were retouched.

Observe the young woman in the middle of page 236 (fig. 3.2, left-hand page, middle row): her parents are identified as from Poland and Estonia. Her hazy gaze is focused on a point slightly above the camera, and she appears slightly more self-confident than the young girl above her. The hair of both this young girl as well as the one above her covers her ear and the left side of the forehead, touching the left eyebrow. Both are marked with a slightly melancholic expression, which, as has been noted, was often associated with the

Jewish gaze. The two girls' profiles face each other, emphasizing the natural relaxed position of the head as leaning slightly forward; their gaze is similar to both the child at the bottom left side of the page and the woman at the bottom left column on 237.

The woman in the middle of the right column on 237, described as a Jewish woman from Germany named Anita Augsburg, and the woman at the bottom of the left column on the same page stand out as exceptions. Their complexion is lighter, their hair curly but blond or light brown (the bottom photograph is burned by overexposure). Their facial expression is self-confident, impressive, powerful and dominating, their hair short and modern and slightly masculine. Both are presented as possessing dominant noses, yet atypically Jewish. Both are recognized as women leaders, and Günther indicated possible presence of a Nordic element. The light falling on their blond hair and across their faces emphasizes their powerful grace. At the bottom of the right column, the photograph of Rosa Luxemburg describes her as a "Jewess from Russia." Photographed from a semiprofile angle, her glance is slightly elevated to a point above the photographer. The texture of her hair looks almost like a drawn picture, but her shoulder, slightly out of focus, betrays that this is a photograph.

Through his use of photographs, we can attempt and infer what kinds of knowledge Günther assumes that his audience possesses. Only following such an assumption can we interpret the photographs as speaking for themselves. The woman appearing in the portrait at the top left column on page 327 is identified as Charlotte Emden (1800–1898), a pianist. Emden, whose picture was cut from a larger painting, was the sister of Heinrich Heine. Similarly, Anita Augsburg (1857–1943), middle of the right column on 237, was the first woman lawyer in Germany and a radical feminist and pacifist. This page was reproduced from Günther's 1930 edition of the book on the Jewish people. Augsburg, who was already from 1923 on a National Socialist list titled "the most hated," was forced into exile in Switzerland in 1933.[19] This series, therefore, could be titled "know thy enemy."

DECONSTRUCTIVE PHOTOGRAPHY: TYPES AND PEOPLES

Analyzing Günther's usage of photographs yields a principal difference between his presentation of the distant past and that of the present. In presenting photographs of his contemporaries, he attempted to demonstrate that these contemporary peoples were deeply mixed, whereas individuals from past periods were depicted as complete types. I will attempt to show that from two different perspectives, that of social history and that of the history of science, Günther's use of photographs could be interpreted in this respect as offering a solution to a problem.

The immediate problem that Günther faced was that no current population, most certainly not any modern European society, could be described as racially homogeneous. Thus, making such a claim would have greatly weakened Günther's scientific authority. Indeed, as Sander Gilman, Ann Stoler, and others have shown, from the end of the nineteenth century, in North America, in Dutch and British colonies, and on the European continent, racial boundaries were becoming increasingly elusive.[20] Some educated and wealthy African Americans could pass for white. Colonial minorities, educated in the language and culture of the colonial power, increasingly could not be distinguished from their European colonizers. In Germany, Jews who from the 1870s were educated and acculturated in unified Germany could often no longer be distinguished in accent, dress, appearances, gesticulations, or behavior from their non-Jewish counterparts. Being German, so to speak, they passed as such. The presumed racial difference in many cases faded to the point of no longer being distinguishable. But alongside these social processes, the belief in racial difference remained steadfast; on the contrary, race was increasingly interiorized and essentialized, conceived as a hidden, camouflaged property. Racial characteristics that had hitherto been held to be visible could no longer easily be perceived and were increasingly conceived as potential rather than actual. These social processes sharpened the need of modern racial theories to concur with complex realities.

Günther, indeed, was not original in his insisting on the differentiation between races and peoples. This methodological distinction clearly addressed this difficulty: races were per definition pure; peoples were a specific configuration of racial mixture. Modern populations such as the German, English, or French peoples were not pure "races" but thoroughly mixed "peoples."[21] His photographic series, indeed, gave concrete and visual expression to this differentiation: statues were used as manifestations of racial types; photographs manifested the modern hybrid condition. This practice reflected the belief that the past was superior to the present, purer, and closer to the natural condition.

This history of science perspective coincides with the social historical one. Following the enthusiastic adoption of Gregor Mendel's laws of inheritance in 1900, which increasingly permeated biological and anthropological discourses, the conceptualization of race shifted toward greater appreciation of genetic variation within racial populations. Many racial writers, including Günther, appropriated from Mendel the belief that genetic characteristics

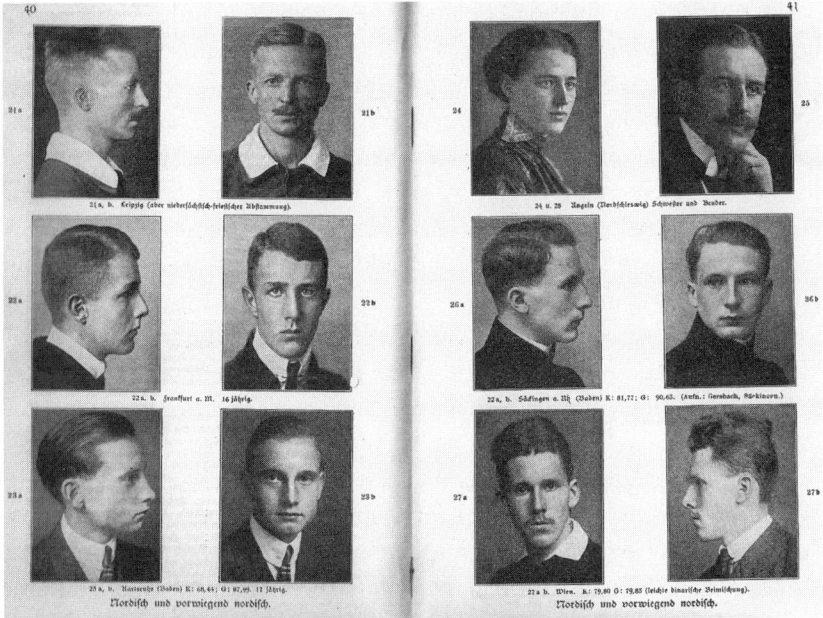

FIGURE 3.3. Racial types and racial components: individuals as carriers of racial attributes. Hans F. K. Günther, *Rassenkunde des deutschen Volkes* (Munich: Lehmann, 1922), 40–41.

could mix and create new combinations but never truly merge. This interpretation of Mendel's theory held races to be stable and even ahistorical entities, and it stood diametrically opposed to Darwinian evolutionist theory, according to which nature itself was unstable, and species and races were historical entities undergoing constant mutation and change. If, throughout history, mixtures did in fact take place between races, these could only bring about mixed individuals—hybrids or racial bastards. If such unions continued over a long period of time and involved significant numbers, they could lead to the creation of a people. Such a population might manifest specific racial patterns, but in strictly racial terms no new types could emerge.[22]

Günther's photographs were analytical in semiotic terms; they did not depict actors acting toward goals but participants who were carriers of attributes (fig. 3.3). Günther's series were not about something that the participants did but rather about the attributes they possessed. The main result of these series is classification.[23] Günther employed photographs naturalistically: they always portray an abundance of detail containing "a multitude of embedded analytical processes," maneuvering between detail and overhaul, where differences would collapse into chaos.[24]

This series is composed of photographs organized in order to create and highlight similarities. Equivalences were created through systematic insistence on symmetrical composition and equal size, distance from each other, and orientation toward the horizontal and vertical axes. Through the plain and neutral background, the removal of geographical considerations, and the use of a frontal, seemingly objective angle, the series gives the impression of timeless classification. Decontextualization is essential for turning the images into typical or generic and against individualization, which is integral to the photographic picture.[25]

Günther's goal in his use of photography was twofold: to convey racial "types," and to deconstruct individuals to their racial components. Günther reiterated the differentiation between race and people on many occasions, but his practice of employing photographs for modern (mixed) peoples and reproductions of statues for (pure) types is a salient illustration of this approach. While Günther's written account left the impression that all recorded history falls under the law of racial unions, combinations, and bastardization, the visual practice sharpened the differentiation between the modern and the premodern periods. This pattern partook in the reactionary logic that I elucidated in the introduction.

In 1876 the Dammann brothers could still display photographs of "racial types" in their *Ethnological Atlas*, although at that time social realities were already rapidly changing. Within these transformed social conditions, racial difference decreasingly corresponded with "pure types" and progressively resisted simplistic schemes of racial classification. Günther, writing almost fifty years later, no longer attempted to claim that modern societies were made up of pure types. In addition, between 1876 and 1922 the conceptualization of race within science and visual scientific practices underwent profound changes. In these transformed conditions Günther spearheaded the use of photographs as a dissecting form of observation not of whole racial types but of far more subtle differences; the merely visible (fig. 3.4).

While Günther's layout principles are appropriated from the discipline of art history, the photographs are designed to extend a Mendelian form of deconstructive vision. In his book *Reassembling the Social*, Bruno Latour gave a very concrete sociological expression to the extension of seeing.[26] Günther's intention was not only to extend the vision of viewers but to thereby transform it.[27] This could be termed a racist-reactionary *Neues Sehen*. One can glean a partial view of aspects of the practical cultivation of seeing from a study conducted by Cristina Grasseni, who studied the case of cattle breeding.[28] Grasseni's account is diametrically opposed to Ann Stoler's interpretation of Stratz

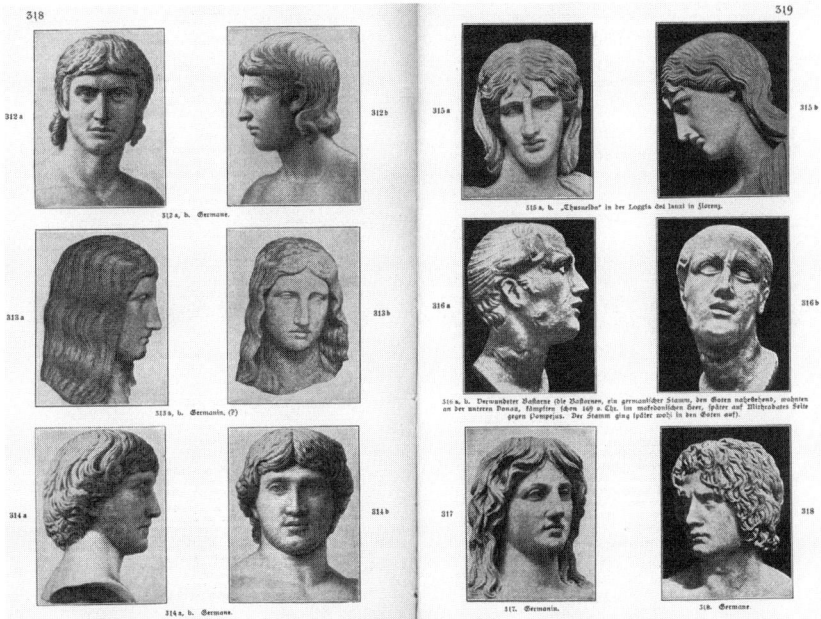

FIGURE 3.4. Types and components: Greek statues as real and ideal. Hans F. K. Günther, *Rassenkunde des deutschen Volkes*, 10th ed. (Munich: Lehmann, 1926), 318–19.

discussed in chapter 1, who argued that it is not seeing but the imagination of seeing that is being cultivated. Yet when speaking of cattle, an entire spectrum of racist considerations (idealistic and antihumanistic, their social and political implications) is absent.

RACIAL *VERFREMDUNGSEFFEKT*

To identify the Jewish gaze, the eyes are sufficient. Günther implied that the eyes are a synecdoche of the type. In terms of anthropological method, Rudolf Martin's *Augentafel*, which was discussed in chapter 1, allowed for the separation of the eyes from the rest of the face (fig. 3.5).

Already in his earliest publications Günther employed photographs for the purpose of estrangement of racial difference. Yet in contrast to Bertillon's usage, as elaborated in chapter 1, Günther employed photography in a more egalitarian manner. He attempted to enable identification of racial differences in social situations, ultimately aiming at creating a racial-political community in which estrangement played a key role.

och' nie beobachtet und bis jetzt nur bei oftischen oder
en Kindern und bei Menschen, die eine oftbaltische Blut-
i konnten. Drews[1]) hat bei Beobachtungen von Lid-
:rischen Kin-
ft stets eine
: Nasenwur-
ünftige For-
raffische Zu-
faltung ent-

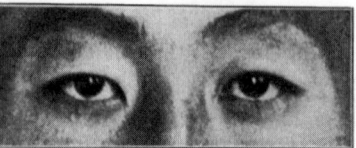

Art der Fal-   Abb. 27}. Lidspalte der mongolischen (inner-
, die im Ge-   asiatischen) Raffe, die Mongolenfalte zeigend.
inthus meist   (Dungane aus Semirjeschtschensk; Aufn.:
)einung auf-   Anthr. Inst., Wien, aus der unten, Fuß-
i Epikanthus   note 2, angeführten Arbeit.)
werden. Sie entsteht aus einer gewissen Schlaffheit

bildet, d. h. der innere Augenwinkel ift
Europa verbreiteten spindelförmigen L
der äußere mehr spitz gezeichnet. Dabei
Längsrichtung leicht nach außen aufwär
                                    der ori
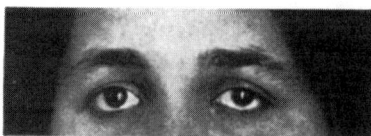   wie ein
                                    wie es
                                    kommt
                                    wie au
Abb. 58. Andeutung von Mandelaugen   eingeb
bei einer Jüdin aus Osteuropa   Braue

FIGURE 3.5. *a*, Hans F. K. Günther, *Rassenkunde des deutschen Volkes* (Munich: Lehmann, 1942), 161; (1926), 139. *b*, Practical redefinition of Rudolf Martin's *Augentafel*, Günther estranges the eyes through their separation from the rest of the face. Hans F. K. Günther, *Rassenkunde des jüdischen Volkes* (Munich: Lehmann, 1930), 70.

The notion of "estrangement," however, has a cultural-political history. A brief reconstruction of its history accentuates the force of the practical epistemological approach presented in the introduction and points to the cultural grounds shared by writers motivated by politically and ideologically opposing views.

Estrangement (not to be confused with the Marxist notion of alienation) can be traced back to Aristotle's *Rhetoric* but was revived by Russian formalists in the 1920s. In fact the term was imposed on them by their Marxist and communist counterparts, who viewed their theories as apolitical, or as uncommitted to the transformation of society. Aristotle viewed estrangement as part of the process of acquiring new knowledge irrespective of its specific content. For Aristotle, the strange was self-evident, and there was no need to estrange the known in order to acquire new knowledge. The perception that acquiring knowledge first required estrangement developed in the twentieth century. Within this new framework the Russian formalists developed the notion of a difference between identifying and seeing. As the identification of something occurs instantaneously, one can no longer see its form. By interrupting perception habits, therefore, one regains the ability to see what has been automatized by habit. If realistic representation renders the thing true to itself—estrangement must run counter to that realistic tendency.

Viktor Shklovsky, perhaps the most prominent theoretician of estrangement, made the renowned observation that the essence of art was to counter habituation, stating that "art exists that one may recover the sensation of life; it exists to make one feel things, to make the stone *stony*. The purpose of art is to impart the sensation of things as they are perceived and not as they are known."[29] The essence of art, according to Shklovsky, was to render the familiar strange and the strange familiar.

Bertolt Brecht introduced the term *Verfremdungseffekt* in 1936, and he is still widely identified with this technique. Brecht's *Verfremdungseffekt*, literally meaning "estrangement effect," was a theatrical device or technique designed to dispel the audience's reality on stage, an anti-illusionist (and antimimetic) device to counteract spectators' emotional involvement and identification with the characters or actions displayed on stage. It was dramaturgically essential that the audience be aware of the fact that "a play is a play," an artifact intended to stimulate thought processes that should then be applied to sociopolitical practice. When such distancing occurs, the action on stage becomes subordinate to a moral or a didactic concept.

Concepts of estrangement were not limited to the left side of the political spectrum. In an article from 1934 on Bruegel's *Macchia*, historian of art Hans Sedlmayr (1896–1984), working in the tradition of Alois Riegl (discussed in chapter 1), elaborated on a form of estrangement far closer to Günther's practice than to Brecht's.[30] Several points Sedlmayr makes in this essay parallel aspects of Günther's photographic practice. Directly tying the notion of alienation to the manifestations of deformation, we can situate Sedlmayr within the "reactionary logic" elaborated in the introduction.

They are all manifestations of life in which the purely human borders on other, "lower" states that threaten, dull, distort, or ape its substance. Primitives—a hollow form of human; the mass—more raw and primitive than the individual man; the deformed—only half human; children—not yet completely human;

the insane—no longer human. These are liminal states of humanity in which and through which the nature of man is cast into doubt. And they are the very subjects . . . to which modern anthropology has turned its attention in recent times, as if it were possible to grasp the nature of humanity precisely in these liminal states, states in which man enters into other realms.[31]

Sedlmayr then turns to a detailed analysis of the ways in which Bruegel's art produces an effect of estrangement (*Entfremdung*). He notes that "the logic of entire portions of a picture breaks down and the objects represented seem strange," while the viewer "is accompanied by the experiences of shock and disturbance" and "sensitive viewers [experience] even of anxiety and something approaching fear."[32] This experience constitutes a form of desta-bilization, as words suddenly lose their meaning. Similarly, Sedlmayr refers to "movement frozen in a snapshot" and notes that the sense of estrangement that photography produces always pertains to the object that captures it.[33]

The final point in Sedlmayr's article that I wish to point to is that of turn-ing the viewer into a suspicious, estranged observer, which toward the end of his article Sedlmayr connects directly with "masking" and concealment.[34] A practical epistemological analysis of Günther's use of photographs reveals that he utilized a similar strategy of estrangement, intended to advance specific racial goals, even before Sedlmayr or Brecht imported and baptized it. It is not clear whether Günther came across the work of Russian literary theorists directly, whether their work reached him mediated through different cultural agents, or whether he developed his practices independently. He was certainly no sympathizer with Brecht's communist ideas, and he was not particularly interested in Shklovsky and others' theories on art. But for his sociopoliti-cal goals he employs as early as in his 1922 *Rassenkunde des deutschen Volkes* practices of estrangement. However, unlike Brecht, Günther never referred to or theorized his use of estrangement, simply practicing it in a subtle manner. The principle of estrangement, therefore, could be put to use for conflicting ideological or political ends.

From a practical epistemological perspective, one can point to at least two different and interrelated forms of estrangement of which Günther makes use: the emphasizing of specific traits by way of serialization, and the disruption of viewer's expectation. Günther's motivation for this usage of estrangement was derived from his belief that the consciousness of "race" had undergone habitu-ation in modern society. Through usage of photographs Günther attempted to dehabituate sensitivity to race by means of estrangement techniques.

Through the careful layout of well-chosen photographs, Günther guided the eye through racial markers. Günther chose photographs of individuals who resembled each other only partially—individuals from diverse cultural, religious, economic, social, or geographical backgrounds—who shared a certain similar trait. The series were therefore frequently organized around the axis of one or two shared traits. The partial similarity between the depicted individuals forced viewers to attempt and identify what they had in common. A series could be organized through a single physical trait—say an aspect of the shape of the nose, the forehead, the eyes, or the skull—but it could also be a more elusive mental trait discernable only in the gaze. The choice of photographs drew a subtle, at times almost invisible imaginary line between the photographed individuals.

The result was an effect of estrangement: the imaginary line connecting between the shared trait called attention to that trait, rendering it visible. This was a practice of defamiliarization, of drawing the observer's attention to a familiar trait through its estrangement. There was a similarity between Günther's practice and that of Bertillon in criminology or Morelli in art history. The latter two performed this estrangement through isolating specific organs, such as the ear, and arranging them in tables. Günther achieved a similar effect without separating the organ from the face.

Through careful choice and organization of photographs, therefore, Günther singles out a trait, bringing it to the surface. This practice was connected closely to Günther's belief that contemporary humans were by and large racially mixed, hybrids. Consequently, Günther employed photographs to defamiliarize not whole types but single traits, sometimes nearly invisible.

This practice is located between the establishment of a racial code and the creation of racial language. The difference between the two is that a code is a set of signs whose meanings are predetermined and fixed; a language is composed of signifiers that never entirely coincide with their signified. These signifiers are in a constant state of slippage, and interpretation cannot be avoided. To follow these definitions, Günther's work was aimed at reducing language to code, eradicating difference or noise, in order to create a fixed racial order in which meanings are transparent, predetermined, and immutable. But ultimately his practice had clear elements of a racial language, of constituting social subjects who must engage in active interpretation of racial markers. Social reality, Günther tirelessly reminded his readers, was racially muddy. In order to raise racial consciousness, it had to first be estranged. For this goal, photographs were his principal tool.

Günther's estrangement practices have what Ian Hacking termed a "looping effect" with regard to ideas, materials, or instruments employed by his predecessors and contemporaries. His work effectively transformed the meaning of elements that had already been published. This can be illustrated through a comparison of Günther's isolation of a pair of Jewish eyes with Martin's *Augentafel* discussed above. Martin's seemingly disinterested "eye table" was an anthropologists' device for classification and measurement. Günther separated an element from a portrait photograph, guiding his readers to identify the Jewish gaze through a pair of eyes alone. Günther's use of the isolated eyes transformed the meaning and the value of Martin's *Augentafel*, placing it in a different interpretative framework, which further accentuated explicit racial asymmetries. As a result, the same instrument gained an entirely different social meaning. In this new set of racial variables, racial difference could be identified not only by the professional anthropologist but by every racially educated (Nordic) person, thereby rendering racial knowledge democratic, concurring with Günther's expressed aim to enhance racial consciousness.

The second form of estrangement, practiced far less frequently, pertains to Günther's play on viewers' expectations. Günther breaks the viewer's implicit expectations by inserting into a series of images one image that inconspicuously disturbed the underlying logic of the series.

In figure 3.1, the portrait at the top left of the left page disrupts the viewers' expectation. Why did Günther insert it there? How does it fit into the series offered on the page? What is the source of the sense of disruption? The person depicted in the photographs appears somewhat like a shrunken grown-up, a grotesque liminal figure, that cannot be classified as a child or a grown-up, as normal or abnormal, as healthy or sick. His elegant suit, tie, hat, and handkerchief accentuate the small and almost entirely shut eyes and the overbig ears. This photograph operates as a teaser. It estranges not a single racial trait but the series as a whole. According to semioticians, in languages written from left to right, when layouts make significant use of the horizontal axis, the elements placed on the left are presented as "given," the elements placed on the right as "new."[35] In this series, the top-left photograph is presented as given, or self-evident. The photographs on the right, therefore, are those to which the observer must pay special attention, not yet known.

The estrangement is evident in the slight, momentary disorientation that the photograph causes the viewer, as the viewer is compelled to ask, what is the relationship between that individual photograph and the rest of the series? This practice allows for various interpretations. In my view this practice enhanced the series as a whole in the sense that it tied together the separate photographs of individuals, thus totalizing the series as a whole.[36]

I will conclude this section by addressing the question of the reason for

Günther's employing estrangement tactics. I argue that this practice was designed to destabilize his audience in order to keep it in suspense. Günther employed these techniques to disrupt the possible habituation that serialization (itself essential) might generate. Günther's aims could be analyzed in light of the distinction between the formalists and the structuralists. If formalists intended to estrange reality through art, structuralists intended to estrange art itself. The difference between formalism and structuralism (and cubism) is important in this respect. Only the latter viewed artistic or literary form, not only reality, as subject to automatization and habituation and in need of estrangement. Hence, for instance, perspective was viewed as a representational form the effect of which gradually diminished as it itself became habit. Günther, albeit for different reasons, is very much sided with the formalists, such as Shklovsky, for whom representational forms themselves can lose their strength.[37]

Following this distinction between formalism and structuralism, the first form of estrangement I identified in Günther's work could be seen to correspond with that of the formalists, while the latter is already bent toward the structuralists. But the fundamental difference between Günther's estrangement practices and both the formalists and the structuralists should not be overlooked. Whereas they were engaged in estranging life or art, Günther was fixed on "race" and estrangement as merely a practical means of enhancing what he viewed as constitutive racial consciousness. Still, Günther was not only employing contemporary cultural tactics but tactics that were associated with his worse political and ideological opponents.

RACIAL ASYMMETRIES

Günther's idea of "race" entails structured and complex asymmetric power relations of the designated classes of a direct or an implied nature.[38] Günther's *Rassenkunde* contained several explicit asymmetries, and the analysis of his photographs unpacks several additional ones. Günther embraced several forms of explicit racial asymmetry on theoretical grounds. These asymmetries touch directly on the relationship between scholarship and ideology in his work. Although on the one hand, he wanted to educate his readers, on the other, he knew he would lose scientific credibility if his work would be considered as mere ideology. Some of Günther's asymmetries would have been considered already at the time of their publication as ideological or propagandistic. For instance, Günther initiated, together with his colleague Eugen Fischer, a photographic contest over the most beautiful Nordic German heads.[39] It would be difficult to imagine him running a similar contest of the most beautiful Jewish head, not merely because he viewed himself as Nordic and held that such contests should be run by members of the same class, but because the very idea of Jewish beauty ran counter to his conception of race.[40] In fact, racial science had been radicalized and politicized between 1900 and the 1920s to such an extent that it would even be difficult to imagine Günther iterating Galton's rather ambiguous verdict of the "beauty" of the Jewish type discussed above. Yet while such propagandistic aspects are important for his interpretation, I am far more interested in those aspects that were akin to then contemporary scientific practices.

In examining his two major publications, *Rassenkunde des deutschen Volkes* and *Rassenkunde des jüdischen Volkes*, we see that the written account glorified the Nordic and vilified the Jews. The deliberate choice of photographs is nonetheless far from simply reflecting those views. The photographs, which could have reduced the complexity of the written account, in fact convoluted it. In this sense, his use of photographs, in contradistinction to his written account, diverged from statistical reasoning.[41] The photographs Günther chose did not depict only stereotypical specimens of Nordic beauty and Jewish ugliness to comply with the written account but constructed patterns of individual portraits that covered a wide spectrum of beauty. These spectrums, however, did not overlap in a Boasian fashion, instead forming patterned clusters of racial characteristics.

Asymmetry was at the core of Günther's conception of race. His positioning of races was consciously asymmetric, and the relationship between the highest race, the rest of human races, and the Jewish people, depicted as the only antirace, according to Günther, was at no point equivalent.[42] The use of photographs indirectly reflected the fundamental asymmetries. There was a tension between the narrow picture Günther would have wanted to show his viewers and far more complex social realities. Günther found himself in a complicated situation—an overly aggressive strategy, removed from social realities, ran the risk of undermining the veracity of his claims.

Some forms of racial asymmetry could be drawn from Günther's use of visual source materials. In the course of discussion of the Nordic race, Günther made use of different reproductions, including drawings by Albrecht Dürer and Rembrandt, icons of German, northern, and Nordic art.[43] Günther made no similar use of visuals by Jewish artists, thereby implying that no equivalent existed. He did, though, employ reproductions of paintings of Jews by Rembrandt.[44] But Rembrandt was a northerner and a Nordic icon, according to Günther, and his pictures not only highlight Rembrandt's Nordic genius but also imply that this racial truth was present and known before the camera or the field of *Rassenkunde*. Finally, it emphasized that a Nordic painter was needed to capture the Jewish racial essence. In other words, he implicitly argued that Jews lacked creative genius to the extent that even a documentation of their own racial character was dependent on non-Jews.

Günther carefully selected his visual material from a vast range that he reviewed and collected. Questions of selection of materials further extend to the decision not to include, for instance, pictures by Mauricio Gottlieb or other Jewish artists as well as a biased selection of past pictures from those at his disposal. Selective choice of source materials contributed to the construction of racial asymmetry.

A more direct source of racial asymmetry derived from Günther's composite notion of racial seeing, or *Anschauung*. *Anschauung* and *Weltanschauung* are central modern philosophical notions. Historian David Naugle does not refer to "race" in his account of *Weltanschauung*, his focus being the relationship between this notion and Christian theology. He demonstrates that for Dilthey, Spengler, Jaspers, or Heidegger, *Weltanschauung* was conceived as intuitive, tied to finite determinacy, contingent on relative knowledge that involved a potential conflict of philosophical systems or perspectives.[45] Far from being a unitary concept, *Weltanschauung* could be, according to Nietzsche, merely an established convention, while Günther, following Chamberlain and others, viewed it as a fixed form of relativism intrinsic to race.[46]

Anschauung and *Weltanschauung*, both containing the word s*chauen*—to see—touch directly on questions of seeing and vision. These terms were not inherently related to race, but certain humanistic branches tied the two together. The notion of a specifically Aryan or Nordic form of seeing did not originate with racial scientists but rather with art historians, including Riegl, Wölfflin, and Hagen. Within the specifically antisemitic and racist discourse, the most prominent and outspoken representative of this concept was Houston Stewart Chamberlain. To appreciate how Günther concretized the notion of a racial form of seeing by means of photographs, it is first necessary to reconstruct the core of Chamberlain's notion of *Anschauung*.

From 1896 onward, Chamberlain developed the idea of a specifically Teutonic *Anschauung*, underpinning it on an explicitly racial basis. Already in *The Foundations of the Nineteenth Century* (1899), he opposed the German word *Anschauung* with the Greek *philosophy*. He translated the former as "intuitive perception" and defined it as wisdom that is never absolute and rests far more on the creative power of seeing than on the power of thinking. Chamberlain exemplified this idea by opposing Rembrandt's *Landscape with Three Trees* with a photograph of the landscape taken from the same spot. The specifically German or Teutonic notion is the opposite of the modern mechanical reproduction of knowledge, the photograph. From Sanskrit, Chamberlain derived that *schauen* is related to *dichten*, which means to invent poetically, "an active exercise of personality." The linguistic metaphors Chamberlain associated with the notion were drawn from vision: *Anschauung* "brightly illuminates" and makes "clearly visible."[47]

Chamberlain developed his comprehensive notion of *Anschauung* following the publication of *Foundations* in his publications on Kant and on Goethe, who he believed epitomized *Anschauung*. Chamberlain increasingly identified with what he viewed as Goethe's interpretation of science and art, giving a racial interpretation to Goethe's statement that "the manner in which a man looks upon the problems of life and of the world, in other words his philosophy, is born with him; it is the necessary result of his way of 'seeing.'" *Anschauung*, according to Chamberlain, was a "new vision" that brought together science, religion, and art in a specifically Teutonic way, merging sensitive observation with powerful conceptualization, sense perception, and creative imagination. *Anschauung* was inherently relative, and therefore partial, a form of seeing that turned "a series of disjointed phenomena into one connected whole."[48]

The stark opposite of *Anschauung* was the Jewish dissecting and sterile form of thought. "Everything," Chamberlain concluded, "tends to make us less able to see and less able to think." Indeed, "in order to understand Kant we must . . . begin by once and for all getting rid of the heavy burden of inherited and indoctrinated Jewish conceptions."[49] The Teutonic *Anschauung*, however, defined as an intuition, was born of incessant observation. This art of systematic observation Chamberlain believed to be a special characteristic of the Teutonic race and of Germanic science in particular.

Chamberlain's notion of *Anschauung*, with its underlying visual metaphor, was adopted by Günther. The latter tied it to additional signifiers he derived from the German language and viewed as specifically Nordic, all containing a visual as well as an idealistic connotation. For instance, in the short preface pages of the photographic competition *Nordic Race German Heads* (1927), Günther referred to the following concepts: *Bild* (image or picture), *Vorbild* (model, example, and type), *Inbild* (ideal), as well as *Rassenbild* (race image) and *Schönheitsbild* (model, but literally image of beauty).[50] Visual and linguistic imageries were closely knotted.

Visual expression of this linguistic ideology can be found in Günther's photographic reproduction of Greek statues (see fig. 3.4). Darkening the background and overexposing the statues sharpens the contrast between the statues and the background, flattening the statues into two-dimensional figures, insinuating their unreal, ideal "Nordic" whiteness.

As the images become more idealistic, the underlying asymmetries become apparent. These asymmetries were usually between the Nordic elements of the German population and the non-Nordic majority and between the racial relativity of beauty and the clear preference of the Nordic. While with regard to Jews, Günther wished to instruct viewers' how to identify minute traits of the racial "other," the photographic contest had far more to do with self-identification.

Within this asymmetric structure Jews were part of the "normal" racial asymmetries noted above, but they also took part in additional, abnormal ones. The Jews, according to Günther, possessed a *Blick*, a gaze that incorporated physiognomic features such as the shape of the eye and a specific form of expression, which is the opposite of *Anschauung*. The overly large eyelid (*Oberlid*), he maintained, drew the attention of observers, as it covered a larger part of the eye than normally found among European races. Basing himself on William Ripley, Günther suggested that Jewish faces give the impression of heaviness, sleepiness, and tiredness. Günther also referred to the claim, which appeared in the *Jewish Encyclopedia*, according to which the Jewish gaze was the source of the Jewish "sly" or "surreptitious eyes" (*verstohlene*) impression. In a footnote he referred to Stratz, developer of a theory of the Jewish gaze (discussed in chap. 1), only to disagree with him, stressing that the Jews' cunning and timid expression combined the mental and the physical.[51]

Günther's opposition of the Nordic *Anschauung* and the Jewish *Blick*, therefore followed Chamberlain's relativistic racial asymmetry. But whereas Chamberlain's idea remained abstract, as he only sporadically incorporated diagrams in his publications and did not use photographs, Günther endowed this idea with concrete visual expression by means of his massive use of photographs. The conceptual continuity between Günther and Chamberlain on race was coupled with the transvaluation of photography. Chamberlain, as we have seen, viewed photography as an expression of sterile mechanical science; Günther appropriated it for the expression of *Anschauung*.

Following an analysis of a number of asymmetries within Günther's work, it becomes clear that these asymmetries generate a multifaceted racial structure made up of explicit (such as *Anschauung*) and implicit (such as his use of "Nordic" art) ideas. I would now like to try and bring these different forms of asymmetry together in order to examine their social implications. I will offer two interpretations, both of which are racially relative, focusing on seeing and on subjectivity.

Already in his first book (1920), before turning to specifically racial writings, Günther complained similarly to Hans Sedlmayr or Belá Balázs (discussed in the introduction) that in many contemporary Europeans, "the ability to observe racial features has been completely lost, has become totally atrophied." Indeed, what was lacking was the specifically Nordic form of observation: "We lack the ability to see in fresh ways, all our perceptions are clouded by an absence of creative observation."[52] Both his criticism and his call for regeneration, therefore, applied only to members of the Nordic race.

The composite asymmetric structure of race was intended not only to regenerate a racial "seeing" that would identify racial differences; it also had a more comprehensive goal: to constitute a racially conscious social subject. One far-reaching illustration of the implications of this can be found in the interpretation of the Nuremberg racial laws of 1935, an explicit case in which scholarly or scientific categories are integrated into the legal order.

Written by government administrators Hans Globke and Wilhelm Stuckart, the document known as the binding interpretation of the Nuremberg laws (*Kommentar zur deutschen Rassengesetzgebung*) explicitly relied on Günther's racial paradigm. In particular, in the section of the document that discusses *Geistesjuden*, the concept of a group of people who are Jewish by spirit, the law argued that the social and spiritual actions of individuals of non-Jewish descent might determine the racial classification of their descendants as racially full Jews. This nonintuitive aspect of racial classification cannot be understood without the asymmetries elaborated by Günther, which highlight interplay of theory and practice, law and consciousness, biology and culture, body and soul. The crux of that legally binding interpretation was not only that race could be determined by conscious action but by the asymmetry between non-Jews and Jews. The addressees of the "call" were the Nordics. The onus of responsibility, activism, and honor was entirely theirs. Answering this racial call was the supreme responsibility of the individual cum racial subject, while the Jews were inherently excluded. In submitting to this call, the racial subject was constituted within a multifaceted asymmetric ideology.

RACIAL PHOTOGRAPHS AND "FAMILY RESEMBLANCE"

There is a tension between Günther's theorizing of race and what can be deduced from his employment of photographs with regard to the status of "type."

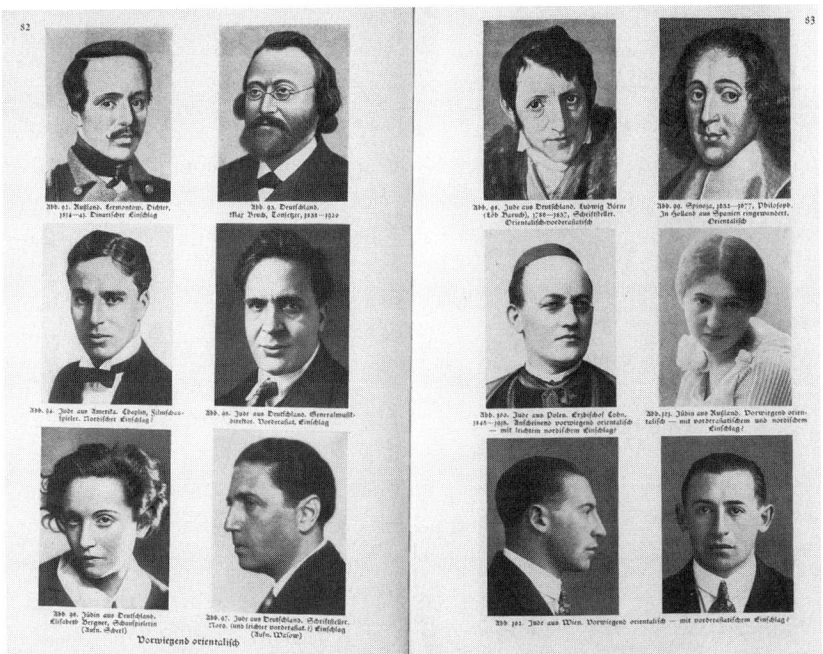

FIGURE 3.6. Lack of easily discernible common denominator. Hans F. K. Günther, *Rassenkunde des jüdischen Volkes*, 82–83.

Günther's theory was based on racial "types" that are characteristics shared by members of the race. In contrast to the Dammann brothers, Günther did not choose photographs of a "typical" specimen to illustrate all the typical traits of the type. His series of photographs did not seek to exhaust respective types, as they lacked an immediately recognizable common denominator possessed by all members of the race. In some of the series, the organization of the photographs created a complicated network of overlapping similarities but often none that could be reduced to any single common denominator. While very different in his motivations, this use of photographs followed in certain respects Ludwig Wittgenstein's idea of "family resemblance." Through an asymmetric comparison, in Jürgen Kocka's terms,[53] I attempt to use Wittgenstein's notion in order to illuminate a somewhat opaque aspect of Günther's photographic practice.

Wittgenstein expresses the idea of family resemblance in the *Brown Book*: "Some of them have the same nose, others the same eyebrows and others again the same way of walking; and these likenesses overlap."[54] In the constructed series in figure 3.6, Günther crisscrosses similarities and differences. Some of the portraits share common characteristics, such as the atypically Jewish straight nose of photographs 94, 96, 97, and 102, while 95 has a far more stereotypical Jewish nose. Photographs 94 and 96 share almond shaped eyes, while 95, 96, 99, and 102 share the more stereotypical round Jewish eyes. The same partial conflation could be pointed to with regard to lips, forehead, cheeks, or gaze. The photographic series creates an elaborate network of partial overlaps, but no single common denominator is shared. While Günther's series escape simple common denominators, they create the impression of shared typical characteristics in the series as a whole. As a Jewish type is presumed, the typical recedes to the background; unstable, it is obscure and concealed.

There is no evidence that Günther came across Wittgenstein's idea or that he was interested in the latter's philosophy; similarly, there is no record that Wittgenstein was aware of Günther's work. Wittgenstein's notion of "family resemblance" was only published posthumously, but it was developed over a number of decades, probably first articulated in writing in 1929 or 1930, further elaborated in 1933–1934, and brought to its final form in the 1940s. There seems to be no direct genealogical link between Wittgenstein's idea and Günther's photographic practice. The history of Wittgenstein's idea, however, is more intimately connected to racial photography than is commonly assumed. Carlo Ginzburg has shown that the connection between seeing, types, composite photography, and racial difference was made by Wittgenstein far earlier.[55]

Based on notes written between 1933 and 1934, the *Blue Book*, published posthumously in 1958, expresses ideas that anticipated Wittgenstein's remarks on family resemblance found in *Philosophical Investigations*, which, although only published in 1953, was ready for publication already in 1946. Wittgenstein's discussion of the typical was closely connected with Galton's photographic method. Wittgenstein wrote: "We are inclined to think that the general idea of a leaf is common to all leaves. (Galtonian composite photograph) This again is connected with the idea that the meaning of a word is an image."[56] The context from which Wittgenstein's concept of family resemblance sprang was contemporaneous with Günther's publications and was directly linked to the former's perception of racial difference and his interpretation of Galton's photographic method. Wittgenstein contemplated similarities that were instantly recognizable but could not be easily formulated by way of common denominators.[57]

In his latest and most famous writing on family resemblance, Wittgenstein omitted Galton and discussed "games" rather than racial traits. His emphasis on seeing, however, remained unchanged.

Consider for example the proceedings that we call "games". I mean board-games, card-games, ball-games, Olympic games, and so on. What is common to them all?—Don't say: "There *must* be something common, or they would not be called 'games' " *but look and see* whether there is anything common to all. For if you look at them you will not see something that is common to *all*, but similarities, relationships, and a whole series of them at that. To repeat: don't think, but look! For if you look at them you will not see something that is common to *all*, but similarities, relationships, and a whole series of them at that. I can think of no better expression to characterize these similarities than "family resemblances" or "family likeness." And I shall say: games form a family.[58]

And the result of this examination is: we see a complex network of similarities overlapping and criss-crossing: sometimes overall similarities, sometimes similarities of detail.

I can think of no better expression to characterize these similarities than "family resemblance"; for the various resemblances between members of a family: build, features, colour of eyes, gait, temperament, etc. etc. overlap and criss-cross in the same way.[59]

While Wittgenstein now focused on games rather than race, the German word he employed, *Verwandschaft*, translated in English as "similarity," in fact means specifically a relationship built on blood.[60] Clearly Wittgenstein's philosophy of language is alien to Günther's racial-visual practices. Yet a basic similarity can be found between Wittgenstein's idea and Günther's practice as they both stem from a similar cultural matrix and share certain concerns regarding "similarities [that] crop up and disappear," which escape formal definition.[61]

Out of the vast repertoire of images that Günther had at his disposal, and given the amount of energy he invested in constructing visual patterns, he could have easily chosen photographs of "perfect" specimens or selected photographs of individuals who manifested a distinct set of traits. Instead, he constructed series that on many occasions lacked easily discernible common denominators. This choice was not self-evident and requires a careful examination.

One possible reason for this choice was rhetorical: to keep viewers in suspense. The strategy had the advantage that it forced viewers to actively examine the series seeking common denominators. Furthermore, the series of photographs of Jews generated a sense of anxiety and alarm precisely because the "typical" image of the Jew could not be localized and remained irreducible to the reader. On the other hand, such elliptic series had the disadvantage that they did not disambiguate types. A second possible reason for this practice is tied to Günther's perception of social realities. Günther had repeatedly stated that all modern peoples were thoroughly mixed; while this did not negate the existence of exemplars of pure types, the majority of living individuals were mixed. Thus his series reflected more precisely the "polluted" state of social reality than his idealized, envisioned state.

There is only a slight similarity between Günther's practice of constructing series that lack common denominators and Wittgenstein's concept of family resemblance. Indeed, Wittgenstein's thoughts on family resemblance appeared in a broader context, as evident already from the opening lines of *Philosophical Investigations*, which included criticism of foundationalism and essentialist philosophies of language. Günther's theory could only count as a caricature of such essentialist philosophies of language. By bringing together Günther's practice and Wittgenstein's idea, my intention was to indicate the link between racial photography and aspects of contemporary culture. The comparison uncovers some of Günther's social and cultural assumptions about the elusive form of racial difference.

Günther's photographic practice of presenting series that lack easily discernible common denominators stands in clear opposition to the earlier racial photographic practices that relied on notions of type. If we wish to attempt to identify from which racial photographic practice it stems, that moment would most probably be Eugen Fischer's study of racial bastards. Fischer, we saw, was the first to shift the emphasis from the study of "type" to the study of racial mixture and to give this shift photographic expression. Günther appropriated Fischer's photographic method for the study of a specific mixed racial community, universalized it for the study of "race" at large, and through constructed series accentuated the elusive nature of racial difference. Through the short comparison with Wittgenstein's notion of family resemblance, however, we see that the *form* of Günther's photographic manipulation reflects wider contemporary cultural and intellectual concerns.

SCIENCE AND PROPAGANDA

Günther's use of photographs closely integrated scientific principles with elements of propaganda. Siegfried Kracauer's apt observations on Nazi propaganda films, *From Caligari to Hitler*, written in the midst of World War II, can be implemented in this analysis of Günther's photographic practices as well.[62] Günther recognized that, to use Kracauer's words, "allusions may reach deeper than assertions" (280), and like Nazi propaganda films, they make a direct appeal to the subconscious (279). Kracauer emphasized the role of repetition in creating patterns (275), a technique similar to propaganda films, and his observation that the use of "visuals in connection with verbal statements is determined by the fact that many propaganda ideas are expressed through pictures alone" applies well to Günther's use of photographs; in instances of this kind, according to Kracauer, the picture does not "illustrate the commentary" but "pursues a course of its own" (279–80).

Thus, if we apply Kracauer's observations to Günther's use of photographs, the latter can be seen to take part in Nazi propaganda. Yet clearly there are differences; Kracauer's analysis examined the use of maps as a dramatic exposition in propaganda films that depict a military strategy that is realized in the following sequence. Günther employed detailed maps, describing racial situations across Germany or Europe, while calling attention to the danger of racial irredentism and pollution.[63] Kracauer's contention that Nazi propaganda films measured time in centuries and not individual life spans can also be applied to Günther, who examined the "time" of races measured in eons. Kracauer observed the use of oppositions in propaganda that it was enough to show one side of an opposition in order to indicate the other side. Hence, on seeing close-ups of "brute Negroes" the "naïve spectator immediately attributes the complimentary ones to Germans."[64] This is only partially applicable to Günther's work. If, as Kracauer claims, "beautiful outdoor shots stress the insoluble ties between primitive people and their natural surroundings,"[65] the photograph of a group of orthodox Jews dressed in black in the white snow highlights the utter strangeness of Jews to the environment and the natural relationship between Nordics and their landscape. This illustrates how Günther's practice involved simultaneously both scientific and propagandistic practices.

Günther's photographic practices were effective because the manipulations he employed were sophisticated, embedded in a complex network of ideas that were already widely circulated and enjoyed a significant amount of social foothold. The aesthetic manipulation was subtle—the series did not simplistically correspond with beauty or ugliness. Günther chose many individuals designated as Jewish who were good looking, whereas not all the individuals he identified as Nordic were equally attractive. Rather than choosing a simplistic propagandistic approach, therefore, racial asymmetry was built into social and cultural codes. In the written account, Günther postulated both a pluralism of races, each with its intrinsic set of values, as well as the superiority of the Nordic race, and his photographic practice was poised between the two. Thus, his series of photographs depicting the Alpine race in *Rassenkunde des deutschen Volkes* indirectly created the impression that members of this race were inferior or aesthetically unattractive compared with Nordics. Unpacking the different manners in which Günther manipulated his reader reveals at least three ways through which the sense of superiority of the "Nordic race" over the other races was generated. These manipulations involve dress and implied social or cultural status, age, and the inclusion or omission of what could be termed "primitive types."

In the first section of his *Rassenkunde des deutschen Volkes*, Günther reproduced from Martin's *Lehrbuch* several photographs of anthropometric measurement instruments, insinuating that his approach to race was similar.[66] But unlike Martin's anthropometric conventions that included the undressing of the photographed subject for measurement purposes, Günther employed a different photographic technique, derived from ethnographic or folklore collections, in which traditional dress is an essential part of the photograph. A close look at Günther's different series reveals that dress is one subtle way in which he conveys his belief in the inequality of the races.

Ironically, Günther instructed his viewers to focus on physical traits rather than on the cultural and social status implied in dress, and thus this manipulation is not easy to detect. Social, economic, and cultural features were implied in the photographic series. Günther's series of European races and the Jewish people portray an uneven spectrum, ranging from well-dressed, clean, and neat individuals on the one hand to poorly dressed individuals, individuals whose dress indicates a low socioeconomic status, on the other. The final image that appears in the *Kleine Rassenkunde des deutschen Volkes*,[67] which was circulated in hundreds of thousands of copies, is of two group photographs of lively youths, some smiling, holding a flag. In *Rassenkunde des jüdischen Volkes*, however, Günther includes photographs larger than the passport size normally employed in his series, taking up nearly a page each, portraying groups of bitterly poor and equally joyful *Ostjuden*.

This photograph (fig. 3.7) portrays *Ostjuden* from Drustopol in Poland, as the caption designates them. They are all dressed in tattered clothes, sitting or standing outside a building and gazing at the cameraman. The men's eyes are almost entirely shut. Two of the men, the old man sitting in the front with his finger pointed and the man standing behind him, possess what Günther's contemporary viewers would perceive as a wicked smile. Observers will be struck by the numerous dark-haired children in the back of the photograph, indicating the intrinsic threat of the endless flood of *Ostjuden*.

In his portrayal of what he terms "European races" Günther was more subtle. These series, with the only exception of *pure* examples of the Nordic race,

scheinend schon lange bestehende ununterbrochene Handelsstraße der Juden von Arabien und Abessinien über Ceylon und Lambri auf Sumatra bis China erwähnt,[1] von der aus wahrscheinlich Abzweigungen auch zu den Handelsstraßen der Juden Persiens und der Kaukasusländer führten.

Bei Vermischungen mit der einheimischen Bevölkerung Vorderasiens muß dauernd der Einschlag vorderasiatischer Raffe in

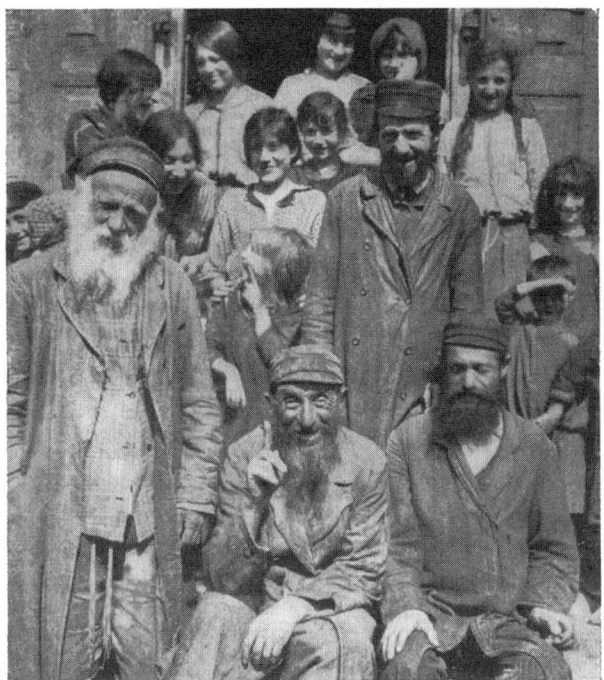

Abb. 183. Oſtjuden aus Druſtopol (Polen)

FIGURE 3.7. Title reads, "Ostjuden from Drustopol (Poland)." Jews as a threat to flood Germany. Hans F. K. Günther, *Rassenkunde des jüdischen Volkes* (Munich: Lehmann, 1930), 183.

contained photographs of the entire spectrum of racial features that made up the type (see fig. 3.3). The photographic layout of the Nordic race shows tidy, shaven or mustached, elegantly dressed young men (and there is a picture of a woman as well.) This spectrum implies that races are heterogeneous "wholes" bearing noble and less noble fruit; the Nordic race alone is intrinsically noble, aristocratic as a whole.

In all the series, with the exception of those depicting the Nordic race, Günther included one or more photographs of individuals who possessed

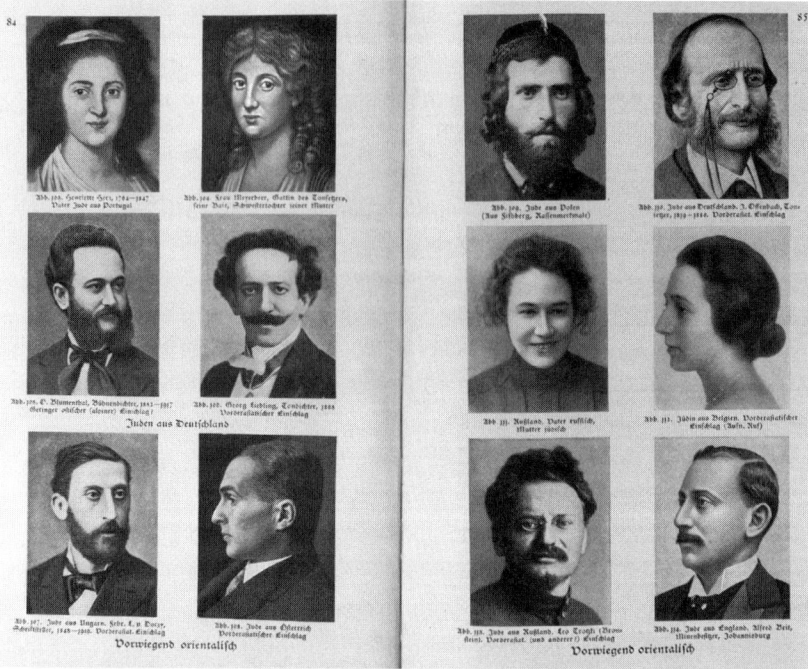

FIGURE 3.8. Construction of similarity through layout. Hans F. K. Günther, *Rassenkunde des jüdischen Volkes* (Munich: Lehmann, 1930), 84–85.

crude physiognomic features. These individuals depict in an exaggerated and almost grotesque form traits that were ascribed to members of the racial class. These individuals function as what I would term "primitive types," that is, they manifest the primordial expression of the type in its most crude expression.[68]

The inclusion of such photographs has an acerbic effect on the rest of the photographs. Günther inserted them there because their presence emphasized similar characteristics that were present, but in a far less accentuated form, in other photographs. Günther therefore implied that in all the photographs, the primitive characteristic existed in protean form. Nordic series, however, did not contain such "primitive" types; the presence of primitive types was expressed in his study of the Jews.

The top two photographs of one photographic layout contrast the better-looking observant young man with the thinner, older man (fig. 3.8). The contrast is further highlighted by the difference in the direction of their gaze. Light falls on the face of the young man from the top left, illuminating the left side (from the viewer's perspective) of the face. His eyes are elevated to a point slightly to the right and above the viewer; his facial impression is serious.

The lighting of the photograph on the right is even, the contrast between the skin color and the background is small, and the quality of this reproduction is poorer. The subject's gaze, slightly lowered, is fixed on a point to the left of the viewer. His thin hair only partially covers the overlarge ears, typical Jewish nose, overly high forehead, partially open eyes, and strained gaze. Both photographs were altered, cut from larger photographs, and the background of the right one was retouched or bleached. This layout constructs a similarity and relationship between the person identified as "a Jew from Poland" (on the left) and "J. [Jacques] Offenbach, a Jew from Germany." Furthermore, the layout constructed a similarity between the two men whereby Günther implies that the individual on the right is the "truth" of the one on the left. The layout leads the viewer to minimize the differences between the two.

Following the notion of "leading lines," Günther's layout of the photographs ensured that the observers' eyes remained focused on the page. Three of the four corner photographs bracketed the series by individuals looking outside in. The fourth corner photograph, at the upper corner on the right page, centered the frame. This was done through the use of the angle of observation: that individual was observed from the center of the layout outward, to the corner.

THE VISIBLE AND THE INVISIBLE

Photography was from its very inception closely related to attempts to study the invisible. In this respect, Günther's photographic studies can be seen as continuing a long line of particularly scientific photographic studies. As noted in the introduction, throughout the history of photography, domains of invisibility have included a wide range of topics, such as spiritualism, astrophysics, and motion imperceptible to the human eye. Günther's studies of the invisible were intimately related to his racial theory. His photographic study of the racially invisible could be seen as the climax of the genealogy of photographic studies of the racial invisible studied in chapter 1.

"Invisible" means more than one thing in Günther's work. The very idea of "invisible" racial difference remains absurd as long as a pre-Mendelian typological notion of race reigns and individuals are seen to be instantiations of types. Günther's attempts to study invisible or almost invisible differences cannot be dissociated from the rise of Mendelian genetics alongside the idea that racial difference could be present but hidden to the eye. The photographic study of invisible racial difference, therefore, was based on a specific notion of race, a specific cultural form of suspicion, and certain social situations.

The fundamental tension between two forms of racial difference is reflected in Günther's photographic practice. The tension between his racial ontology and his appreciation of the modern condition lay at the core of his competing photographic aims. Günther wanted to demonstrate the self-evidence of racial difference alongside its fundamentally concealed nature. Günther viewed racial hybrids, in particular those made up of what he conceived as greatly different races, as monstrous. From his photographic practice it is possible to draw two different forms of the representations of hybrids: monstrosities either instantly discernible or monstrosities cunningly hidden. As racial mixture ran counter to the natural order, Europe was seen to have become a bastardized continent. The lighter and more easily discernible form of monstrosity was that in which two racial essences cohabited in one individual.

A second form of monstrosity, directly associated with Jews, was more elusive, more difficult to isolate visually, and also perceived as far more dangerous. Günther believed Jews manifested a specific immediately recognizable racial difference alongside a difference that was fundamentally hidden and elusive.

Precisely because this second form of difference verged on the invisible, photography was Günther's prime tool for its study, as he believed photographs conveyed what words could not express. Photographs betrayed traces of the invisible, subtle visual signs of that which was concealed. Günther sought hidden differences on the photographic surface; he was not interested in the identification of racial elements and their ascription to one of the respective parents, as in Eugen Fischer's method. His goal was not simply to portray visual racial difference in any traditional sense but rather to depict a form of visual unconscious.

The organizing logic of the series of photographs is frequently difficult to decipher. But when looking closely, one can sometimes identify a subtle, almost unobservable trait that the eye was guided to observe. This practice closely followed Kracauer's statement with regard to Nazi propaganda newsreels, which he described as directly addressing the visual unconscious, deriving their power from the fact that they were independent of the written word and that they superseded it. Günther's series indicated the presence of invisible racial difference rather than attempting to visualize it or make it fully explicit. This aspect of Günther's practice is suggestive of notions of racial difference as disguised and hidden.

Günther merged the tradition of careful observation of minute racial differences developed by Stratz, Salaman, and Fischer with the specifically antisemitic intellectual genealogy of the Jews' camouflaged hidden difference developed by Houston Stewart Chamberlain, Alfred Rosenberg, and Adolf

Hitler. Chamberlain wrote of the Jew, "he conceals himself, he slips through the fingers like an eel. . . . One doesn't recognize him."[69] Günther gave this view a concrete photographic expression. At the same time, he remained loyal to the Mendelian view that priority must be accorded to the invisible (recessive) essence over the visible phenotype.

RACIAL IMAGINATION AS A TOTALIZING EFFECT

In previous sections of this chapter I have examined the various ways in which imagination as a racial attribute was interconnected with visual perception and embedded in a hierarchy of racial asymmetries. In concluding this chapter, I wish to point to an additional and more evasive social aspect of racial imagination. This chapter showed that Günther's use of photographs was guided by epistemological assumptions regarding how photographs could enhance the imagination: the more one sees the more one imagines, and the more one imagines the more one sees. These assumptions stand in direct opposition to the intuition according to which there is a negative correspondence between seeing and imagining. I now wish to address one specific social implication of this belief.

Kracauer offered the following sensitive observation: "Any creative process approaches a moment when only one additional experience is needed to integrate all elements into a whole."[70] Günther's paradigm was geared toward such a totalizing moment, and photography was essential for that "additional experience."

Günther brought together Simmel's and Wölfflin's interpretation of seeing as a constitutive activity of the subject, Chamberlain's antisemitic notion of a specific Teutonic *Anschauung*, and scientific conceptions of the photograph as an immediate, transparent medium. These conceptions, as Richard Gray has convincingly argued, were crucial in instructing the readers how to see, teaching them racial observation and a bottom-up form of surveillance.[71] An additional and more ambitious dimension has to do with racial imagination. This kind of photographic imagination had a totalizing social effect. It could be viewed as a nodal point that was the intersection of different registers.

Several points made in the introduction are crucial for understanding the totalizing effect of racial imagination in Günther's work and the role of photography therein. As noted, imagination, as a prerational capacity, was covertly politicized in the philosophical discourse of the final years of the Weimar Republic. Photography had the capacity to combine the actual with the potential. Imagination itself was conceived as relative to race. By the use of imagination, one can transform a deficient object into its ideal and vice versa. Photographs could be conceived as a powerful instrument in this regard.

We have seen that Günther's use of photographs deconstructed individuals in a Mendelian fashion to what he believed were their racial components. This allowed him to address the "impossibility" of the actual world around him: the racial inconsistencies of a tall long-skulled Jew or a short round-skulled Nordic; Nordics who had a Jewish nose and Jews who possessed light blue Nordic eyes. At this point, in light of the threatening chaos of racial differences, imagination was arguably the strongest instrument in extricating, classifying, and separating the potential from actual, the *Sollen* from the *Sein*.

The only attempt to analyze the role of the imagination in social life, as I have noted in the introduction, was done by Cornelius Castoriadis, who developed a theory about the role of the imagination in the constitution of society. Castoriadis noted that the answers that a society gives to the questions that it raises are ultimately irrational and are shaped by the imagination. Hence these can be considered answers only in a metaphorical sense.

Analyzing Günther's employment of photographs, I have become increasingly convinced that their ultimate function—over classifying visualization of racial difference, or, through constructed series, the formation of meaning—was a totalizing effect. Demonstrating the "reactionary logic" discussed in the introduction, the totalizing effect was to elevate and defend what he conceived to be the natural order and return it to its original and natural state. The totalizing effect, however, was also the ability to serialize: to move from the part to the whole or vice versa, to privatize a universal category, or to universalize a particular instance. The totalizing effect established an invisible bond; through an imaginary act it created an imagined racial community. At that moment various aspects of his theory converged and penetrated social life. Günther's theory, in Castoriadis's terms, was an answer to social questions, and the photographs, in particular, were crucial in engendering a totalizing effect. But the totalizing effect could also be conceived as a moment of disclosure: the bond of the racial identity—transcending social and geographical differences—became real. This was a point of conversion akin to the moment in which an observer recognizes that he or she is facing not individual trees but a forest but more significant because it marked the assimilation of the observer into the class observed.

*

Günther's use of photographs was more opaque than his written account. The photographs were closer to the racial objects than the written word and ultimately transformed the reader into a viewer and attempted to draw the viewer into experiencing race firsthand. Günther raised photographic matrices of the kind addressed in the previous chapter to an altogether new level of sophistication. Günther employed the photographs as if they enjoyed an indexical quality, in Charles Peirce's terms, but in fact he employed photographs to construct meaning through careful layout strategies and a system of suggestive subtitles. Thus nomenclature is classification; classification is ontology. I have attempted to deconstruct some of the manipulations he exercised on his audience, particularly in terms of the layout of his photographs, and have demonstrated how these sought to teach people to see not only others but to identify themselves through the photographs. Employing a practical epistemological perspective, I have demonstrated how aspects of his photographic practices were reflective of wider cultural considerations, from estrangement, normally considered with the opposite side of the cultural map, and Wittgenstein's notion of "family resemblance." I attempted to show that Günther's photographic practice ultimately aimed at the imagination. The reason was that it was there, he believed, that the Nordic racial community was constituted and held together. While politically, ideologically, and in terms of endorsement of the National Socialist idea of Nordic superiority, very close to Günther, the photographic method of Ludwig Ferdinand Clauß, discussed in the next chapter, takes us to different notions of race and versions of science. According an equally central role to the imagination but based on different photographic techniques and notions of photographic serialization, we will see Clauß's attempt to transform the racial photograph from a form of evidence into a self-referential sign.

Racial Photographs as "Thought Experiments": The Photographic Method of Ludwig Ferdinand Clauß

Disjunctions between the visible and the invisible are at the heart of Ludwig Ferdinand Clauß's racial photographic method. Clauß repeatedly emphasized that his aim was to "teach people to see,"[1] but attempting to understand what he meant by "seeing" in the Weimar and Nazi periods and the work that he wished photography to do in this respect leads us from science to philosophy, from measurement to phenomenological description, and from representation to the imagination. The general practical epistemological approach employed in this study reveals Clauß's method to be the last original development in racial photography, with the redefinition of the racial photograph as a self-referential sign. But with Clauß the practical epistemological interpretive framework developed in this study is also stretched to its limit.

Even the simplest biographical facts about Ludwig Ferdinand Clauß confront the historian with great interpretive difficulties. In the 1930s, Clauß initiated a legal procedure within the Nazi party, to which he belonged, in an attempt to be allowed to continue his professional cooperation with his long-time Jewish assistant, Margaret Lande.[2] In a legal twist, the procedure was then turned against Clauß, leading to his being expelled from the party and banned from teaching. In an exceptional act of courage, Clauß then hid Lande in his country mansion, thereby saving her from certain deportation and death, and for this act of courage, Yad Vashem recognized him as one of the Righteous Among the Nations shortly after his death. It was only following the intervention of German historian Benno Müller-Hill several decades later, citing the disgrace of including a leading Nazi race theoretician whose prominence was second only to Günther, that the title was revoked.[3]

Clauß, who saved the life of his Jewish assistant, was among other things also a close friend and scientific collaborator of Bruno Beger's, who was a medical doctor and physical anthropologist who played a central role in Himmler's *Ahnenerbe* project and who stands out even within the most horrific chapters of murderous racial science. Beger planned and carried out the creation of a collection of Jewish skulls. To that end he wrote a memorandum, which Himmler approved, about how to select and measure Jews while still alive, then kill them without damaging their skulls, and finally employ particular chemicals and techniques to remove the skin, flesh, and tissues from their bodies. Beger carried out his program in spite of significant technical problems caused by the wartime situation: he handpicked his victims in Auschwitz, measured them (and caused some of them to undergo extraordinarily cruel medical experiments while they were still alive), had them sent to the camp at Natzweiler to be properly killed, and had their bodies put into specially designed containers with acid. After their skin, flesh, and tissues had been fully removed, he sent their skulls and skeletons to research institutions. Clauß housed Beger in his country mansion (probably at the same time as he was hiding Lande), and they remained close friends and associates after the war. With funding from the *Deutsche Forschungsgemeinschaft* (DFG; German Research Foundation), they traveled together to areas of the Middle East, a trip that formed the basis of Clauß's last major publication.[4]

It is not only Clauß's biography but also, and equally, his work that challenges the historian so severely. One has only to look at the extremeness of his racial determinism juxtaposed with the lucidity of his philosophical prose, his insistence that he is participating in a rational discourse, or the philosophical seriousness with which he puts forth absurd contentions.[5] While it would be easier to go to one or the other of two possible extremes—either viewing his work as mere pseudoscience or else taking it seriously while smoothing out the ideological and political elements—this chapter adopts a messier interpretive strategy that insists on their integration or fusion.[6] This strategy is not without tensions or contradictions, for instance between Clauß's explicit philosophical arguments and their (sometimes opposite) political implications, or between photographs in his economy of demonstration and their meaning along wider contemporary coordinates. I address these tensions in the three parts that make up this chapter: I begin with a concrete analysis of the use of photography in two "cases," Jews and the landscape; in the second part, moving to a more abstract register, I analyze the philosophical framework—Gestalt psychology and phenomenology—in which Clauß's photographic practice was embedded; and the third part shows the intertwining of the elements of his own scientific economy and strains of the wider National Socialist culture.

Rasse und Seele (Race and Soul), Clauß's principal publication, first published in 1925, comprises two parts, both founded on the discussion of photographs. The first part consists of a series of chapters in which Clauß meditates on photographs of racial types of man. The second part of the book, "Grundfragen der Rassenseelenkunde" (Fundamental questions for the study of racial soul), attempts to develop a synthetic framework for the study of what Clauß terms *psychoanthropology*. Here Clauß briefly discusses the photographic method of arresting movement in still photographs taken at short intervals,[7] but on an ethnographic level: talking about responses to requests for people to allow themselves to be photographed, or getting the "right moment" for a photograph to capture the expressive response to a question.[8] He does not touch on the fundamental status of photography as scientific evidence. Reconstructing his use of photographs in the two following "cases" brings to light their critical role in his argumentation.

Jews

The opening lines of the book, focusing on the performance human (*Leistungsmensch*) type of the Nordic race, refer the reader to a series of photographs that are integrated into the body of the text. This structure recurs in all chapters, including those in which Clauß discusses Jews: the chapter on the revelation man (literally "human being": *Offenbarungsmensch*) of the desert race (*wüstenländischer Rasse*) and the redemption man (*Erlösungsmensch*) of the Near Eastern race (*vorderasiatischer Rasse*).[9] This discursive form immediately engages the reader with what Clauß believes to be the particulars of racial gestalt without, however, addressing questions concerning the definition of race, the number of existing races or types in the world, why certain types are selected for discussion, or the selection of individuals for photographs.

Three photographs of a well-groomed young Arab man, dressed in a collarless white button-front shirt and traditional white Arab turban, against a white background, open the discussion of the revelation man. The angles differ from those found in standard anthropological photographs: one is a semiprofile portrait, and the other two are taken from more or less frontal angles. None of the photographs follow the physical-anthropological convention of neutralizing the facial expressions; on the contrary, the photographs show the young man smiling lightly while posing for the camera. Clauß states in his preface that virtually all the photographs in the book are his own and that they were selected out of several thousand in his possession.[10] He almost always employs series of photographs taken in one single sequence. The reason for this, as he explains in the second part of the book, is to arrest and capture movement, which is the essence of racial style.

ClauB forces a sense of estrangement on the viewer by selecting low angles and by reproducing partly blurred photographs, indicating that the subject was in motion. He also reproduces photographs that would normally be judged as deficient in other ways, for instance, photographs of individuals with gaping mouths.[11] In general, he employs similar methods for the study of various types, but the chapter on the Nordic race has only one open-mouthed subject. Gaping mouths in photographs indicate low intelligence or culture; such differences in photographic treatment, which are sometimes rather subtle, remain unacknowledged in the book. To discuss the "redemption man" type, ClauB uses six additional photographic series, ranging from one single photograph of a young Bedouin man, taken from close up and a lower angle,[12] to a series of six photographs of a young Jewish Yemenite woman,[13] as examples. Three of the photographs of the young Yemenite, taken against a gray background from angles varying from three-quarter profile to full frontal, show her with her eyes shut or closing. ClauB pays special attention to the form of the eyes when they are shut as a component of racial style, and his discussion moves freely from observations of the photographs to observations of the type.

For the revelation man, all knowledge of the world stems from revelation, the pinnacle of his value system.[14] In his world, all is God given—all happenings come from God; all that exists is predestined by God and has the status, ultimately, of booty (*Beute*).[15] If a person of this kind runs across a defenseless person in the middle of the desert, for example, that person is God-given booty. God is permanent and stable but by definition unreachable; all else is fluid, fleeting, and momentary. Social and cultural forms flow from this basic structure, determining how individuals of this kind respond to the camera.[16] ClauB derives general observations concerning the type from the photographs and ruminates on the photographs based on the gestalt of the type.

ClauB has a particular interest in the study of the gaze: he contrasts a young Arab's gaze with that of the Nordic performance man.[17] The performance man creates the object of his gaze by way of distance between subject and object— ClauB repeatedly plays on the literal meaning of the German word *Gegenstand*, which means "thing" or "object" but the literal meaning of which is "stand against." This distance is the defining feature of Nordic objectivity. The Nordic man, stabilizing the object in front of him, judges it on the basis of a clear differentiation between himself and the object. Studying the gaze of the young

Arab man, Clauß determines that this type lacks the hard objectivity (*sachliche Wesenshärte*) of the Nordic race. This type of man, belonging to the children of the moment (*Kinder des Augenblicks*), cannot fix his gaze on an object because his internal world fluctuates and objects, for him, are as a result fluid and unstable. Clauß's observations about the photographs cannot be separated from the observations he makes about types. The observations could be characterized as being on a continuum between phenomenological, ethnographic, or cultural observations at one end and stereotypical, racist, and at times ridiculous generalizations at the other.[18]

Jews are textually and photographically integrated into Clauß's discussion of the revelation man, although with that type Clauß, for all practical purposes, has Muslim Arabs (in particular, Bedouins) in mind, who are educated to follow their racial form. For understanding Jews, the redemption man type is far more important. Originally, Clauß contends, these two types had nothing in common. When he discusses Jews in terms of redemption man, then, he endorses and gives a phenomenological interpretation to von Luschan's theory of the Hittite (as distinct from Semitic) origin of modern Jews.

Clauß's discussion of the redemption man type is structured around photographs of six individuals, but most of the chapter centers on photographs of a man Clauß describes as a Kurdish Jewish porter in Jerusalem (fig. 4.1).[19] Where von Luschan excluded Jews from his photographs in order to include them in the category of modern Europeans, Clauß photographs Jews in order to exclude them from that category.

The photographic angles range from a full profile of the man as he takes a drag on a cigarette to four almost identical frontal photographs which show the man's face with different degrees of what looks like a soft, faint smile. All the photographs center on his head. They emphasize the man's Arabic turban, covering roughly a third of the area of the photograph, and the close range accentuates the man's bulky nose. Unlike the Turkish Armenian's photograph analyzed in chapter 2, this man's skull shape cannot be observed. All the photographs are taken from close up, outside, against a neutral off-white background; his eyes are open, though obscured by the lighting conditions, in all but one of the photographs (and in that one his eyes are shut).

The porter incompletely manifests his type, according to Clauß, because he fails to exhibit the particular Jewish form of religious learning (*Rasse und Seele: Eine Einführung in den Sinn der leiblichen Gestalt*, 88). But this attribute, this failure to manifest a basic part of the type, is in fact essential to the type and does not express a failure on Clauß's part to select a proper example. In

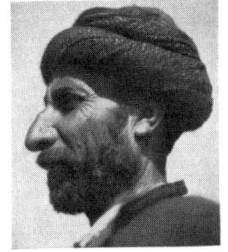

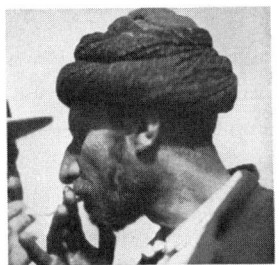

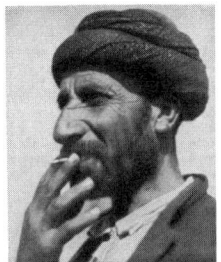

Bild 62: Jüdischer Lastträger aus Kurdistan.
Erlösungsmensch, vorderasiatische Rasse. Lastender Stoff.

Bild 63: Derselbe. Zu S. 77.

Bild 64: Derselbe. Zu S. 77.

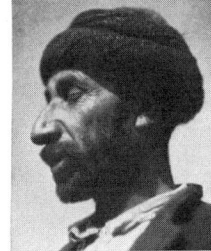

5. Der Erlösungsmensch   81

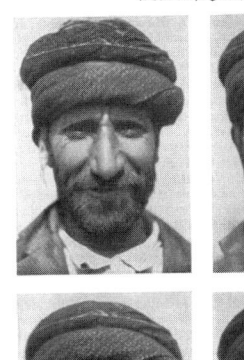

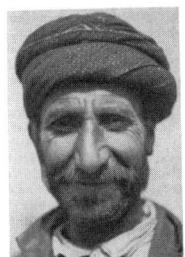

Bild 65: Derselbe. Ein gewöhnlicher Anlaß (vgl. S. 87) kann unversehens
den Ausdruck einer vergeistigten Heiligkeit hervorrufen. Dadurch entsteht ein
für diese Menschenart kennzeichnender Widerspruch zum Verhaftetsein im Stoffe.

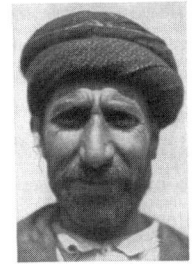

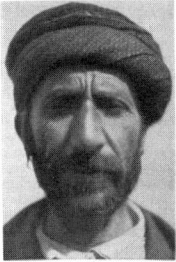

FIGURE 4.1. Through breaking and freezing the movements of an individual Clauß wishes to derive the essential mental characteristic of the type. Ludwig Ferdinand Clauß, *Rasse und Seele: Eine Einführung in den Sinn der leiblichen Gestalt* (Munich: Gutenberg, 1937), 78–81, photographs 62–69.

Bild 66—69: Derselbe. Ein Versuch zu schlichter Erheiterung (Bild 66/67)
mißlingt: ein immer bereitliegender Mißmut gegenüber dem Einfach-Leben-
digen bricht durch (Bild 68/69). Er entspringt aus dem Zwiespalt zwischen
„Fleisch" und „Geist", der zum Wesen des Erlösungsmenschen gehört.
Clauß. Rasse und Seele.   6

the deepest sense, the failure to meet its own racial gestalt is the source of this type's difficulties and what makes modern Jews an acute "problem" (99).

The complexity of the redemption man stands as the exact opposite of the simplicity of the Nordic performance man. For the former type, nothing is simple. Its basic phenomenological gestalt, which Clauß reads from the photographs, is structured around spirit (*Geist*), around the word, as an attempt to free oneself from flesh (*Fleisch*), against which spirit is defined. Redemption, for this man, can only be reached through the spirit and by overcoming the flesh (85), but redemption is always and by necessity doomed to failure, as there will always be a remainder of the flesh (80–81).

Discussing, again, the photographs of the Kurdish Jewish porter in Jerusalem, Clauß states that much derives from this fundamental structure. Thus, although this man lacks Jewish learning, he views any aspect of life that does not contribute to spiritualization (*Vergeistigung*) as sin—a fact that Clauß derives from this man's gaze. This individual can neither accomplish his gestalt nor free himself from it (88). But this form of life is life defying: only a priest can reach spirituality through the word, through the acquisition of enormous amounts of religious knowledge, and through overcoming the flesh (99). Clauß, in a discussion of a photographic series of a Greek monk, suggests that this Jewish structure is at the heart of Christianity.

Violence, according to Clauß, is an individual attribute, not a racial one. But if for the Nordic performance man body (*Leib*) and soul (*Seele*) are inseparable, then, according to Clauß, spiritualization is inherently connected to violence because the attempt to liberate the spirit from the flesh is violence (82). A more serious problem stems from the fact that members of the redemption man type attempt to flee their own structure. This only leads to substitute forms (Clauß names Freud's psychoanalysis as an example). This type wants to see slaves, so it creates abstract forms of domination (Clauß names money; 97–98), and these forms of domination serve as revenge against any expression of life, because this same gestalt defines such domination as sin. While Clauß does not restrict the type to Jews, they are its most prominent example.[20] (If all this appears blatantly antisemitic, it should be noted that when Clauß's book appeared in print, reviews in Hebrew by Jewish authors were not unanimous in their response to his interpretation. Some praised Clauß for his fresh method, describing his treatment of Jews as being among the most beautiful descriptions of Jews found in literature.[21])

Both when accomplishing their gestalt and when attempting to break away from it, members of this type are doomed to disharmony with themselves and with others. In this sense, Clauß states, this type is a pressing problem, which is accentuated all across Europe by the infiltration of seemingly "raceless" individuals.[22] Clauß reiterates these statements in an article that appeared in 1941 (the year that the legal procedure was turned against him and the same time in which the persecution and murder of European Jews was escalating).

In his biography of Clauß, Peter Weingart more or less dismisses this latter article as a self-serving device. It probably was that, but Clauß's ideas remain consistent in it. The context of the article ("Woran erkennt man den Juden?" [How does one recognize the Jew?])[23] is the unfolding of the external "solution" to the problem, of which Clauß seems to be well informed.

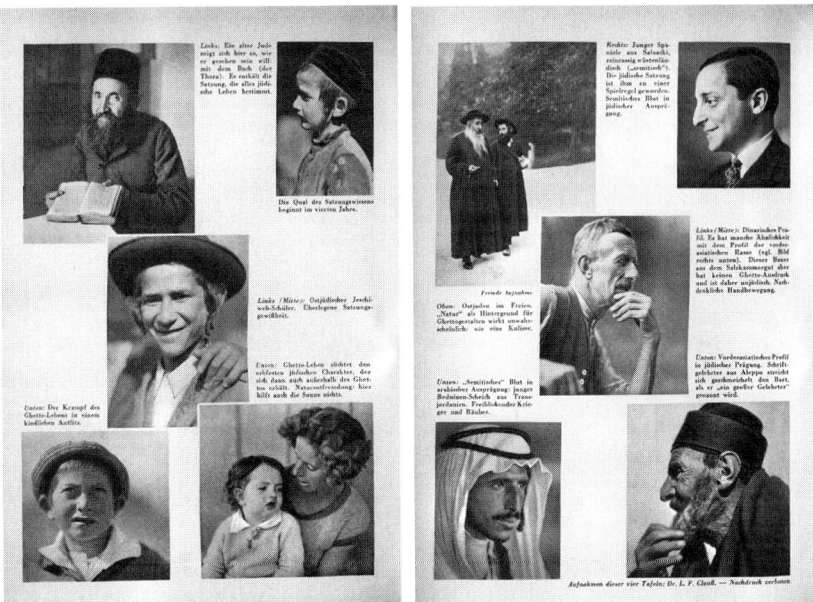

FIGURE 4.2. Racial essence of Jews has eroded to the extent that they are almost unidentifiable. Ludwig Ferdinand Clauß, "Woran erkennt man den Juden?," *Wir und die Welt* 11 (1940), 458.

Everyone knows or believes they know what a Jew looks like. The more the Jewish question [*Judenfrage*] is externally solved, the less the Jew is present in German life; in a word, the more invisible [*unsichtbarer*] the Jew becomes for the German eye, the less will the eye encounter real experience and the more will the representation of Jews rely on racial books and caricatures. A generation [*Geschlecht*] will emerge for whom no real experience of the Jew will exist but from books.[24]

Inasmuch as an academic publication can partake in persecution and advocate murder, Clauß's article does this. His use of photographs in the article deals more clearly with allegedly invisible, hidden difference than it does in the book; the hidden difference, he contends, is more dangerous than the easily discernible one and requires more knowledge and detecting skill. It is the gaze, more than any anatomical features, that is the key to identifying the Jew (fig. 4.2).

Clauß demonstrates this by comparing Jewish to non-Jewish individuals who bear "Jewish" traits. He then discusses Jews who lack Jewish physical traits but can be identified through their gaze. This gaze does not directly face its object—in fact it has no fixed object—but squints, as its goal (*Ziel*) is constantly fading (*ständig im Schwinden*).[25] Even describing this gaze is hard because Indo-Germanic languages lack a word to designate it.

Clauß thinks to demonstrate this with photographs of a Sephardic Jew. In line with much of Nazi propaganda, Clauß inverts the real historical roles of aggressor and victim, justifying in practice the violence inflicted on Jews because the latter are the source of dehumanization:

> The written constitution, the word, the letter are from God. . . . Because this code was created for Jews alone, it applies only to them and only they should know it. It draws a line around Jewish life. What lies beyond is godless and against God [*gottwidrig*]. The people of Israel are the people of God. . . . All others are *gojim*, and the Talmud teaches that they should be treated as animals.[26]

As the Jews of Poland and Russia were incarcerated in ghettos, Clauß vowed that they could be identified by their ghetto expression. The word *ghetto* was invented in the sixteenth century, Clauß noted, but he claimed that its real history was longer and what made Jews discernible: in truth it was invented by Jews, who conceived of themselves as the slaves (*Knechte*) of God,[27] and the Jewish gaze was the outcome of the ghetto and of the spirituality of this type.[28]

The basic use of photographs in the article is the same as in the book: Clauß reads the attributes of the type through the analysis of photographic series of individuals. In the article, however, he emphasizes hidden dimensions found in the gaze, and that hiddenness also serves as justification for the eradication of the Jews. The racial imprint has eroded to such an extent that they are almost unidentifiable (see fig. 2),[29] and this camouflage is part of their strategy as "destroyers of culture," to use Hitler's phrase.[30] Their form is now irreversibly compromised.[31] The German external solution correlates positively with the utter degeneration of the Jews.

Landscape

Clauß's views on "seeing" continue ideas of the kind we encountered with Wölfflin and Simmel (in chap. 1), but whereas Wölfflin and Simmel, as we saw, expressed reservations about the photographic medium, Clauß may be credited with being the first to use photographs to advance the idea that "seeing" is constituted by the imagination. Clauß opens his discussion of the landscape by warning the reader that when speaking of the notion of "lived spaces" (*erlebte Räume*), one is necessarily called on to employ the words of poetry rather than science. It is the words of poetry that are needed to describe the connection between the "inner life of races" and the relationship between racial soul (*Rassenseele*) and designed spaces (*gestaltete Räume*).[32]

All forms of seeing are creative, Clauß claims, extending Simmel's concept of "creative" interpretation.[33] But whereas the neo-Kantian interpretation views the constitutive side as subjective and universal, Clauß is interested in "specific differences." To see is to establish or set meaning (to be "meaning-setting," *bedeutungsetzend*): "race is form and forming."[34]

Starting from this theoretical axiom, Clauß addresses the "Nordic man," who sees in terms of distance and whose seeing creates borders (is

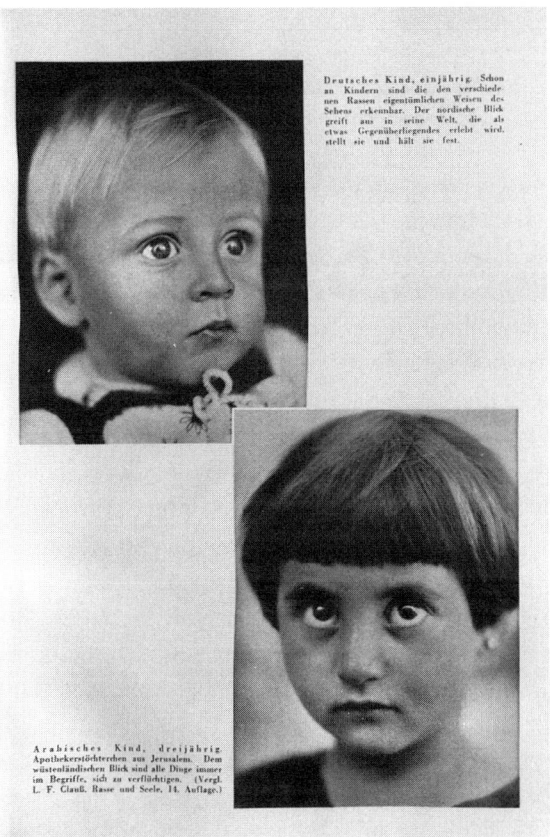

FIGURE 4.3. Gaze as racially determined in infants or toddlers. Ludwig Ferdinand Clauß, "Rasse im Raum," *Wir und die Welt* 1 (1940), 9.

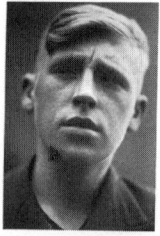
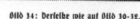
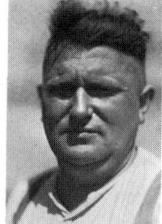
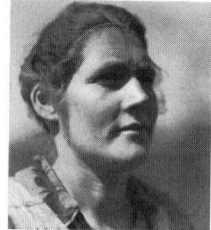
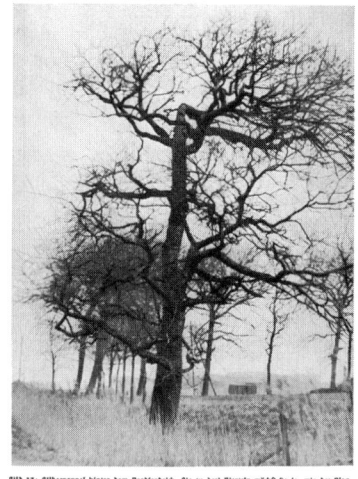

FIGURE 4.4. Landscape reflects the inner form of respective types. Ludwig Ferdinand Clauß, *Rasse und Charakter: Das lebendige Antlitz* (Frankfurt: Diesterweg, 1938), 70–71.

"border-creating" [*grenzenschaffend*]).[35] Elsewhere, Clauß employs landscape and architectural photographs in order to demonstrate different racial styles. Here, however, he employs photographs of individuals accompanied by short descriptions (fig. 4.3).[36] He begins by analyzing the photographs of two young children. At the top left corner of the page there is a photograph of a blond German one-year-old child with his lips pressed shut and his eyes wide open, his gaze fixed on the distance ahead of him. The caption states,

> It is already possible to recognize the different races' specific forms of seeing in children. The Nordic gaze reaches out to grasp the world, which is experienced as something opposite [*etwas Gegenüberliegendes*], positions it, and then holds onto it [*stellt sie und hält sie fest*].

The bottom right corner of the same page shows a photograph of a three-year-old girl, identified as the Arab daughter of a pharmacist from Jerusalem. The girl has straight dark hair; her lips are pressed shut, just like the boy's; her eyes are wide open, and her gaze is fixed on a closer point, in front of and above her. The caption reads, "To the desert gaze, all things are always about to evaporate."[37] By placing her photograph at the bottom of the page, Clauß accentuates the position of the reader as looking down at the girl.[38]

On other occasions, Clauß places human and landscape photographs together, suggesting that landscape reflects the inner form of various types (fig. 4.4). Here, for instance, Clauß suggests that close observation allows us to see that the posture of the people shown on the left-hand page of the layout and the lines of the tree on the facing page (shaped to accommodate the North Sea storms) share the same form.

Detecting a dependent relationship between race and environment could be understood as a form of geographical determinism, but Clauß argues for an opposite causation: every race has its inner landscape (*innere Landschaft*) independent of the actual one it happens to inhabit. The gaze appropriates, transforms, and molds the landscape to its inner form.[39] Nonetheless, every space contains the contours that determine and constrain the ways in which it can be formed.[40] Nordic space is particularly appropriate for Nordic seeing. But this only seems like a tautology, Clauß states, because we name races according to the geographical areas in which they formed their seeing. Landscapes, he emphasizes, in contrast are transferable (*Räume können wandern*). Through this conception of the landscape as molded by race, the basic asymmetry that characterizes Clauß's racial typology, though it is only implicit at this stage, becomes visible. With its powerful gaze, the German *Volk* is predisposed to appropriate distant areas and mold them to its inner form.[41]

PART TWO: EPISTEMOLOGY AND METHODOLOGY

Clauß through Arendt: The Status of Examples

Let us now move to a different register of discussion. Clauß's major publications are all constructed of a form of textual exposition of photographic series. The series are always of one single individual at a time, such as the Jewish porter discussed above, but the individual is not a specimen in the anthropological tradition: Clauß shows no interest in any measurements, and the photograph is not an illustration. Rather, they are "examples." But what does "example" mean? And what are they examples of? Approaching Clauß's photographic practice from these questions leads us fairly directly to the role of the imagination, rather than perception, and to a philosophical, rather than scientific, tradition. All this can be further elucidated by bringing Hannah Arendt's short seminar on the imagination into the discussion.

Galton's attempt to generate visual statistical racial means, Martin's attempt to generate racial statistics, and von Luschan's attempt to capture racial essences share a basic feature despite the important differences between these ventures: in all of them, photography is used in order to represent or reproduce what they believe are racial universals. In one way or another, photography is used to subsume the particular under a general rule. Even a brief examination of Clauß's publications is sufficient to make it clear that his photographic series of individuals follow a different logic—the logic of particulars or, in the language of Arendt's analysis of Kant, the logic of "exemplary validity," in which the rule is derived from the particular. Analyzing Clauß's photographic practice using the concepts developed in Arendt's short seminar, given in the New School for Social Research in the fall of 1970, reveals his series to be tightly tied to this notion of the "example" and underscores the specific role of the imagination therein.[42]

The point that Arendt makes in the seminar is that according to Kant, the same faculty that provides schemata for cognition, namely imagination, also provides examples for judgment.[43] In the main body of the seminar, Arendt reconstructs the role of imagination in the first edition of Kant's *Critique of Pure Reason*. According to Kant, if there are two branches of experience and knowledge—those two branches being intuition (sensibility) and concepts (understanding)—then imagination is the faculty that provides the synthesis between the two, and it does so (this is Arendt quoting Kant) by "providing an image for a concept."[44]

Arendt concludes her seminar by succinctly summarizing several decisive points, which are crucial for analyzing Clauß's photographs as "examples" (I am paraphrasing her but retaining her use of quotation marks). First, the perception of a particular table contains the idea, or schema, of "table" as such.[45] Hence, no perception is possible without imagination. Second, the schema "table" is valid for all particular tables. Third, without schemas we would be surrounded by a manifold of objects, but no knowledge or communication would be possible.[46] In other words, what makes something particularly communicable is that (*a*) in perceiving the particular thing we have a "schema" in the back of our minds, and that (*b*) this schematic shape is in the back of many different people's minds (because even disagreements presuppose that we are at least talking about the same thing). Fourth, the *Critique of Judgment* deals with reflective judgments as distinguished from determinant ones. Determinant judgments subsume the particular under a general rule; reflective judgments "derive" the rule from the particular. In the schema table, one actually "perceives" some "universal" in the particular. Arendt stresses that Kant hints at the distinction between these two forms of judgment in the *Critique of Pure Reason*. Fifth, we need imagination in order to recognize sameness in the manifold. Imagination, therefore, is a condition of all knowledge and determines sensibility a priori. Finally, Arendt notes that in the *Critique of*

Judgment there is an analogy to the "schema": the example. Examples play a role in both reflective and determinant judgments, which is to say whenever we are concerned with particulars. The example is the particular that contains or is supposed to contain a concept or general rule.[47]

Schema here, as a kind of archetype or idea of something through which its particular instances or instantiations can be identified, has a different meaning than the one we used in chapter 2. By way of a textual elaboration on photographs of individuals, following the logic of particulars, Clauß attempts to reveal features essential to the type, which can be recovered from the particulars because they preexist in the imagination.

The photographic series of the Kurdish Jewish porter in Jerusalem opens Clauß's section on the redemption man type, of the Near Eastern and Armenoid race. What follows are twelve pages of meditations on the photographs. Scrutinizing the series of eight photographs of the same individual, all taken within short time intervals of each other—in other words, breaking down and freezing the components of one sequence of this individual's movements—Clauß attempts to derive the essential mental characteristic of his racial type (see fig. 4.1). The photographs, then (and as pointed out earlier), are not meant to be illustrations in the physical-anthropological or ethnographic tradition but are used as an "example," a particular that contains a concept or a general rule.

Clauß opens his analysis by looking at such things as the way the turban sits on the head of the man, the form of the line the turban creates with his forehead, or the relationship created between the man's beard and his nose by the line connecting them. In all of this, Clauß looks at the individual as a "whole," making no attempt to separate between the physical-anthropological, expressive-physiognomic, and cultural-environmental dimensions; on the contrary, in fact, he studies their interconnectedness. By the end of the second paragraph of analysis, Clauß has already drawn from his observations the unity of the type's expressive possibilities (*Geschlossenheit der Ausdrucksmöglichkeiten*)and determined that this type has, by nature, nothing to do with elegance, which flows from the blood.[48] The Jewish porter, Clauß determines, lacks Nordic distance, and from his gaze, although the porter received no religious education, Clauß deduces the Jewish form of learning, marked by the "wish to know" (*Wissenwollen*) found in the yeshivas.

Through contemplative analysis of the same photographs to the meaning of being Jewish, Clauß adds his deduction that being Jewish is a process of becoming Jewish by way of undertaking to acquire religious learning.[49] He claims that the porter's strained look results from guilt (*Schuld*), a guilt that tends toward violence, over having failed to acquire Jewish religious knowledge.[50] Clauß's additional observations fuse a form of phenomenological observation with antisemitic stereotypes, but his entire discussion moves, crucially, from the particular to the rule that Clauß believes to derive from the close study of that particular. What appears, from the perspective of the scientific methodologies studied in chapter 1 (and irrespective of our view of its veracity), as a form of circular logic, however, is based on a specific conception of perception as tightly tied to imagination. Clauß also differs from the scientists of race in that for him, the example captures the typical but remains distinct from the normal in terms of statistical frequency.

Clauß devotes his energy to elaborating on the examples, but he never defines the relationship between photography and race. If we apply Arendt's short analysis of Kant to the imagination, however, their distinct roles can be defined:

> Intuition [sensibility] always *gives* us something particular; the concept [understanding] makes this particular *known* to us. If I say: "this table," it is as though intuition says "this" and understanding adds: table. "This" relates only to this specific item; "table" identifies and makes the object communicable.[51]

In this analogy, the "this" is the photograph and the "table" is race. In other words, the photograph is the particular, which intuition or sensibility "gives." "This" (the photograph) relates only to this specific item; "race" identifies and makes the object communicable. We can reformulate Kant's phrase, quoted above via Arendt, about imagination "providing an image for a concept," in the following manner: the imagination produces the synthesis between sensibility and understanding by "providing the photograph (image) for race (a concept)"—the photograph presents to the imagination what is absent from self-perception.

The "mechanism" could be described in this way: the photograph connects an object with the imagination, which transforms the manifold ("the totality of unorganized experience as it is presented in sense") into a representation. Specifically, the photograph connects the Jewish porter with the imagination, transforming the porter into a racial representation of the Jews. The imagination activates the understanding, which begins to supply concepts that relate the image to the redemption man type, creating the synthesis of sensibility and understanding. Instead of reducing the object to a particular case, the concepts bounce the mental materials back to the imagination as part of a circular, continuous process. Rather than ending the discussion of the concept of the

"Jew," this photographic and textual discussion renders it incomplete. The imagination uses such incomplete attempts at synthesis for mutual feedback and indefinite activity.[52]

The fact that I am applying Arendt's terms to Clauß's racial photographic practice should not be misunderstood to mean that I am discussing her judgment of Clauß's method or practice. For one thing, Arendt, unlike Clauß, was attempting (as was Alessandro Ferrara, more recently and drawing on Arendt) to develop a nonfoundationalist view of validity. My analysis only aims to show the deep ambivalence, or even duality, in many chapters of the history of race and photography reviewed in this study. On the one hand, it can be shown to be consistent and accord with a certain strain of philosophy or science. On the other hand, it cannot but be viewed as a caricature of knowledge. But it is a caricature of a specific kind of knowledge: Clauß's practice presupposes a Kantian conception of the productivity of the mind, as understanding is not merely receptive but productive and based on a relativist outlook on race.

It is not unimportant that the analysis of the epistemological status of Clauß's use of photographic examples draws on Arendt's elaboration on the validity of the exemplar in the context of Kantian *political* philosophy given that Clauß's use of examples to fire the imagination of his reader-viewers has a political dimension and ramifications. In the introduction, it will be recalled, I employed the metaphor of the trees and the forest for the imagination. Ferrara quotes Paul Ricoeur's image of the "trail of fire issuing from itself" to describe how an entire forest can be set on fire just by catching one tree after another.[53] This metaphor, mediating between imagination, which is always individual, and the social, which is about many individuals, applies very well to Clauß's photographic method invoking the imagination of his reader-viewers. Our practical epistemological analysis has enabled us to reveal Clauß's photographs as "examples" and to define the photograph as the "this," "race" as the concept, and, thus, the synthesis of the two through the imagination as what gives photography the status of scientific evidence.

"Thought Experiments"

All of Clauß's key terms and concepts—including "seeing" (*Sehen*), "soul" (*Seele*), and "form" or "gestalt" (*Gestalt*)—are drawn from Gestalt psychology or phenomenology. Clauß reiterates that race is gestalt, or form ("Rasse ist Gestalt"), and, as the practical epistemological analysis will show, developed the use of the photograph as a form of racial "thought experiment."[54]

The previous section pointed to the Kantian presuppositions in Clauß's practice, but the core of Clauß's philosophical and psychological beliefs extend out of intellectual traditions that are *reacting* to neo-Kantianism.[55] In particular, as Mitchell Ash has shown, these traditions are reacting to the notion that the unification of impressions was achieved first by a survey of the manifold and then by its summation, and to the belief that access to phenomena was inherently mediated by and modeled on the human categories.

Max Wertheimer, one of the central figures in the development of Gestalt theory, drew on examples such as the difference between a stick broken in two and an arrow broken in two to demonstrate that certain objects possess a form that is not synthesized from component sensations.[56] Objects that possess gestalt qualities are not mere collections of properties; furthermore, psychologists and philosophers gradually came to conceive of gestalts not as psychological constructs but as real, obliterating the difference between the phenomena and the perceiving subjects.

In the reinterpretation of the ontological status of phenomena as possessing inherent gestalt, humans were redefined: humans were specially equipped to perceive gestalts. With regard to the relationship between seeing and knowing, Clauß continued the Gestalt tradition, in which Wolfgang Köhler called people to "learn to see," that is, to transcend mere visual perception (seeing what *is*) toward something deeper and more comprehensive (seeing the essence *in* what is).[57]

Clauß's use of examples as part of the demonstration is also modeled on the Gestalt tradition. In experimental psychology, illustrations have a didactic or educational function but are, scientifically speaking, secondary and derivative. Gestalt writers, meanwhile, bring forth examples of forms that are instantly recognized, that are natural "units of thought."[58] Gestalt writers search for "good" or "pure" examples with which to structure their arguments: invariant forms or structures, demonstrated through well-chosen and carefully formulated examples, that are seen to be universal, constant, and integral to the argument. Some of the most famous examples in the Gestalt tradition are "thought experiments,"[59] and Gestalt writers are especially keen on visual thought experiments such as Edgar Rubin's "figure and ground," in which the viewer sees, alternately, either two profiles or a container.[60] Clauß also followed the Gestalt tradition in turning readers into active, participating observers and photographs into demonstrative evidence.

The role of the photograph in Clauß's practice is to demonstrate a racial gestalt.[61] Clauß takes the photograph of the German (Nordic) farmer from *Rasse und Seele*, abstracts from it the figure, in the form of a single line, and then, gradually, modifies individual parts of the line, a single feature at a time: the

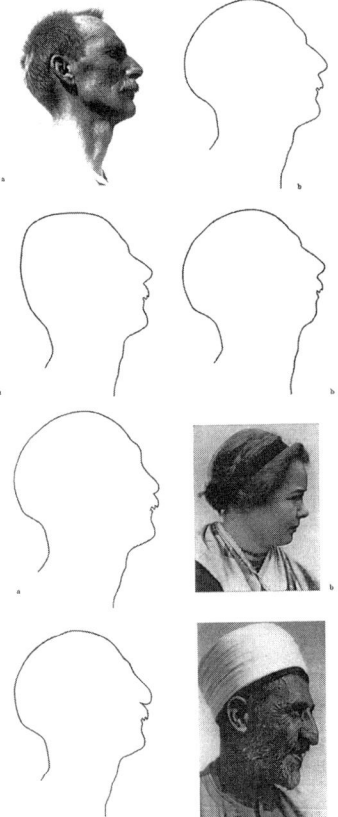

FIGURE 4.5. Photographically based demonstration of Nordic Gestalt through its aberration. Ludwig Ferdinand Clauß, *Rassenseele und Einzelmesch: Lichtbildervotrag* (Munich: Lehmann, 1938), slides 3–5.

nose, the chin, the back of the head, the angle of the head. He accompanies this with a close description, over several pages, of the modifications, suggesting their consequences for the form as a whole: "Dennoch—etwas stimmt nicht." ("However—something isn't right").[62] His intention in distorting the line of the figure is to gradually damage the Nordic gestalt by infusing elements foreign to it until his reader-viewers are forced to perceive that the internal form present in the original figure has been transformed into that of another race. Through this "thought experiment," Clauß believes himself to have proved the reality of the Nordic gestalt (fig. 4.5). In this version of the thought experiment (there are several others as well), Clauß displays two columns of photographs, *a* and *b*. The photograph of the Nordic farmer he places at the top of column *a*.

He then places its abstraction at the top of column *b*. The reader is requested to compare column *a* with column *b*. In column *b* he then carries out alterations in the abstracted figure. Note the minor alterations of the back of the head and the chin in the abstracted figure beneath. By placing a photograph of a different racial gestalt beneath the altered figure, Clauß suggests that following the alterations, the Nordic gestalt is transformed to the extent that it is damaged, and it approaches the gestalt of a different race.

Clauß reproduced this experiment in numerous publications over several years, both during the Nazi period and in his postwar publications. "The experiment [*Versuch*] was tried on many people, of different educational backgrounds, and always led to the same outcome," Clauß wrote in his last major postwar publication. And "after the second modification was carried out observers responded with a painful sense of disappointment."[63] Furthermore, observers always experienced this modification as "*a violation of a rule*: the rule of this gestalt" (*Verletzung eines Gesetzes*: des Gesetzes dieser Gestalt).[64]

In a variant of the thought experiment included in his postwar publication, Clauß places a series of photographs of middle-aged or old men on facing pages. What he claims they share is their attitude toward knowledge, as expressed in their strained expressions, emanating from their responsibility as "guardians of a flame" ("Hüter einer Flamme").[65] The text states that these are photographs of professors, and that in his experiments with numerous viewers, he has confirmed that they instantly identified all but one of the men in the photographs as German professors. The one whom they identified as different was, indeed, an American professor (fig. 4.6). Similarly to the American professor, "A Jewish wise man," Clauß adds in the same vein, lacks the expression of Socratic undoing or doom (*Verhängnis*). He then reproduces a photograph of Martin Buber to highlight the difference between the gaze of the Jewish professor and that of the German professors.

Why, in the first thought experiment mentioned above, did Clauß abstract a graphic line from the photograph rather than altering the photograph itself? Retouching practices existed and were readily accessible to him. My speculation is that this reflected his commitment to photography as a transparent reproduction of the real, which retouching would have been taken to undermine. Clauß's thought experiment was a form of projection, and while one is tempted to invalidate it as pseudoscientific racist propaganda, it is important to stress that in this sense it is no different from virtually *all* Gestalt thought experiments.

Like most authors discussed in this study, Clauß never elaborated on how he selected the individuals for his photographs. As with Günther, his selection

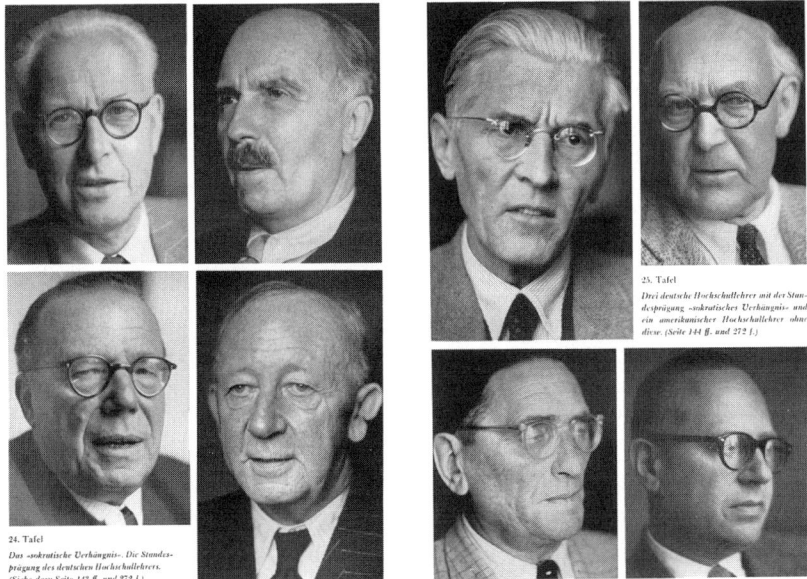

FIGURE 4.6. Photographic "thought experiment": The gaze of the American professor stands out. Ludwig Ferdinand Clauß, *Die Seele des Andern: Wege zum Verstehen im Abend- und Morgenlande* (Baden Baden: Grimm, 1958), between 128 and 129.

cannot be easily reduced to propagandistic terms of beauty. While Clauß's selections unquestionably work to deepen stereotypes of "races" rather than attempting to deconstruct or defixate them, the selections do not always easily conform to idealized stereotypes of racial beauty. The German farmer who is the object of analysis representing the Nordic type may be more beautiful than the Jewish porter from Jerusalem, but is far from an ideal Nordic beauty.

Clauß lacked any medical or physical-anthropological education. We can assume, but have no way to corroborate, that his selections involved considerations of both availability for and consent to being photographed, as well as correspondence with Clauß's preconceptions, along with other considerations that were specifically photographic. Konrad Lorenz spoke about "filmable" animals,[66] and we can only contrast Lorenz's recognition that certain specimens were more photographable than others with Clauß's lack of any such discussion. This was probably related to the fact that the specimens were humans.[67] For one thing, a discussion of the "filmability" of animal specimens does not necessarily discredit their typicality, whereas with humans, because of their undeniable individuality, it presumably would. This evasion serves to naturalize the categories under discussion and their representation.

Photographs as Objects for the Imagination

Several aspects of Clauß's racial photographic method and practice only become understandable on the basis of Husserl's phenomenology. The few publications that touch on the relationship between Clauß and Husserl attempt to avoid contaminating the phenomenological movement with racial ideas.[68] But Clauß was Husserl's assistant (*Mitarbeiter*) in Freiburg between 1917 and 1921, regularly participated in Husserl's seminar, and was considered one of his senior students, invited to contribute an article to Husserl's seventieth-birthday festschrift in 1929.[69] In this context, two issues are particularly relevant and important.[70]

The first issue concerns the possibility of studying others. Husserl's theory is grounded in a commonality between ontological form and the notion of empathy (*Einfühlung*).[71] Empathy allows the individual to experience other human beings as living organisms with their own forms of humanity through the projection of the self into the position or situation of the other.[72] Clauß restricts empathy to members of the same race; but if empathy only functions between members of the same race, how can the racially other be studied? To answer this question, Clauß introduces into his photographic method the notions of mimesis and *Mitleben* (something akin to vital experience; the literal translation is "living with"). *Mitleben* is based not on commonality of form but on imitation by way of which one attempts to dispense of one's own form and approximate that of the other as a precondition for the projection of the self into the position or situation of the racially other.[73] Hence, when Clauß learned Arabic, converted to Islam, and immersed himself in the life of a Bedouin tribe in Transjordan, he was practicing his solution to the problem of the study of the racially different (fig. 4.7). In certain senses, one could liken this method to cultural anthropological ideas about the study of distant cultures using their own language, ideas, and myths. And yet this comparison makes clear the implicit difference with regard to photography: Franz Boas famously had himself photographed acting out Native American religious rites; Clauß did not do anything comparable because that would have undermined the understanding of race as a natural status and implied that the photographs were mediated or constructed.

The second issue concerns the investigation of hybridization. Clauß was unambiguously a "purist," yet his books very often study what he viewed as aberrations of racial style.[74] It took me a long time to realize that this is not out of interest in hybridity but is, rather, the result of bringing Husserl's notion of "eidetic variation" together with the notion of style, which Clauß appropriates

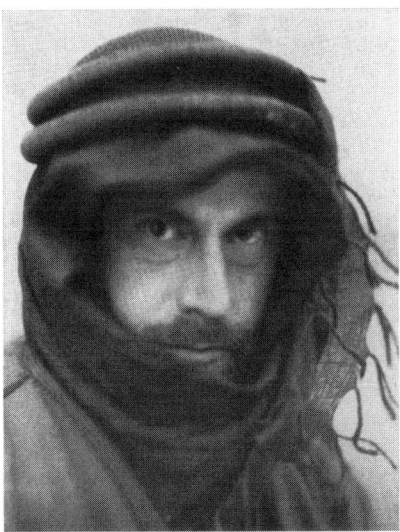

Mhammad Farîd (L. F. Clauß)

FIGURE 4.7. Clauß as Bedouin: *Mitleben* as solution to fundamental racial difference. Ludwig Ferdinand Clauß, *Als Beduine unter Beduinen* (Freiburg im Breisgau: Herder, 1933), between 140 and 141.

from art historians. In art history, style is often defined as "the presence of a common formal denominator in the visual production of a period."[75] Clauß often repeated his definition of the psyche that expresses itself in the animate body as "that which is governed by a style."[76] Applying style to race, Clauß basically studies racial phenomena in art historical terms. By *imagining* variations in the properties of photographed subjects, Clauß attempts to differentiate between their essential and their accidental features.

Not only Clauß's terminology but his method and practice as well, when observed closely, can be seen to merge phenomenology and art history. Differentiating between the essential and the accidental as a way to capture the essence of a phenomenon is an approach that comes from phenomenology, while the concept of style is based on art historical assumptions that "the style forbids certain moves and recommends others as effective."[77]

Knowing this, the philosophical foundation for three photographs of a woman that Clauß reproduces in the methodological part of *Rasse und Seele* becomes comprehensible. He claims that the first photograph demonstrates her Nordic features when she is still; the second reveals Ostic features around her eyes when she smiles; and in the third, Clauß claims, the upper part of her

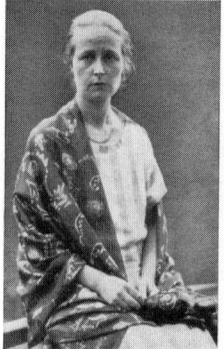

9. Stilwechsel im Ausdruck   127

Stilwechsel im Ausdruck.

Norddeutsche Künstlerin.

Bild 98: In unbewegter Haltung
tritt nur das Nordische hervor.

Bild 99: Das Lächeln weckt ostische
Züge um die Augen.

Bild 100: Die obere Gesichtshälfte
lacht ostisch, die untere nordisch.

Bild 98

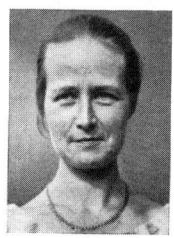

Bild 99

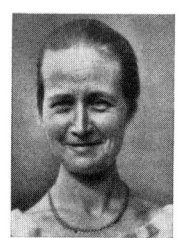

Bild 100

FIGURE 4.8. Study of switches in racial expression. Ludwig Ferdinand Clauß, *Rasse und Seele: Eine Einführung in den Sinn der leiblichen Gestalt* (Munich: Gutenberg, 1937), 127.

face laughs in Ostic style while the bottom part expresses a Nordic style.[78] The study of hybridization in photographs is an *instrument* for the study of coherence in (racial) style (fig. 4.8).

This form of interrogation of photographs directly involves Husserl's later theory of the imagination. In the three books of *Ideas*, Husserl argues that each domain of objects can be correlated to a form of *Anschauung* (discussed at length in chap. 3).[79] Seeing, according to Husserl, is the primordial mode of consciousness. Immediate seeing, rather than merely sensual seeing, forms the ultimate source of validity for rational assertions. Through "imaginative variation," Husserl delineates those features that are essential to the thing, leading to the *eidos*, the essence of the thing under scrutiny.

In his early work, *Logical Investigations* (1900–1901), Husserl describes imagination as being a form of quasi perception. In perception, the object appears to us as itself present,[80] while in fantasy the object appears as if it were there, a presentation of a nonpresent object.[81] In his later work, however, Husserl fundamentally transforms his view, determining that perception has no advantage over the imagination.[82] Both are seen as parallel acts of consciousness, each with its own evidential force: perception provides evidence in the realm of the transcendent experience of actual objects, fantasy in the realm of possibilities. Clauß, then, employs photography to *present* racial essences, not to *represent* facts of natural science.[83] While from a modern perspective the attempt to enhance the imagination by providing it with visual information may seem counterintuitive, in the sphere of religion, and ever since Lessing's *Laocoon* (1766), a positive correlation has been established between the two.[84]

Clauß subordinates the imagination to racial relativism and (not unlike Wölfflin, as noted in chap. 1) claims that "every people has its own law of seeing and will."[85] Clauß applies this view practically, with a specially created educational "kit" for schools, encouraging students and pupils to photograph as a means to perfect their investigative gaze.[86] But in the context of an epistemological history of racial photography, the transformation of the photograph is even more important. If Husserl defines evidence as "an experiencing . . . of something itself," a self-evident form of experience,[87] Clauß transforms the racial photograph from a medium of representation, a form of evidence (*Beweis*), into the thing itself (*Evidenz*). In this sense photographs are immediate encapsulations of racial essences, self-evident racial signs.

PART THREE: PHOTOGRAPHY AND NATIONAL SOCIALIST CULTURE

Up to this point I have approached Clauß's racial photography from within his work and methodology. I would now like to move the discussion to its intersections with wider visual and cultural patterns and codes of meaning. In fact, approached from this perspective, a major apparent contradiction in Clauß's work becomes visible. Clauß follows Paul Schultze-Naumburg, as we will see, in defining photography as a mechanical form of reproduction. This guarantees its objectivity and scientific legitimacy but debars it from qualifying as genuine art. Unlike Rudolf Martin, Eugen Fischer, or Felix von Luschan, however, Clauß intersperses his publications with photographs that he himself took and that emulate art. What should we make of this apparent contradiction, and what can we learn from it more generally about this chapter of racial photography? Using several examples, ranging from the cultural definition of photography and proceeding then through signifiers to particular photographs, I will move from Clauß's deliberations to their intersection with wider National Socialist frameworks of meaning.

Let me start with the definition of photography in the context of Clauß's version of science. Clauß reflected on the epistemological foundations of his science only once, in his inauguration speech at the University of Berlin, which he gave on November 16, 1936, dressed in a brown Nazi Party shirt.[88] (He also discussed the status of photography once, in his major postwar publication in 1959.) Drawing on Husserl's critique of the experimental sciences (but omitting any mention of Husserl himself), and basing his classification on Kant, Clauß contrasted two kinds of truth and two practices of science in order to situate his version of racial scholarship as on a higher level than competing ones.

Following Leibniz's work, Clauß stated, all knowledge (*Erkentnisse*) was split into two, and the differentiation between "truths of reason" (*Vernunftwahrheiten*) and "factual truths" (*Tatsachenwahrheiten*) was established. The former are derived directly from reason; the latter are mediated by experience (*erfahrungsbedürftig*) and verification (*Bestätigung*), as the weight of truth is carried solely by facts.[89] These two forms of truth call for two forms of demonstration and proof. Truths of reason, unlike empirical truths, need no verification but follow a pure intellectual conception (*reine geistige Anschauung*) and can be demonstrated with *Schauhilfen* (literally "show devices"). Different forms of demonstration are tied to the different forms of science between which Kant differentiates

"Nature," according to Kant, is bipartite, divided into objects pertaining to the external senses and those pertaining to the internal sense.[90] A rational doctrine of nature is truly a science only when it is underlain with natural laws, cognized a priori, rather than from experience.[91] A merely empirical science, such as chemistry, according to Kant, is only a "systematic art."[92]

The polemical intentions of this classification become more visible as Clauß approaches the discussion of race. Although it proceeds empirically through the accumulation of cases, the racial science of experimental psychology and of natural science ultimately comes down to individual phenomena. Implicitly criticizing the models of Galton and of Martin, Clauß is opposed to any racial photograph that relies on a statistical understanding of race. Joining a longstanding tradition of criticism of statistics, he contrasts the "truth of style" (*Stilwahrheit*) with the "heaping together of cases," as statistics is only capable of depicting regularities from a descriptive, external point, whereas his method is closer to the rational, a priori model.[93] Both modes are legitimate, but they are not on a par with each other.

Gestalt, according to Clauß, may not be equivalent to pure reason, but it is closer than statistics to the rationally a priori. When Clauß states that gestalt can be presented to the mind through aberrations departing from the gestalt or that the force of photographic examples stems from their being self-evident,[94] we recognize issues discussed above, but we note that Clauß is also providing philosophical justification for Paul Schultze-Naumburg's contention concerning examples.

Schultze-Naumburg belonged to an older generation of nationalist, and later Nazi, intellectuals and ideologues; he called for Germans to be provided with "visual judgment" in order to know what was "beautiful," "good," and "practically useful for the preservation of the future of their race by means of examples and counterexamples,"[95] and he mentored both Clauß and Günther. Clauß's scientific practice partakes of and helps to constitute the National Socialist visual, moral, and political culture of examples and counterexamples as didactic, ideological, and political instruments.

Clauß's most direct and elaborate discussion of photography can be found in an appendix to his last book, *Die Seele des Andern* (The soul of the other), in 1959.[96] In an echo of Simmel's interpretation of art (discussed in chap. 1), Clauß contrasts photographs with art objects and the photographer's vision with that of the artist. The artist's vision, unlike the photographer's, "leaps" into the seen, determining its meaning and form.[97] The weakness of photography is also its advantage: it represents the other free of interpretation.[98] Artworks generate reality; photographs reproduce it.[99] Now, while this understanding of photography has a long history in the history of science, Clauß almost certainly draws it from Schultze-Naumburg, who justified this view on racial grounds and whose understanding of it became National Socialist doctrine on photography.[100] Schultze-Naumburg explains why photography can never qualify as true art: "The choice of a subject is, of course, common to both photography and art, but the technical image simply mirrors the subject. In painting, the situation is quite different, for here what is reproduced is not the subject of the picture but the artist himself."[101] Crucially, the differentiation between the two is directly related to the reproduction of racial types. In the next chapter we will encounter an opposite conception of photography, one according to which the photographer is an artist who manipulates light on two surfaces, the photographed object and the film or glass surface, at the same time. The internal scientific controversy over the status of photography cannot be separated from the wider cultural and political controversy.[102]

While in his science, Clauß defines photography as a mechanical means of representing reality that cannot qualify as art, Clauß's photographic practice,

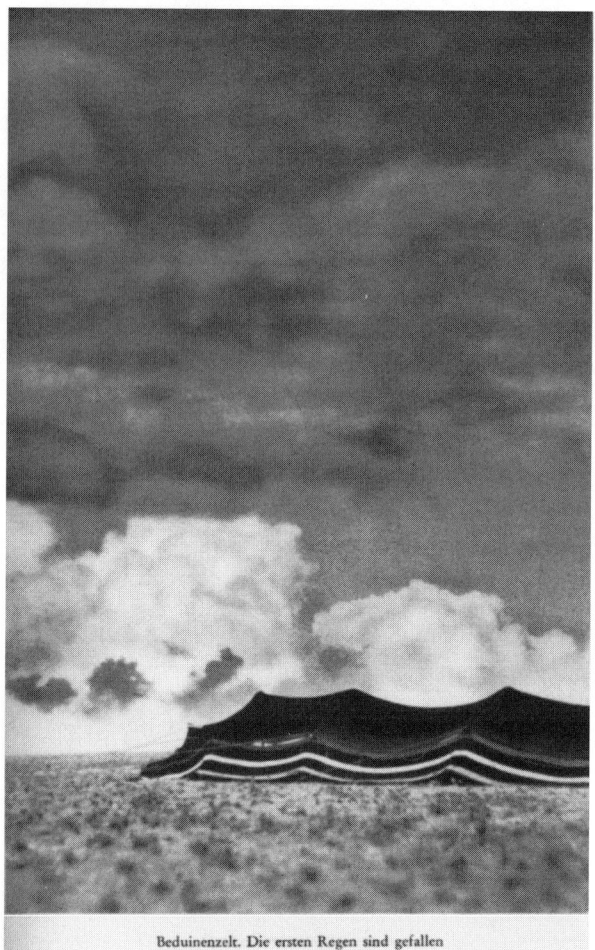

Beduinenzelt. Die ersten Regen sind gefallen

FIGURE 4.9. Pictorialist photography. Ludwig Ferdinand Clauß, *Als Beduine unter Be-duinen* (Freiburg im Breisgau: Herder, 1933), between 12 and 13.

his pictorialist photographs of people, architecture, and landscapes, emulate art (fig. 4.9). This contradiction (his failure to follow through on his definition and to avoid artistic photographs) is a form of blind spot stemming from his commitment to National Socialist cultural doctrine, according to which photography could at best emulate art but without ever qualifying as true art.[103]

The intertwinement of Clauß's photography with wider cultural and visual patterns can be observed in the type of the Nordic man, which Clauß defines as "achievement" or "performance" (*Leistung*). As Eric Michaud has shown in *The Cult of Art in Nazi Germany*, *Leistung* was one of the central keywords in

National Socialist language, culture, and mythology: "Just as there was a *Füh-rerprinzip* there was also a performance principle (*Leistungsprinzip*), which constituted one of the bases of 'practical' National Socialism."[104] Through the concept of *Leistung*, "work" assimilated creative work, but it was also this concept that determined inclusion in and exclusion from the national or ethnic community (*Volksgemeinschaft*).

Clauß's performance man, then, intertwines horizontally with cultural and political National Socialist patterns and codes, both visual and linguistic, and this intertwinement also has particular visual aspects. This young athletic man was photographed from behind and slightly to the side. Completely naked, his body is tanned and erect (fig. 4.10). As he stands on what looks like beach sand, holding what appears to be a metal ball, the light falls on him from the left, making his body shine. The photograph, which seems to have been cut out of a larger photograph, emphasizes his muscled masculinity. He is relaxed and yet poised and concentrated. His eyes are fixed on a distant point ahead

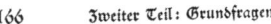

166   Zweiter Teil: Grundfragen

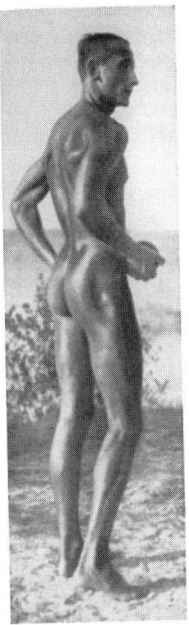

Bild 159: Nordischer
Wuchs, männlich.

FIGURE 4.10. Male form of Nordic type of man. Ludwig Ferdinand Clauß, *Rasse und Seele: Eine Einführung in den Sinn der leiblichen Gestalt* (Munich: Gutenberg, 1937), 166.

167

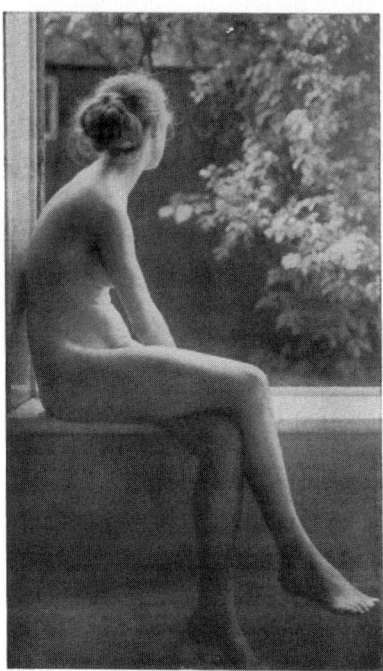

Mit Erl. d. Verl. d. Schönheit
Bild 160: Nordischer Wuchs, weiblich.

FIGURE 4.11. Female form of Nordic type of man. Ludwig Ferdinand Clauß, *Rasse und Seele: Eine Einführung in den Sinn der leiblichen Gestalt* (Munich: Gutenberg, 1937), 167.

of him, manifesting Clauß's concept of the distance that defines the gaze of the Nordic performance man. The horizontal lines that cut through the background of the photograph, caused by the shade falling on the sand just above the man's knees and the horizon line just above his waist, emphasize his perfect, Hellenic proportions. The title states: *Nordischer Wuchs, männlich* (Nordic stature, male).[105] The photograph manifests Clauß's ideal performance man: natural lightness, grace in movement, inner harmony, and harmony in relationship with the environment. For this type, work is creative work and leisure. Through the notion of *Leistung* and the visual representation of that notion, Clauß's scientific economy becomes one with the wider National Socialist culture.

The counterpart to the photograph we have just discussed, on the facing page, depicts the female Nordic type and functions similarly (fig. 4.11). The photograph shows a young light-skinned woman in the nude. She is photographed almost in full profile; her legs are crossed, the back of one of her shoulders is toward the camera, and we see her long slender neck and only the back of her head as she looks out of the window into a garden. The light falls onto the left side of her body from outside the window, and her neck, back, arm, and leg almost glow from the light reflecting on them. Although she is naked, her breasts and genitals remain out of sight. The photograph, rather than being a pornographic image intended to sexually arouse male viewers, expresses sublime and unreachable beauty. The woman is at ease with her naturally graceful body, manifesting the attributes of Nordic beauty. Her idealized body suggests the beauty of a Greek statue or goddess (thereby interacting with photographs of Greek statues and with statues and photographs of statues by Georg Kolbe, Arno Breker, and Josef Thorak). Like other expressions of National Socialist art, the photograph resonates within the direct line that traces the continuity from classical to Nordic beauty. We can see the Nordic woman as Daphne and Germany as the bearer of classical values. In its pictorialist language, showing the soft contours of her image in shades of gray, this photograph extends core National Socialist doctrines concerning the visual arts, not only in its commitment to naive figurative art and the expression of pure, clean, unpolluted life in its natural form but even in such a motif as the use of the window as a framing device.[106]

Before we turn to what is in one respect a parallel history and in another a different universe, namely racial photography in Palestine, I would like to end here by returning to the 1959 photographic thought experiment discussed above in which observers were called to distinguish the non-German professor from the Germans based on the strained gazes of the Germans. This form of gaze, according to Clauß, was a motif that expressed the unbroken (racial) genealogy of scientific knowledge from Hellenic Greece to the West German present. The gaze marks critical knowledge, a knowledge characterized by the willingness to criticize the foundations of knowledge and to accept nothing as given.[107] The gaze signifies that the people who have it are the bearers of the torch of knowledge.[108] Here, the meeting between the *signifier* and its visual *image*—another landmark of National Socialist imagery—illustrates the intertwinement of Clauß's photographic practices with some of the main visual images and codes of National Socialism well into West German history.

One of the major problems we face in examining Clauß's work is the fact that the condemnation of racist and antisemitic discourses has been so complete and successful that the contemporary viewer is no longer able to "see" what Clauß, using a particular photographic method, set of techniques, and practice, wanted to teach his viewers to see. Such forms of seeing have lost their conceptual viability and their experiential force. To acknowledge that many observers, within both the academy and the wider public, once had access to these forms of seeing does not bestow validity or moral value on them; the historicization implied in such an acknowledgment is merely a necessary precondition for a historical judgment of the role of photography in advancing these forms of seeing.

CHAPTER 5

Racial Photography in Palestine

From around 1918 one finds in the writings of numerous Jewish and non-Jewish authors the contention that Palestine is a mosaic of Jewish communities from all around the world and, as such, an exceptional site for anthropological observation. Between 1918 and the 1940s, the focus of anthropological studies of Jews, including the use of photography, at least partially moved from Europe to Palestine. Conceptually, racial photography in Palestine was not novel but derived from one particular European tradition, that of Felix von Luschan. In the previous two chapters, I discussed the photography-race nexus in two different but ideologically and politically closely related cases in Weimar and Nazi Germany. In terms of the background of the authors discussed in this chapter, this chapter runs parallel to the previous two. It consists in a practical epistemological exploration of racial photography from the bottom up in a historical context that is both connected and parallel to them. By bringing to the surface "small tensions" involving, for example, racial type photographs of one's own family members or one's own research assistant, or false identifications of individuals in publications, my aim is not only to show how the social and political meaning of the same photographic techniques or practices was redefined in the new context but, more generally, question some of the core assumptions of the current framework of interpretation of racial photography.

A close study of the place of race and photography in the work of Arthur Ruppin and Erich Brauer occupies the main part of the chapter. Ruppin and Brauer, both of German-Jewish background, employed photography in writings on race that were both researched and written in Palestine. The chapter then shifts perspective, from scholars of race to their subjects, examining the "case" of Yemenite Jews, arguably the most racialized group in the context of Jewish Palestine. This shift in perspective points more prominently to the complicated relationship between scientific and artistic discourses. The final part of the chapter turns back to those who crafted representations, and by examining the work of two prominent photographers and artists whose work touched on racial discourse, Ephraim Moses Lilien and Helmar Lerski, I further explore the two-way flow—and, indeed, also fluid boundaries—that existed between art and science.

One of the aims of this chapter is to show that the relocation of racial photography constituted a social context in which science and politics interacted differently than in Europe. To this end, a brief historical elucidation of the crucial tensions involved is required. With the gradual emergence of a modern Jewish society and polity in Palestine (before the establishment of Israel in 1948), photography and race were redefined along two axes: first, the discussion of differences between various Jewries acquired new, politically concrete meaning; and second, the figure of "the Jew" was redefined and reconfigured in a complex way along its direct or implicit opposition to the figures of both "the Aryan," on the one hand, and "the Arab," on the other. The interpretation of the genesis and meaning of this new context is not, however, free from difficulties.

The only existing theoretical frame of interpretation for racial photography in Palestine is the postcolonial one. According to the standard deployment of this form of analysis, small incidents such as those noted above are illustrations of a "bigger picture" of the appropriation of European colonial forms of racialization and "othering"; that is to say, racializing and othering European categories that were applied to Jews in Europe were relocated to Palestine, where European Jews now applied them to non-European Jews and to Arabs, even toward family members or research assistants.[1]

Methodologically, the weakness of this approach is that it merely replicates already existing categories of analysis, thereby failing to identify the novel context of historical interpretation. Historically, the weakness of this framework is that the dichotomies on which it is based simplify convoluted historical subjects; for instance, the relationship between race and nation.[2] Jewish nationalism, and in particular Zionism, included Jews within the same national project who were considered to be of various different racial backgrounds. At the same time, Arabs who were viewed as racially close to part of Jewry were excluded from the national project. While it is not possible to completely separate the racial and the national here, it is also not helpful to collapse the two into each other, as deployment of the postcolonial framework would appear to necessitate. Even if the line between the two is elusive and often impossible to fix, for an epistemological history of racial photography it is methodologically critical to insist on the difference between the two.

Avoiding reductive or deterministic explanations, the interpretation that I offer in this chapter focuses on the racial and not on the national aspects of this history; and for this reason it is necessarily partial. The authors addressed in this chapter repeatedly contended that Palestine, with its mosaic of Jewish communities from all around the world, was now an exceptional site for anthropological observation. Arguably the most important tension in this respect is between the forward-looking national project, with its aspiration to create a "new Jew," and the backward-looking authors wishing to study Jewish racial types before they mixed and disappeared forever. Looking at materials from the bottom up and attempting to bring to the surface the necessary assumptions without which photographic practices would not be intelligible, I am interested in the tentative, hard to detect moments in which existing practices or ideas generate a new context of meaning, which, in this one particular historical context, redefined the interface between science and politics.

PART ONE: ARTHUR RUPPIN, RACE, AND PHOTOGRAPHY

Of the various Jewish authors who studied race and employed photography in Palestine in the first half of the twentieth century, there is little question that Arthur Ruppin is the most prominent, and his career graphically illustrates some of the complexities involved in the transfer of ideas and practices from Europe to Palestine. Recently, Ruppin's idea of "race" has been the subject of growing scholarly interest, but while it is recognized that he employed photographs, they have not hitherto been the subject of significant study. In what follows, I will first address several empirical hurdles that must be surmounted in any discussion of Ruppin and photography. The central part of the section, however, consists of an intellectual experiment in the application of the practical epistemological approach: focusing on the almost identical German and Hebrew editions of the only book in which Ruppin made use of photographs, I will show the significance of various practical aspects of the publications, such as size, layout, or the reproduction quality of the photographs. As photography formed part of Ruppin's multilayered model for the study of contemporary Jewry, I will then attempt to analyze the relationship in his work between photography and statistics, the latter on which his model as a whole was built. Finally, Ruppin's appropriation of photography contributed to and was informed by the generation of a new social and political context in Palestine. But here the discrepancy between the order his science attempted to confer on the chaos of social differences and his own identity in particular proved to be problematic. I will conclude by noting some of the ironies that Ruppin's own case involved with regard to social visibility.

Ruppin's initial encounter with photography, although it left no recoverable photographic evidence, was when, as a young person, he traveled to eastern Europe at the beginning of the twentieth century to encounter, study, and photograph "true" Jews firsthand. When he returned to Germany he established in Berlin the Bureau for Jewish Statistics and Demography, which he directed between 1902 and 1907 and which did *not* collect photographs as part of its mission to assemble statistical information about contemporary Jewry. Ruppin moved to Palestine in 1908 to establish the Palestine Office, which was responsible for the purchase of land for Jewish settlement (a sphere in which he imported ideas from Germany's conflicted eastern regions, in which he had grown up).[3] Ruppin only employed photography for professional purposes for the first time in the late 1910s, and his interest in photography peaked in the 1920s, culminating in a major publication in 1930.

Missing Links in Ruppin's Photographic Career

Any attempt to reconstruct Ruppin's photographic career faces certain empirical lacunae. As just noted, it is known that Ruppin had shown interest in photography from as early as his sojourn in eastern Europe, and in private communication in the early 1920s he expressed an intention to publish a racial photographic atlas of Jewish types in Palestine. One of his files in the Central Zionist Archive (CZA) in Jerusalem contains numerous photographs of Jews clipped from various English and Yiddish language newspapers. But the book never appeared. In fact Ruppin only made use of photographs in his 1930 book *Sociology of the Jews*. This book is not the atlas, and the relationship between his plan for an atlas and this book is not clear. Until recently it was known that the CZA, in which Ruppin's entire *Nachlass* is found, possesses several boxes of Ruppin's glass-plate photographs. But there was no inventory of the photographs, and access to them was prohibited because of their fragile state. There are thus advantages to the fact that the present study took much longer than planned: in the summer of 2012 the glass plates were scanned, and the images were made available to the public. I will have something to say below about the relationship between the photographs in this archive and those reproduced in the book, and in general, it is possible to draw some information about Ruppin's photographic career from these photographs. Nevertheless, and as will become clear, the archive leaves unanswered more questions than it answers.

The Two Editions of *Sociology of the Jews*

Ruppin's use of photographs is tied directly to his discussion of race, and it is not accidental that his only book to contain photographs was also the one with the most comprehensive treatment of race. I have conducted research into Ruppin's ideas on race before,[4] but approaching it here by way of his use of photography is illuminating because it brings into focus and makes much more apparent Ruppin's debt to von Luschan. Ruppin cited von Luschan on virtually every occasion he discussed race, invited him to open the first issue of his journal, and evidently saw himself as in some way a disciple.[5] In particular, Ruppin credited von Luschan with freeing the study of race from its false association with the study of language groups, thereby allowing a breakthrough in the study of modern Jews.[6]

Ruppin followed von Luschan's Jews as a mixed race (JMR) idea, but by the time Ruppin wrote his book, undermining the Semitic definition of Jews was old news. Furthermore, with his rejection of assimilation and endorsement of Zionism, Ruppin was not bound by von Luschan's political and ideological convictions. Nevertheless, by the time he published *The Sociology of the Jews* in 1930, Ruppin was also swimming against the tide, not because he, Ruppin, had changed his direction but rather because, outside of Weimar and Nazi Germany, the tide itself had turned. By the early 1920s the study of "race" had become politically polarized, and an earlier spectrum of ideas had increasingly been reduced to political or ideological stances. The same Ruppin who established the Berlin office for countering antisemitic allegations in what came to be known as the "statistics war," and who from the outset of his career had embraced a soft form of racial determinism, would now increasingly find himself aligned with antisemitic authors, whose time and energy he spent countering.[7] In fact, Ruppin first included photographs in his work precisely at the moment that they disappeared from the publications of opponents of racial determinism and antisemitism.[8]

But no less important is the fact that with the materialization of a national Jewish society in Palestine and the political changes in central Europe, the implications of some of his ideas changed quite dramatically.[9] For example, a statement about harsh conditions of existence that increased the pressure of selection among Ashkenazi Jews (in the context of relations between Jewish and non-Jewish Europeans in Europe), which pressure was largely absent among Arabs or Sephardic Jews, came to function as a wedge between Jewish populations and between European Jews and Arabs, whose coexistence was, by the 1920s, not confined to the pages of a book.[10] In a word, Ruppin's ideas were fairly static, and the transformation in the meaning of his thought resulted rather from the shifting of ground and context.

From this perspective, the two editions of *The Sociology of Jews*—appearing in German and in Hebrew in, respectively, Germany and Palestine—are marked by deep ambivalence. It is not possible to situate Ruppin in one single linguistic and geographical context, and his statements, at least in retrospect, appear somewhat outdated for their time. Ruppin introduced his ideas about race with hesitation: he prefaces the first volume with the qualification and explanation that race is not one of his research subjects, and his decision to include a section on the topic is because it is necessary for the discussion of the source of diversion of Jews from the social patterns of their respective environments as externally forced or as flowing from their racial essence. His assertion that his discussion is not aimed at providing a final answer to this question but is only an attempt to clarify concepts reads as an attempt to pull back what he was at the same time putting forward.

With regard to the employment of photographs, Ruppin voices no similar hesitation. He also does not accompany their use with any methodological elucidation. The comparison of the Hebrew and the German editions, therefore, offers a remarkable site for a practical epistemological perspective. Both editions are based on realistic and deictic assumptions. Photographs in both editions are exclusively devoted to the discussion of race and reproduced at the end of the first volume. But there are important differences between the editions.

The appendix of the Hebrew edition is made up of a 166 photographs on eleven pages, moving from reproductions of ancient monuments to standardized squared frontal photographs of persons (sometimes accompanied by a profile angle). Drawn from several kinds of resources, their source is not disclosed: some are anthropological sources, as can be discerned from their mug shot style (the frontal and profile angles, on even background, and the uncomplimentary angle below the face displaying a disproportionately large nose); some are privately produced studio portraits (men in dark elegant suits, white shirt, and black tie—light falling softly on their faces from above, persons endowed with dignified appearance); and some were cut from larger photographs. With few exceptions, each page is made up of a photographic matrix of four on four and consecutively numbered. The reproduction quality varies, but it is on the whole low. There are discrepancies between the appendix numbers and references in the text (photographs are accompanied with Hebrew letters, whereas the text uses Arabic numerals; the text refers to numbers 121–123, whereas the appendix ends with 116).[11]

Ruppin's discussion of the visual materials is minimal. He exploits the Jewish religious interdiction on the creation of images to explain the absence of Jewish images of ancient Jews (and to reproduce images of non-Jews).[12] He also notes cynically Günther's claim that the Emorites were blond-haired and blue-eyed Nordics that it was based on pictures found in Egyptian tombs the distinctive characteristics of which had been proved to be chemically caused by the dark and damp conditions.

Although drawing on the idea of the JMR, the photographs in the appendix of the Hebrew edition derive from a different repertoire. Similarly to von Luschan, Fishberg, Feist, and Günther, Ruppin breaks down types into their constitutive components by displaying photographs of non-Jews in order to render visible certain features that Jews allegedly share with them. Here (similarly to Boas), Ruppin is insinuating that the "Jewish type" is not specifically Jewish. But readers who have not read the relevant passages in the book will not know that they are observing photographs of non-Jews. In contrast to Boas, Ruppin did not wish to demonstrate the Jews' assimilation but rather to corroborate their difference from northern Europeans. Ruppin divides what he takes to be the three racial components of the Jewish people: the Aramaic (von Luschan's Hittites), the Arabian (Bedouins), and the southern European (the Philistines).[13] Ruppin's discussion alludes to various physical features associated with these respective types and refers readers to a specific monumental reproduction, implying that the discussed features are observable in reproductions. In terms of the tension between projection and cultivation of vision, discussed above, this case corresponds to the former.

Ruppin makes use of a variety of representational forms, discussing ancient monuments, for example, as if they were archeological mug shots. Placing of a first century El Faiyûm portrait (discussed in chap. 2) alongside a painted portrait of a contemporary Jewish child from Jerusalem, and because of the poor reproduction of both and consequent lack of detail, he suggestively increases their apparent resemblance.[14] But an equally necessary premise for his statement is Ruppin's implicit assumption that photographs (and paintings) are neutral, objective renditions of a real referent. In other words, sculptured reliefs and portraits are understood as photographs, that is, transparent representations of referents.

Ruppin's discussion of the complexity of inheritance and the fact that physiognomic features differ even among brothers is particularly important in the comparison between the editions. This is because the reproductions to which Ruppin refers are so small that they do not allow anything to be seen. A different picture emerges from the German edition, however, because there *one sees much more.*

The reproductions in the German edition are larger, technically superior, and printed on better quality paper. Instead of the rigid matrix of the Hebrew edition, the photographs here appear in two columns, which are spread over sixteen rather than eleven pages. The source of virtually all of the photographs is specified and accompanied with detailed subtitles. The underlying credits of this edition in fact make clear that Ruppin himself produced most of the photographs. Ruppin's photographs express his debt to von Luschan in more than one sense.

In terms of photographic style, Ruppin here extended von Luschan's type photographs, which were free from scientific control in terms of distance, camera type or lens, lighting conditions, background, or angle but that nevertheless, by way of the selection of individuals for photography and by their photographic treatment, tended to create patterns of similarities. Similarly to von Luschan, Ruppin took photographs with strong light, from a low angle, and short distance. As an effect, faces appear like mug shots with little facial detail, and in particular noses are enlarged and distorted. Ruppin's photographs thus belong to von Luschan's type photographic genre rather than Martin's anthropometric tradition.

Without stating so openly, Ruppin also followed von Luschan in another very important way. The credits do not specify when, where, or under what circumstances the photographs were made. But it is possible to deduce from the credits that while (unlike von Luschan) Ruppin reproduced photographs of Jews, he only photographed non-Jews. In a complex way, because the social conditions were quite different in this instance, Ruppin extended von Luschan's inhibition on photographing Jews as white Europeans to Jews as a whole.

I want now to make concrete by way of several examples the general observation that in the German edition one sees more. Photographs (fig. 5.1) in the Hebrew edition show two men whose features are virtually illegible and who are referred to at the point in the text at which Ruppin states that even brothers can differ phenotypically. The German reproductions of these two particular photographs are larger, their technical quality in terms of reproduction is higher, and their titles state that the subject of photograph 66 is a Jew

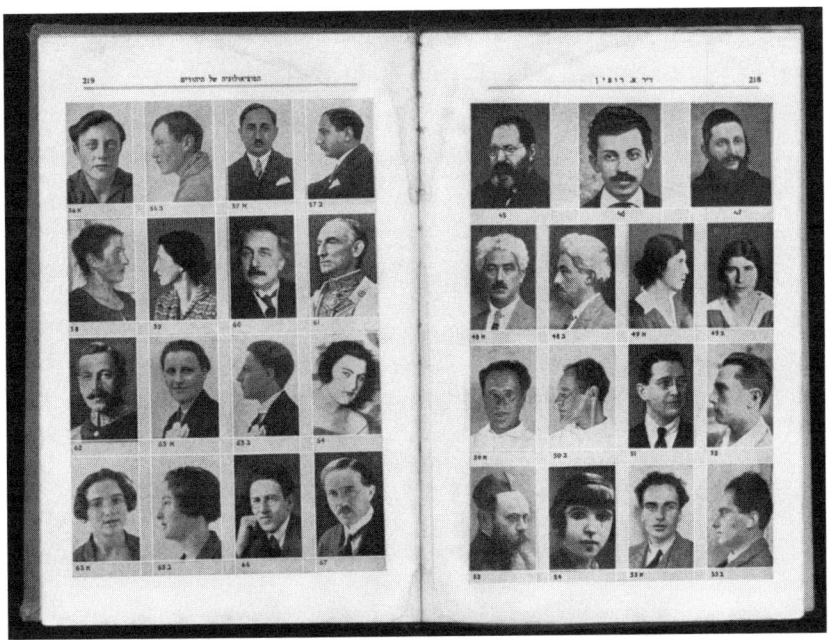

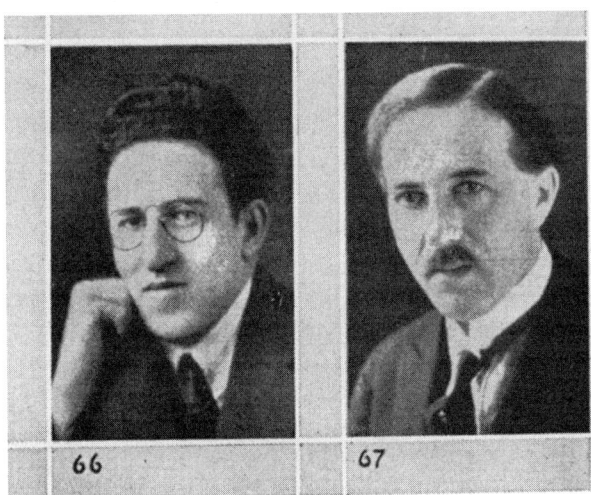

FIGURE 5.1. *a*, Practical epistemology of small differences: titles, page layout, size, and quality of reproductions. *b*, Detail from Hebrew version of the book. Hebrew version of Arthur Ruppin, *Sociology of the Jews* (Tel-Aviv: Stiebel, 1934), 66–67.

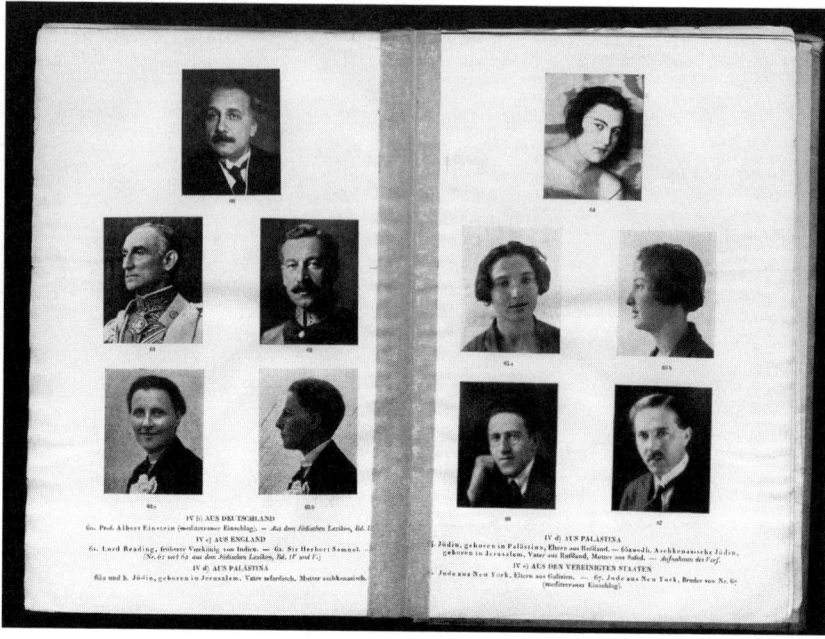

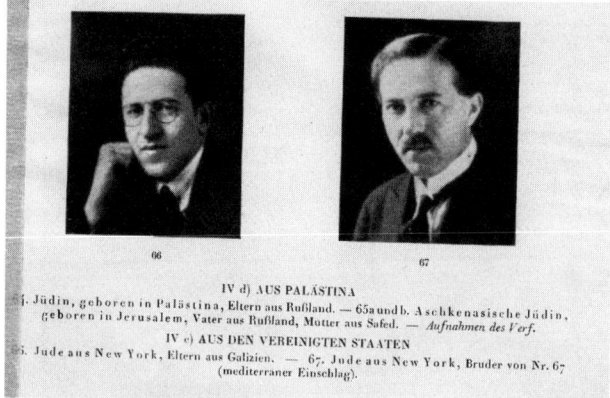

FIGURE 5.2. *a*, Practical epistemology of small differences: titles, page layout, size, and quality of reproductions. *b*, Detail from German version of the book. German version of Arthur Ruppin, *Sociology of the Jews* (Berlin: Jüdischer Verlag, 1930), 66–67.

from New York born to parents from Galicia and the subject of photograph 67 is his brother (in brackets Ruppin states "Mediterranean touch"; fig. 5.2). No source is disclosed, and we have no way to authenticate his statement of their social background. The differences between the individuals can be taken to corroborate his claim. On closer observation, however, similarities between both facial structures and source of light can be discerned (the light in both photographs is falling from above and slightly to the left, leaving the right side of the face darker; they seem to have been taken in similar conditions). In comparison with the German edition, the poorer reproduction quality of the Hebrew edition generates less of what Roland Barthes called a "reality effect" because fewer contingent features are legible.[15]

If photographic faults sometimes enhance the "reality effect," another "fault"—only observable in the German edition—touches on implicit assumptions about types and race. In chapter 2 I argued that it is crucial for understanding the effect of racial photographs in publications that viewers believed that they represented a fraction of the actual reproductions found in archives. In one series of images (fig. 5.3), there appears a bearded man in a white shirt against the background of a white fence, seemingly shot from two angles. The German edition describes the subject of both photographs as a Jew from Fez (fig. 5.4). If one looks carefully at the German edition, one can see that these are not photographs of one individual from two standard angles but

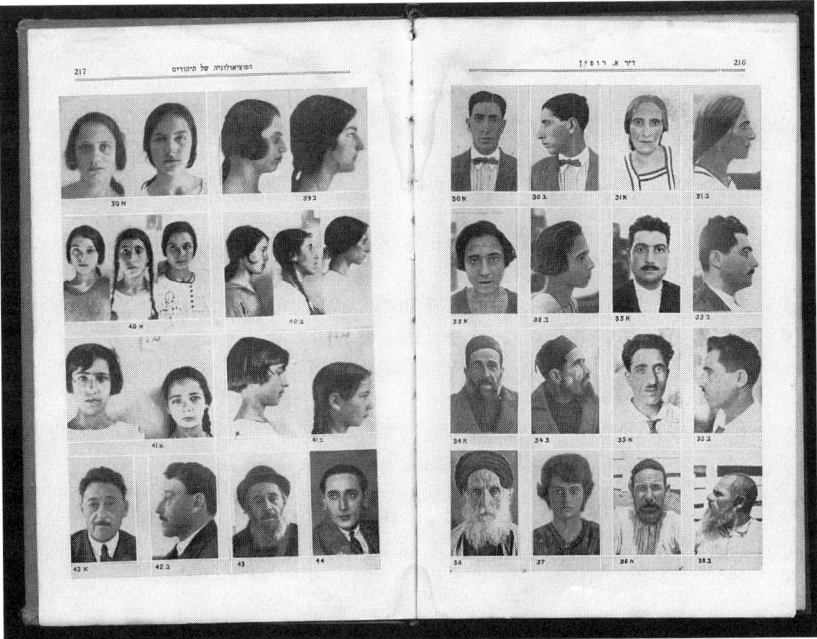

FIGURE 5.3. Practical epistemology of small differences: creation of assumptions through layout, titles, and angles. Hebrew version of Arthur Ruppin, *Sociology of the Jews* (Tel-Aviv: Stiebel, 1934), 38*a*, 39*b*.

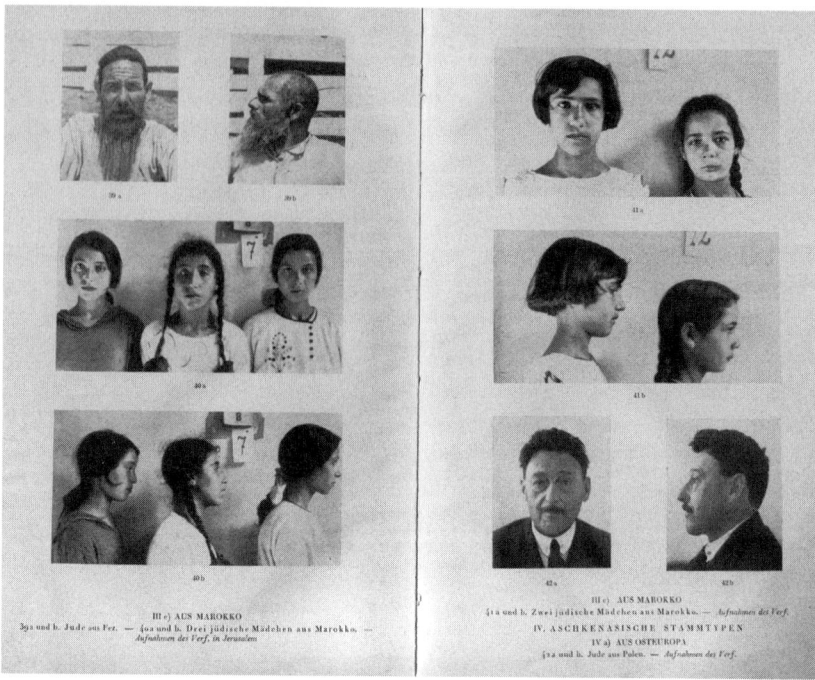

FIGURE 5.4. Practical epistemology of small differences: creation of assumptions through layout, titles, and angles. German version of Arthur Ruppin, *Sociology of the Jews* (Berlin: Jüdischer Verlag, 1930), *39a*, *39b*.

two individuals, one photographed frontally the other from the profile. The hair of the man in full profile is shorter and receding, his beard is longer and pointed, his straight shirt does not have the folds of the other's. The angles, layout, and above all the title make it difficult to recognize that these are not two photographs of one specimen of a type but rather two specimens of one type. This instance illustrates beautifully what Margaret Olin terms "mistaken identification," that is, the power of photographs to generate (false) referential relationships. Because of the size of the reproductions and their quality, this is not discernible in the Hebrew edition.

Was Ruppin's mistake deliberate? For our purposes, what is important is that identifying the mistaken identification enables us to retrieve the conditions that rendered it invisible to readers. Unearthing them illuminates the intertwining of assumptions about the photograph and assumptions about types. Most importantly, the mistake could only go unnoticed as a result of two separate but here interlaced assumptions: that the specimen manifested the type, and that the photograph authentically represented it.

By unearthing mundane differences between the two editions, I have tried to show how affects were tied to practical dimensions of reproduction. Photographs in both editions of Ruppin's book partake in an objectifying scientific ethos that we no longer share today. But materials found in his archive possibly cast a question mark over some of our assumptions about the othering function of racial photographs.

Knowing that in 1976 Ruppin's widow donated to the archive photographic materials from her late husband's estate, I expected that the negatives of the book would be found there.[16] With few exceptions, however, that expectation was not met.[17] Based on the earliest photographs in the archive—dating to Ruppin's 1907 visit to Palestine, which include pictures of himself, landscapes, and a small number of ethnographic photographs—it is possible to pinpoint the appropriation of "type" photographs to the early 1920s.[18] Possibly even more importantly, the archive mixes private and professional photographs. We saw that the photographs produced by Ruppin were of non-Jews. Two photographs found in his archive, both from 1925, seem to reflect the opposite end of the spectrum. These are type photographs, from frontal and profile angles, of a woman identified as Sophie Ruppin.[19] Today we view type photographs as objectifying human beings, as a form of unacceptable reduction of human individuality to racial categories. We take von Luschan's and Ruppin's refraining from photographing white Europeans as implicit recognition of this fact. But if the only way we currently have to understand a racial photograph of one's own family member is as either a contradiction or as internalization and extension of practices of "othering," Ruppin's archive seems to call our assumptions into question.

Photography and Statistics

Similarly to von Luschan, Ruppin's photographs were aimed less at classifying than at illustrating what he took to be natural (racial) classifications. In practice, however, they unquestionably contributed to and in some cases partook in the creation of social classifications that they had been intended merely to illustrate.[20] In this, it might be thought, photography was not fundamentally different from Ruppin's statistical classifications, which were at the core of his model. In this section I briefly compare the two. This comparison aims first at addressing the relative status of photography in Ruppin's scientific economy, second at determining whether the epistemological status of photography and statistics differs, and third at determining whether photography possesses any epistemological specificity.

Both photography and statistics are embedded in Ruppin's wider scientific framework. Remote from philosophical notions of relativism (such as "style," "perspective," or *Weltanschauung*), the introduction of both he viewed as part of the substantiation of empirical, objective standards.[21] That is to say that Ruppin viewed photography as well as statistics not as a form of science but as auxiliary scientific instruments. Ruppin's scientific model as a whole was based on statistical models of measurement. In this sense his model agreed with Martin's physical-anthropological version of science, which was based on statistical measurement, rather than the practice of von Luschan, whom as we saw rejected statistical models for the study of race. In this sense Ruppin's model of science was mixed. The comparison between statistics and photography in Ruppin pertains more to practical questions of classification of what he took to be the whole into its natural segments or classes.

There is a certain similarity between Ruppin's assumptions concerning the capacity of, respectively, photography and statistics to represent social reality. Photographs of well-selected individuals illustrate the type of the group, while a properly segmented class of data represents the class as a whole. Furthermore, the respective medium (photography or statistics) is, according to Ruppin, neutral and transparent in its representation of the real. Crucially, in both cases Ruppin seldom felt the necessity to justify his classification or selection processes. For example, Ruppin did not justify the selection of girls supposedly exemplifying the Yemenite type or patterns of crime of German Jews and non-Jews, and there is no record of his working process, that is, how these segmentations or individuals were selected.[22] Indeed, Ruppin never discussed the status of photography or statistics in general, and this was because his scientific ethos did not require him to do so. *Not* discussing certain basic epistemological instruments is thus a principal feature of this history.

Epistemologically, then, photography and statistics were both conceived by Ruppin as transparent mediums that represented reality without meddling with it. Nevertheless, and as noted above, statistics stood at the core of his model from the outset while photography was incorporated only in his 1930 book. Beyond this difference, beyond the question of (statistical or photographic) classification and sampling, there is an epistemological difference.

Historians and philosophers of science have shown how the statistical method gained an autonomous status in the nineteenth century when statistics were no longer reducible to underlying causes but seen as the explanation itself. Ruppin's statistics of race gained this autonomy when it was no longer believed to be needed to explain it by subjecting it to underlying causes: in statistics on birthrate, mixed marriages, professional dispersion, or crime patterns, the statistics *were* the racial explanation.

We saw in the previous chapter that photographs were increasingly integrated into scientific economies of demonstration. But in Ruppin's economy of demonstration photography remained essentially illustrative. Conceptually, photography never reached the autonomy of statistics. Furthermore, photography never came close to statistics in terms of elucidation, sophistication, prominence, or the degree to which it could be scientifically manipulated (e.g., through the study of correlations). Statistics remained the dominant sociological and demographic model for the study of contemporary Jewry, whereas photography disappeared.

The Ironies of Visibility

Ruppin's study of race, including his use of photography, was part of his attempt to classify, understand, and regulate the chaos of social reality. If, as could certainly be argued, Ruppin was attempting to secure for himself a natural geographical, racial, and national place (in Palestine), somewhat ironically, his attempt bounced back at him. Ruppin never managed to overcome the ambivalences of his place. Haim Weizmann, later the first president of the state of Israel, who deeply admired Ruppin, described his first encounter with him in the following way:

> When I was introduced to him for the first time in Haifa I was somewhat taken aback. I saw before me a young German—I would almost have said a Prussian— correct, reserved, very formal, seemingly quite remote from Jewish and Zionist problems. I was told that he was an assessor, or assistant judge, that he had had a successful business career, that he had come out to Palestine in the spring of 1907, and spent several months there studying the land. All that one perceived on first meeting Ruppin was a German statistician and student of economics, but beneath that cool exterior there was a passionate attachment to his people, and to the building up of Palestine. I learned this in the course of the years.[23]

Ruppin moved from Germany to Palestine because, he believed, Jews were out of place in Germany. In his "natural" home, however, he found it exceptionally hard to master the Hebrew language, and he continued to write in German. Twenty years after his arrival in Palestine, in the late 1920s, his lectures in the newly established Hebrew University, which he was required to give in

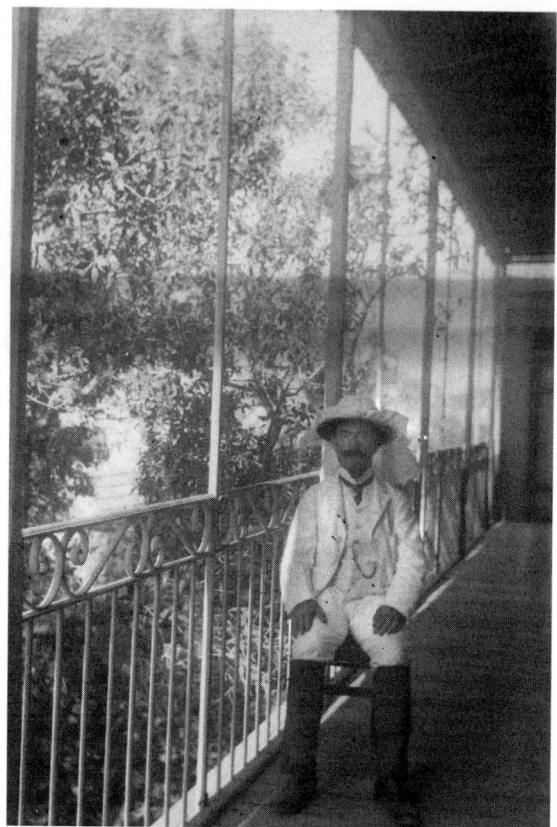

FIGURE 5.5. Arthur Ruppin on his first visit to Palestine. Central Zionist Archive GNAR\
3310038.

Hebrew, were written in transliteration, or so the legend goes. Audio record-
ings of Ruppin speaking Hebrew are marked by an extraordinarily heavy German accent. He was "the German" in a Zionist leadership made up of eastern
European Jews.

In the photograph in figure 5.5, found in the Central Zionist Archives and
dated to 1907, a year before Ruppin emigrated to Palestine, colonial associa-
tions cannot be avoided. He is photographed seated on a balcony overlooking
a cultivated garden and dressed in a spotless white suit and hat. His body is
stiff, and his facial expression is stern. Ruppin's stance says "here I am a European, out of my natural habitat"—quite distant from Zionism's language of
return to a land from which Jews were physically expelled for centuries but
which remained, nonetheless, their natural home. Social reality was no doubt

messy, but Ruppin had no conceptual space for the hybrid. His biography was far messier than the order he attempted to impose on reality. Moving between continents, languages, and national categories, Ruppin could not secure the unproblematic "place" to which he aspired.

PART TWO: FOLLOWING THE PROTOCOL: THE JERUSALEM PHOTOGRAPHS OF ERICH BRAUER

In the history of science, Erich Brauer, the second most important scholar who studied race and employed photography in Jewish Palestine, occupies a place far less prominent than that of Ruppin. However, a study of Brauer is not only necessary for mapping the "race"-photography nexus in Palestine, but in certain respects it is precisely because of his lesser status that Brauer is advantageous for recognizing, first, how certain ideas, practices, and techniques became part of the scientific protocol, and, second, how their meaning and significance were redefined with their relocation to Palestine. Furthermore, from an epistemological perspective Brauer offers a unique advantage: Brauer's estate, including a large number of photographs, is located in one archive housed in the Israel Museum (in Jerusalem). This provides the exceptional possibility of comparing the photographs he selected for publication with his entire collection, and I will close my analysis with several observations drawn from this archive on the relationship between the scientifically "good" and the aesthetically "beautiful" photograph. To begin with, however, I will analyze several key moments in his writing on race and his use of photography in his traveling between Germany and Palestine.

With regard to race and photography, Brauer to a great extent worked within the coordinates set by von Luschan. His claims for originality did not consist in his use of photography or in his notion of race, and his work therefore allows us to point to the emergence and solidification of certain patterns in Palestine.

Brauer's photographic work is consistent with the subjects of chapter 2 in technique and style. Rather than rely on the archives of others, however, Brauer (like Ruppin) produced the photographs himself. Another point, then, has less to do with the photographs as representations than, in the context of an emerging society, with the production of photographs as a complex but asymmetric form of social relations. A closely related point is that Brauer photographed groups that, while not exclusively Jewish, consisted only of non-Europeans. The combined result, as I will show, was the extension of von Luschan's core idea and core photographic practices of the Jews that, while redefining them socially, he nevertheless created an image of Yemenite, Kurdish, or Afghani Jews as allochronic. This was achieved through the selection of individuals for photographs, in particular with regard to hairstyle and gear, through the selection of photographs for publication, and most importantly through deft deployment of titles.

Brauer's Academic Career

Erich Brauer has been to a certain extent rediscovered in the last decade, and today he is recognized as one of the pioneers of Jewish folklore and anthropology in prestate Israel; this despite the fact that his work was not continued after his death and did not serve as the basis for the later development of Israeli anthropology.[24] Brauer, who was born in Berlin in 1895, was to some extent a marginalized figure already during his lifetime. In part this was due to a medical condition—following an illness he suffered as a child he was left a hunchback. But it was also because he never obtained a permanent academic position. As a Jew in Germany he had had only a slim chance of obtaining an academic position, and when he finally settled in Jerusalem, after the Nazi party secured control over Germany, Jerusalem was flooded by individuals with the highest German academic credentials who could not be absorbed by the small newly established Hebrew University. Up to his death, then, Brauer conducted his research on temporary fellowships and grants.[25]

Brauer's early work touched on race but made no use of photography and had nothing directly to do with anything Jewish. He started his academic education in Leipzig on the eve of World War I, and subsequently moved between several universities, including, in 1917, a period in Berlin with von Luschan. His doctoral dissertation was submitted in 1923 to the University of Leipzig and was written under the supervision of the distinguished Africanist and director of the *Völkerkunde Institut*, Karl Weulle (1864–1926), on the religion of the Herero.[26] His first trip to Palestine was made on behalf of the Museum für Völkerkunde zu Leipzig in 1925 in order to collect the ethnological objects of the Arab community. He returned to Jerusalem in 1927, conducting fieldwork on the Jews of Buchara who had settled there. In 1931 he returned to Leipzig, and in 1934 he published his book *Ethnologie der Jemenitischen Juden* (Ethnology of Yemenite Jews), still the most cited book on the subject outside the Hebrew language. Around this time, however, his supervisor Weulle was succeeded by the antisemitic Otto Reche, which ended any prospective employment for Brauer at Leipzig. But in 1932 Brauer was offered a teaching position at the University of Berlin by one of his former teachers, Diedrich Hermann

Westermann. The negotiation over the position ended abruptly, however, as the Nazi party ascended to power. Brauer therefore returned again to Palestine, this time for good. Brauer's passages between Germany and Palestine, therefore, took place on grounds that changed dramatically in 1933.[27]

If we attempt to pull together the threads of Brauer's career as they touched on race, photography, and Jews, we realize that the three only came together in the late 1920s. His dissertation did not touch directly on Jews or use photographs. It did, however, focus on the religion of the Herero from the perspective of the Hamitic theory, and as such it certainly touched on race. Nevertheless, aside from the motto that Brauer took from James Frazer, which speaks of the study of "lower races," race plays little part in the work; the bulk of the text was composed of an ethnological study of religion that built on English and German scholarship, in particular that of Carl Meinhof and Charles Seligman. Brauer believed he had corroborated the Hamitic thesis with regard to elements of the Herero religion.[28] While Brauer did not continue his work on the Herero, basic strains of his scholarship remained the same.

Brauer was little inclined to theory. Raphael Patai reports that on being asked to what school of thought he belonged, Brauer replied that he was only interested in facts. Irrespective of the specific subject he was studying, his implicit notion of "fact" was based on positivist assumptions. He was nevertheless influenced by the *Kulturkreise* theory, which emphasized the mutual relationship between cultures. When he moved to the Jewish context, then, he documented the mutual relationships between Jews and their respective environments. Standing in a certain tension to that practice, however, he nonetheless assumed that the different Jewries were part of some Jewish "whole." But unlike Ruppin, Brauer never attempted to elucidate the relationship between the components and the whole, and to a great extent his monographs were based on the assumption that each such diasporic Jewish group was a separate component with more or less permanent boundaries. As with Ruppin, Brauer viewed the Jerusalem of the 1920s as an exceptional site of opportunity for the study of the (Jewish) world at large. Only with his first studies of Jewish populations in the second half of the twenties did Brauer employ photography, thereby bringing Jews, race, and photography together for the first time.

Brauer provided general statements on visual culture and race in two short exhibitions at the Hebrew University Club on the representation of women in Pietr Bruegel and Albrecht Dürer and with a show in 1939 of one hundred photographs of Jews of the East—that the perspective was shifting is evident in that *East* no longer meant *Ostjuden* but Jews from Muslim or Arab societies. In an unpublished four-page document that most likely accompanied this last exhibition, Brauer noted the absence of knowledge about the peoples among whom Jews live (he used the present tense), reiterated von Luschan's stance about the Jews being made up of Oriental and Armenoid races, and emphasized that the Jews of the East should be prioritized for study because they were closer to the original Jews (*yehudim rishonim*). The visual materials through which he discussed the traits of Kurdish and Yemenite Jews—paintings by Rembrandt, photographs of Kurdish Jews, Hittite reliefs—were the same materials as those studied in the previous chapter. In this short text Brauer followed the typological model, writing of individuals as "exemplars" and as "representative" of their race. His major publications on Yemenite and Kurdish Jews, both conducted in Jerusalem before the exhibition, afford us a more complex picture with regard to race and photography.[29] It is to questions that pertain to the differences between these books that we now turn.

The Yemenite Book

Brauer's discussion of Yemenite racial features occurs in his third chapter, which is devoted to Yemenite physical anthropology. He opens with an account of a Saturday afternoon on which he is standing in Jaffa Street, Jerusalem, observing the various people walking on the street. Yemenite Jews, he says, with their slender dark forms, stand out ("diese dunklen, kleinen, schmalgliedrigen Gestalten aus dem bunten Bilde der übrigen herauskennen"). He complements this account of everyday observation with statistics that point to the smaller average height of the Yemenites as compared with Sephardi Jews, their slender form, their skin color, their long-shaped skulls, and the particular forms of their noses, foreheads, irises, lips and mouths, and hair.[30] At the end of the book are eight tables of photographic reproductions, each containing four to six reproductions, to which Brauer refers throughout this discussion.

By and large, Brauer's discussion proceeds by way of reference to photographs or photographic tables. For instance, he states that "the lips are not infrequently full (table I, 5; VI, 4). The mouth is sometimes broad (table I, 1)."[31] The photographs lack standardization: they vary in terms of angle, light distance from the lens, and facial expression.

Brauer contends that Yemenite Jews do not comprise one pure type, an alleged fact that he explains by the processes of mixture that this community has undergone over the course of its long history. But two African types do stand out particularly clearly in the features of the Yemenites, he insists: the Negroid and the Hamitic. In conclusion, Brauer mocks Niebuhr's impressionistic observation of similarity between Yemenite and Polish Jews, commenting, "it seems that Niebuhr did not encounter too many Polish Jews in his life, because anthropologically there is nothing that connects Ashkenazi Jews with Yemenite ones."[32] The miniscule size of the photographic reproductions, however, ensure that even the keenest observer is unable to find visual verification of this assertion. Significantly, however, Brauer does emphasize the differences between their respective skull shapes. He refers to photographs in his discussion, but their role is secondary to the deep structure, which draws on what he believes are anatomical statistical facts. Brauer agrees with von Luschan that the long-skulled Yemenite differs from the short-skulled *Urtype* Jew. Yemenite Jews are Arabs. In a statement that reads somewhat ironically, coming as it does from an exiled German Jew, Brauer transplants the self-perception of many German Jews as Germans whose Jewish identity is merely religious to the Yemenites. Stating that "the Yemenite Jew . . . is an Arab of the Jewish faith" (*Araber jüdischer Konfession*),[33] Brauer implied that the Jewish national project was aligned with religion rather than race.

In a strange way, though, Brauer also echoed the most radical German branches of writing on race. Race, he states, is not merely a matter of skull or body form but pertains to attachment (*Verbundensein*) and the will to be a member of a race. *This* turns Yemenite Jews into a racial group, within the wider Arab environment, and permits their inclusion within the Jewish whole. The idea of racial consciousness as a racial attribute, as developed by Günther, was discussed in the final part of chapter 3. Here we need simply note that Brauer employed the idea in order to enable Yemenite Jews to distinguish themselves from their Arab environment and join the Jewish national project. The form of justification, however, indicates the penetration of elements of *Rassenkunde*'s holistic interpretation of race within the scientific discourse in Palestine.

Two Versions of the Book on Kurdish Jews

Brauer died suddenly and prematurely in 1942, leaving behind an unpublished monograph on the Jews of Kurdistan. With regard to race and photography, there are significant differences between Brauer's 1934 book on the Yemenite Jews and the two posthumously published versions of his book on the Jews of Kurdistan that Patai translated, edited, and published, first in Hebrew in 1947, and then in English in 1993.[34] As the overall structure of the book is similar to the Yemenite book, the absence of a section on physical anthropology stands out, and several interpretative questions cannot be avoided. We know Brauer possessed some knowledge about the physical-anthropological features of

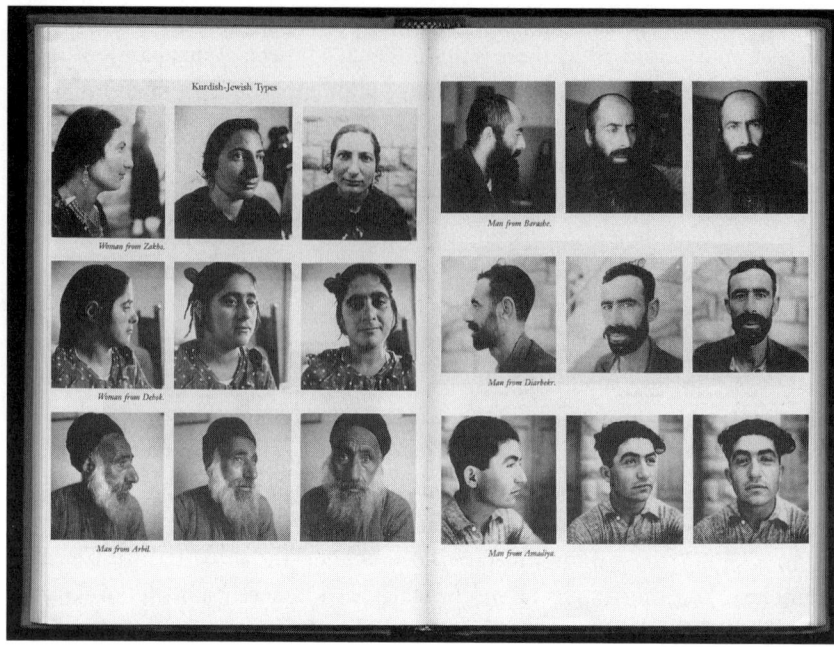

FIGURE 5.6. Erich Brauer, *The Jews of Kurdistan*, ed. Raphael Patai (Detroit, MI: Wayne State University, 1993), 202.

Kurdish Jews. A posthumous Hebrew article that appeared before the book in Patai's journal *Edot* (Tribes) on birth rites among the Jews of Kurdistan was accompanied by a three-photograph series of a Kurdish Jew (in accordance with Martin's post–World War I convention) with facial and skull indexes.[35] With regard to the absence of photographs in the book, we can speculate that the cost of their reproduction in 1947 Palestine could not be met. In any case, however, we cannot know what exact role Patai played here.

From the correspondence between Patai and Brauer's sister, we know that Patai asked for photographs, which she then provided. Lacking a physical-anthropological chapter, the English version is nonetheless accompanied by thirteen pages of photographic reproductions of which five are designated "Kurdish-Jewish Types." The photographic plates are placed in the middle of the book, but in contradistinction to the Yemenite book, there are no reference to the photographs in the body of the text. The photographic plates form a series of rather poor quality reproductions of individuals photographed by Brauer at three angles: frontal, full, and half profile. These are von Luschan–style type photographs. Photographed from very close, the noses are distorted—heavy noses almost stick out of the frame (fig. 5.6). Taken from a low angle, the faces possess a menacing appearance. Some series are accompanied by subtitles describing the origin of the respective individual and identifying the person by name. In terms of photographic style, then, there is no break between the Kurdish and the Yemenite book.

The "Good Example"

What qualified as a "good example"? We cannot know whether the archive possesses all of Brauer's photographs, but as the unpublished amount is much larger than the published, we have an exceptional opportunity to compare published and unpublished photographs. Brauer, his sister, and Patai left no direct evidence concerning the selection process, but from the comparison of the published with the unpublished photographs, it is possible to deduce what was considered a good scientific example.

In terms of style, there is a great amount of conformity in Brauer's photographs. Similarly to von Luschan, we can assume that Brauer's selection of individuals for photographs tended to cluster around certain types and hence created certain patterns of resemblances. Some differences pertain to the physiognomy of individuals, but the more important differences pertain to indications of social status. Differences between the archive and the published materials pertain to dress, glasses, and hairstyle. In the previous chapter I discussed at length the example from a philosophical point of view. Here, however, it is sufficient to note that by *good* I mean the (unstated) characteristics that qualify an image as exemplifying what Brauer wishes it to exemplify; by *beautiful* I mean whether the image pleases the eye in aesthetic terms. With regard to both criteria, there are clear discrepancies between the published and the unpublished photographs.

The archive possesses series that are in technical photographic terms more successful and depict individuals who are aesthetically more beautiful than those reproduced in the publications. Put negatively, Brauer found the less beautiful and the technically more deficient to be more typical and therefore more suitable as scientific example. In publications Brauer also eliminated photographs that contained indications of modern society, and even more crucially, photographs that generated intimacy or empathy with the photographed or otherwise engaged the observer. Whereas the selection of photographs could have minimized the distance between viewers and photographic subjects, it actually did the opposite. The joint effect of all this was to endow subjects with an allochronic sense of meaning, as if they belonged to a different time and place.[36]

The photographs of the radiant sixteen-year-old girl identified as Yehudit

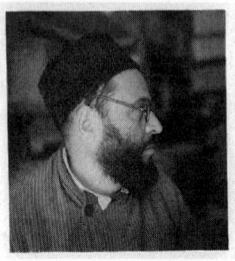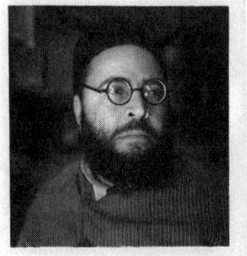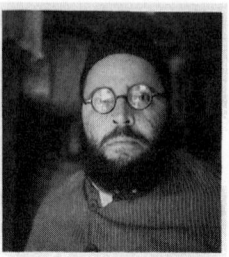

FIGURE 5.7. Three angled series of Eliyahu Cohen. Erich Brauer, unpublished, from Brauer's estate at the Information Center for Jewish Art and Culture The Israel Museum, Jerusalem.

Amar, to provide one example, are far superior to the coarse series Brauer reproduced in any of his books. Her eyes instantly seize the observer, who cannot avoid her personality. Perhaps the best series in the archive is of the forty-five-year-old Eliyahu Cohen, dated to 1938 and identified as an Afghan Jewish businessman (fig. 5.7). In the frontal angle the light reflects on his spectacles. This photographic "fault," however, only emphasizes his personality as a modern, educated, gracious middle-class man whose personality transcends the anthropological framing of the photograph. Brauer kept out of his publications photographs with these kinds of qualities, choosing rather photographs that were strict and even harsh and which, following in von Luschan's photographic tradition, did not engender empathy or any sense of intimacy with the photographed.

Brauer's unpublished photographs of Yemenite Jews portray them in European gear, with short, fashionable haircuts and with round, modern glasses; photographic traits (such as the entrances of British Mandate Jerusalem stone buildings in the background) connect them to 1930s Jerusalem.[37] The titles found on the unpublished photographs distance the photographed from their time and place. As if an attribute of the subjects, the museum cards that accompany the photographs state "Sana" (Yemen) or "Herat" (Afghanistan), omitting their actual date and place of production. The "good example," then, is one from which residues of the time, place, and modern context of production have been removed or minimized.

Photography as Social Practice

Within the internal Jewish context, and when compared with the photographic practices discussed in chapter 2, the most important transformation brought about by the shift of work to Palestine is that Ruppin and Brauer produced their photographs themselves. This shift in production was coupled with an intensification of selectivity.

I have already analyzed their selection of individuals within what they perceived as a natural class. But multiple other forms of asymmetry and selectivity were involved in Ruppin's and Brauer's photographic practices regarding whom they photographed (and whom not) and whose photographs they reproduced (and whose not). Ruppin, similarly to von Luschan, pretty much photographed only non-Jews, but he reproduced photographs of Jews taken by others. Brauer photographed only non-European Jews. If von Luschan excluded Jews from being photographed in order to include them in the class of white (non-Jewish) Europeans, it could be argued that Ruppin extended this logic. Brauer only photographed Jews from the non-European diasporas he studied. Semiotically, it is possible to argue that for both of them European Jews (i.e., their own class) were known, and there was no need to study them or any interest in so doing.[38] But this selectivity had implications for the construction of the composite Jewish "we" in Palestine, in which the national project encompassed various racial components. Rather than undermine borders between members of the same emerging society, these photographic practices both naturalized and further corroborated them.

If from our contemporary perspective the socially oppressive aspects of these photographic practices seem obvious, it is helpful to end this section by reintroducing incidents that fail to meet our assumptions. In his 1924 recommendation letter for Brauer, his former adviser, Weulle, stated favorably, "Brauer possesses the specific spiritual sharpness of his race in abundance" (Brauer besitzt die spezifische Schärfe des Geistes seiner Rasse in ganz ausgeprägten Mass).[39] In Jerusalem, not many years later, Brauer photographed his assistant Hai Avraham Shabati from three angles: frontal, full, and semiprofile (fig. 5.8). Was there a negative discriminatory undertone to Weulle's recommendation? Did Brauer attempt to "other" his assistant? For brushing history against the grain, such moments are crucial precisely because they do *not* fit our formed assumptions. The unsettling quality of these incidents, I believe, is related to their liminal quality, to the blurring of spheres we have become accustomed to conceive as separate, such as a scientific category and the personal sphere of relationships. This is worth remembering as we move into the next section to the interplay and mutual flow between the scientific discourse and the artistic field through discussion of arguably the most racialized group in prestate Jewish Palestine, Yemenite Jews.

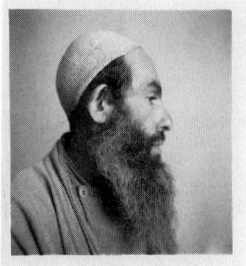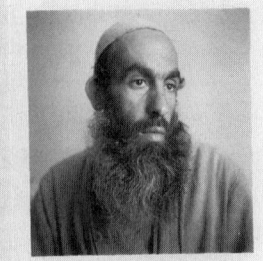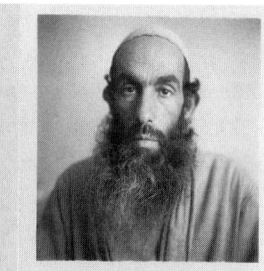

FIGURE 5.8. Three-angled series of Hai Avraham Shabati. Erich Brauer, unpublished, from Brauer's estate at The Israel Museum, Jerusalem. Information Center for Jewish Art and Culture, identified by Ann and Izidor Polck.

PART THREE: FROM SCIENCE TO ART—THE "CASE" OF YEMENITE JEWS

Yemenite Jews have been the subject of increasing critical historical interest of late.[40] A recent exhibition focusing on the photography of Yemenite Jews in the Museum of the Land of Israel in Tel Aviv in 2012 by photographer, curator, and historian of photography Guy Raz revealed that virtually all important photographers operating in Palestine in the first decades of the twentieth century dedicated at least some of their energy to representing Yemenite Jews (as did virtually all painters as well).[41] By focusing on their artistic rendition, this section explores how art photography drew on but also undercut the anthropological discourse.

It is crucial to grasp here that within the racial discourse Yemenite Jews in this period became a question and to some extent even a problem. Study of the Jews' racial constitution was based on several assumptions, two of which clashed with regard to modern European Jewry and Yemenite Jews: first, that Jews were racially defined, and second, that the Jewish type persevered over time and space. From these two assumptions it followed that Jews across the world shared a basic structure. Indeed, a foundational study by Richard Andree insisted on a great similarity between the physical form of Ashkenazi and Yemenite Jews.[42] Nevertheless, as early as the 1880s authors began to note that the skull shape of European Jews tended to the brachycephalic while the supposedly Semitic skull, as manifested by the desert Bedouins—and Yemenite Jews—was identified as dolichocephalic. Scholars committed to a racial account of the Jews thus faced a contradiction. What was the (racial) relationship between European and Yemenite Jews? Which of the two reflected the ancient

Jewish type? How could the problem be harmonized if embracing the one excluded the other?

Over the following decades several attempts to grapple with this problem were made, none of which succeeded in providing a fully satisfactory answer. On the whole, scientists tended to approach this contradiction as probably the result of incomplete scientific theorizing and insufficient empirical data.[43] Von Luschan, as we have seen, set out on a new path by redefining the racial constitution of the Jews as mixed from several components, a redefinition that allowed for the separation of European Jews from Yemenite Jews. But while this explained the structure of the former, it could not explain the divergence of the latter. Samuel Weissenberg, whom in chapter two we saw negated von Luschan's explanation, insisted that the original type of the ancient Jews was Semitic but that following a long process of racial penetration by Slavic elements, the European Jewish skull type had been transformed into its current brachycephalic form. Sephardic Jews (into which Yemenites were now assimilated), Weissenberg argued, remained dolichocephalic because, unlike the Ashkenazi branch, they had mixed with dolichocephalic Mediterranean populations.[44] Weissenberg studied 200 Yemenite Jews in Jerusalem and concluded that between them and the Jews of Europe there was no relationship whatsoever. The Yemenites maintained the Semitic type in its fullest purity; in aesthetic terms clearly superior to European Jews, their long and thin faces were marked by fine and subtle features.[45] Weissenberg remained well aware that if Yemenite Jews correspond to the original Jewish type, "European Jews are not Jewish."[46] But from the same racial argument Ruppin and Brauer chose to draw the opposite conclusion—that Yemenite Jews were (racially) not Jewish. Based on the assumption that the Jewish type was permanent from ancient times to the present, Ruppin developed a methodological distinction between "original Jewish types" and "special types," his prime example of the latter being Yemenite Jews.[47] Brauer, as we have seen, referred to Yemenite Jews as "Arabs of the Jewish confession," that is, Jews by religion but not by race. We saw in chapter 2 that the separation of various variables (race, language, religion) does not destabilize the referent under discussion; the variables in fact remain intact. Yemenites were integral to the Zionist national project, but the racial question remained unsettled.

Almost all descriptions of Yemenite Jews emphasized their long skulls, dark skin and eye color, and small stature, their high degree of literacy, the presence of Talmudic learning among them (their books, it was noted, had arrived directly from Lithuania), and their high intelligence. Documenting several photographic series consisting of frontal, profile, and semiprofile angles, Guy Raz was surprised to find out that photographers touched on the anthropological discourse.[48] Focusing on only two photographers, E. M. Lilien and Helmar Lerski, I show they did so in strikingly different ways.

Lilien's Idealization

E. M. Lilien (1874–1925), "the first Zionist artist," visited Palestine three times: in 1906, 1910, and 1914. But his three-angled photographic series of Yemenite Jews dates to his first visit in 1906.[49] In addition to his famous photograph of Theodore Herzl (on the balcony of the Three Star Hotel on the Rhine), Lilien was well known during his lifetime for his illustrations and engravings. Only very recently has the full extent of his use of photography become known. While famous during his life, interest in his work declined sharply after his death, the result of the rise of the Nazis in Germany as well as the fact that his art was influenced by the *Jugendstil*, which declined after the first decade of the twentieth century while he remained out of touch with the major revolutions in art. By the late 1920s, his photographic/realistic etchings no longer matched the spirit of the times.[50]

Lilien would eventually marry a wealthy German Jewess, settle in Braunschweig, and gradually distance himself from the Zionist idea and institutions. Before that date, however, he was swept up by Zionism in its earliest phase, and his art at this time was intended to advance its cause, to endow Zionism with images as well as with a philosophical and aesthetic vision. Lilien's depictions of the Jordan River, Jerusalem's mountains, or "Jewish heads" were thus deliberately idealized. His idealization of Yemenite Jews was thus an integral part of his general outlook. The specific form it took was the depiction of beautiful, allochronic, biblical figures.

Photographer and curator Micha Bar-Am has shown that photography was more central for Lilien than had been thought; in fact, Lilien employed photography quite extensively throughout his career but minimized or even hid this fact.[51] Lilien's photographs, however, were not only professionally executed with regard to their technical quality but also affected his etchings and illustrations in several specific ways.[52] In the first decades of the twentieth century photography was often perceived as a documentary medium in relation to the fine arts. Like many artists, Lilien viewed the camera as a mere auxiliary tool, preparatory for the artists' plastic work. In his letters to his wife, which were collected and published in 1985, he mentions photography numerous times, discussing his extensive photography in Palestine, both of scenery as well as of "Jewish heads." Nevertheless, in these letters he "warns" her that photographs cannot reflect the color and the true characteristics of the real object.[53] His rather amusing observation that the animals in the Berlin Zoo are so used to the camera that they perform for it indicates his understanding of photography as a reciprocal medium.[54] In these letters he also discusses his practice of drawing from photographs:[55] photography, he insists, is central to his working process and form of observation, but it is not an end in itself.

In photographing Yemenites, then, Lilien was not interested in documentation. Viewing the camera as an apparatus subservient to the plastic arts, he believed photographs were free from those prescriptions and acculturation processes that affected his work as a painter and sculptor.[56]

Lilien does not discuss "race" in any elaborate way in his letters to his wife. His Zionist outlook, however, clearly hinges on the conviction that "we are a people [*Volk*], a race [*Rasse*]. All of us, rich and poor Jews, educated and uneducated, have *one* past. . . . The Jew is my brother. . . . The Jew cannot become French, German, or English."[57] These words were written in Berlin on June 27, 1905, and they demonstrate that Lilien is unaware of the anthropologists' differentiation between "people" and "race." What he means is that the Jews are a people (a biological component included) rather than a religious community and that bonds of solidarity hold them together. He is enthusiastic about the "wonderful Jewish heads," the beauty of which he believes to be but faintly reflected in the photographs he sends to his wife.[58] If his representations of Yemenite Jews depict them as an "other," this is an internal other, part and parcel of the same Jewish people or race to which he commits his own belonging.

Lilien's negatives, today held mainly by Tel Aviv Museum of Art, were found only many years after his death, almost by chance, and probably reflect only a fraction of his work. But they are sufficient to allow the observation that the photographs of Yemenites embody a point of view and form of observation. They idealize Yemenites as "authentic," beautiful bearers of ancient Hebrew heritage. Lilien respects his Yemenite subjects, and his photographs magnify the beauty he ascribes to them in his letters. The racial angles he employs have an affinity with their depiction as ancient Hebrew Semites, fusing myth, fantasy, and the shock he encountered from the state of the land and its inhabitants on his first visit there.[59]

The three-angled series is not a standard anthropological series (fig. 5.9). The photographs of the adult man dressed in striped jacket and buttoned collarless shirt with an oriental hat, sidelocks, and thin beard, taken against a stone wall, possesses a distinct sense of movement. In the frontal photograph the eyes of the subject cannot be seen because he is looking downward; the profile is achieved by the man turning his head slightly backward, as the photographer is standing behind his shoulder; only in the semiprofile angle, centering on the face, are the eyes visible. They look directly at the photographer and beyond him, the gaze expressing transcendence and distance. This same sense of almost terrible beauty, transcendence, and distance is induced in Lilien's other portraits of old Yemenite Jews. Photographed frontally or from their pro-

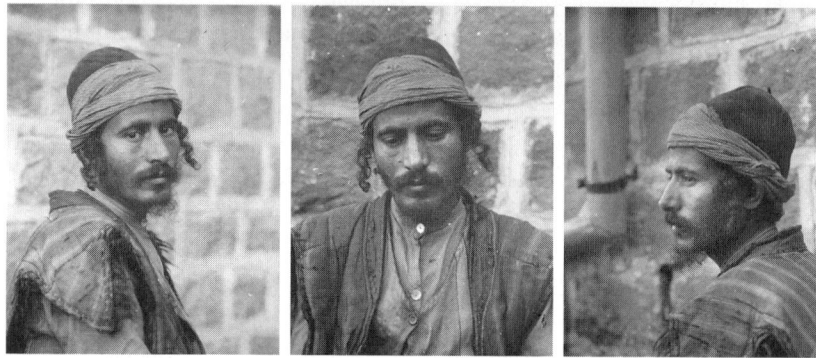

FIGURE 5.9. Three-angled series of Yemenite man. E. M. Lilien, from Lilien's estate at the Israel Museum, Jerusalem.

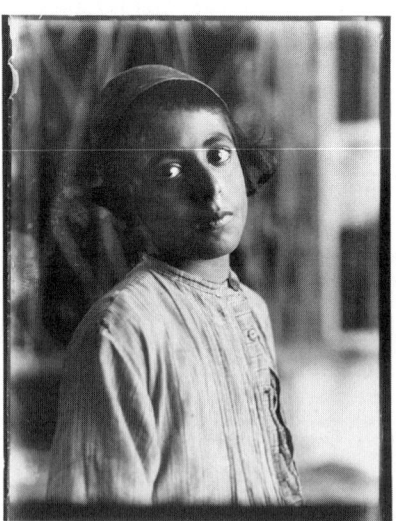 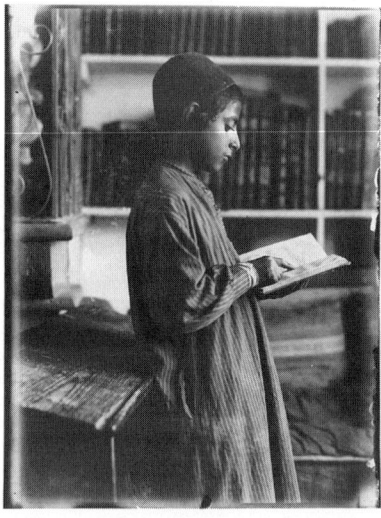

FIGURE 5.10. Yemenite Child in Palestine, 1906. E. M. Lilien, from the estate of Lilien at Tel-Aviv Museum of Art.

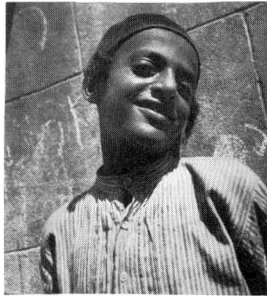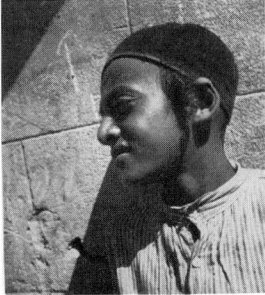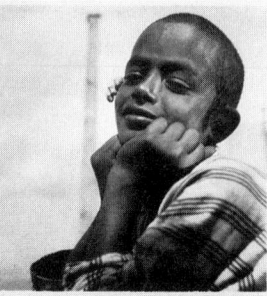

FIGURE 5.11. Three-angled series of Yemenite child. Photograph by Herlmar Lerski, from Guy Raz, *A Yemenite Portrait: Jewish Orientalism in Local Photography* (Tel Aviv: Eretz Israel Museum, Yoav maron Collection, 2012).

file, holding the Talmud or reading it, they are dark and different. The light is falling from above on their foreheads, and the pictures thus express interiority and piety. It is possible to compare Lilien's photographic series of the Yemenite child with that of Lerski, who we will turn to next (fig. 5.10). Lilien's shows the child in a room against a background of shelves of religious books, from the profile, and in reading the sacred book his eyes follow his finger, light falling on his beautiful wide open dark eyes, contrasting between them and their white surrounding. Lerski's ostensibly similar study, by contrast, minimizes the distance between the photographed subject and the viewer: shown from three angles, from slightly beneath his face and to the side, the smiling child returns with light playfulness that which he received from the photographer (fig. 5.11). The photographs of Yemenite Jews, produced by Lilien and Lerski, differ profoundly. Their contrasting understandings of the photographic medium were as important as their different anthropologies.

Helmar Lerski's Universal Photographic Language

Between 1931 and 1948, during which period Helmar Lerski lived in Palestine, he, too, photographed Yemenite Jews. But if Lilien's photographs depicted them as beautiful Jewish "others," Lerski applied to Yemenite Jews his unique but universal photographic language. Situating these Yemenite photographs in the larger framework of Lerski's photographic and social or political philosophy, one is led to see an undermining of the representation of Yemenites as "other," an anti-indexical statement about photography, and an untangling of identity from visuality.

Guy Raz chose to serialize Lerski's portraits of Yemenites next to those of Lilien and Brauer, the result being that what was serialized was the *Yemenite* subject. Such juxtaposition generates the assumption that all three focused on the Yemenite *referent*. The perception arising from serialization, as argued in the previous chapter, is deepened by the shared medium irrespective of possible differences in authorial intention. Furthermore, in the text accompanying the exhibition, Raz frames the entire history of the photographs of Yemenites in prestate Palestine in a Saidian framework and claims that Yemenites replaced Arabs as the conduit of the biblical Jew in the service of Zionist ideology. In this general framework, photographs of Yemenite Jews were necessarily highly ambivalent "others." But if we situate Lerski's photographs of Yemenites with his earlier photographic projects and his conception of photography, a quite different interpretation emerges.

Lerski's career was marked by several rather unexpected moves. Born Israel Shmuklerski to Polish Jewish immigrants in Strasbourg (then Germany) in 1871, he spent most of his childhood in Zürich, Switzerland. At the age of twenty two, after having trained to become a bank clerk, he immigrated to the United States and began a career as an actor in the German-speaking theater. In 1909, however, he gave up on the stage and opened a photography studio with his first wife; based on a unique method that involved the use of mirrors, he won recognition almost instantly among American critics and scholars. But in the fall of 1915, in the midst of World War I, he relocated to Berlin, where he worked intensively in Berlin's theater and cinema.[60] In 1931, now over sixty years old, he relocated again, this time to Tel Aviv, which he left to return to Zürich in 1948, a month before the State of Israel was proclaimed. What is important for our purposes is simply the fact that before his arrival in Palestine, Lerski developed an idiosyncratic photographic method that was coupled with a humanistic conception of man, both of which remained constant during his sojourn in Palestine.

The unique photographic method developed by Lerski, it could be argued, strongly undermined its deictic assumptions, assumptions shared by all the scientists or authors addressed in this study. If the unstated assumption of the latter was that a well-executed photograph expresses a real relationship between the photograph and the referent that it depicts, Lerski unsettled this assumption by way of his manipulation of natural light. Purposely disobeying the rule of "proper" studio portraits, portions of the faces of some of those he photographed were in shadow while others were overexposed by strong light. This technique produced images that did not agree with the conventions of realistic photographic portraits and in practice undermined dependent and equally conventional ideas of "natural" lighting. Ontologically, Lerski's photography was not about a referent being truthfully depicted by a mechanical process but an artist manipulating light on the surface of a photographed object.[61]

Potentially, this notion of photography could be applied to any object, but in practice Lerski focused almost exclusively on human faces. His classic study *Köpfe des Alltags* (Everyday heads) appeared in Germany in 1931 (a year after August Sander's *Antlitz der Zeit*, Erich Retzlaff's *Das Antlitz des Alters*, and also Günther's book on the Jews) and treated individuals of various social backgrounds with the same photographic technique. Faces fill almost the entire frame, and any social contextual clues that normally indicate social status were removed from the background. Names were not disclosed. But in a separate list the reader can find out the social status of the photographed, who range from beggars to factory workers. The beggar's expressive portraits, for instance, could easily be taken for those of a theatre celebrity or a poet.[62] Lerksi coupled a conception of photography as two-pronged "surfaces of projection" (*Projektionsflächer*) with a humanistic perception of man, his motto being "In jedem Menschen steckt alles" (everything can be found in every person). This vision was expressed most fully in his project *Verwandlungen durch Licht* (Metamorphose through light), which he carried out in Palestine, in which 175 photographs of one single individual portrayed him so differently that on many occasions it is not possible to see that one is looking at the same face.[63]

Working at the same time as August Sander, Erna Lendavi-Dricksen, Günther, and Clauß, who all, in different ways, photographically touched on social, national, or racial types, Lerski's work was interpreted in more than one way. Lerski's second major project, which he began in 1930 (before he left Germany), and which was termed "Jewish Heads" and later "Palestinian Heads" (now including Arabs), never materialized. It is clear that the context of this project was the rising flood of racist, antisemitic ideas and visual images in Germany. According to Florian Ebner, Lerski's photographic philosophy remained constant in these years. But I believe that Ebner's interpretation of the project misses Lerski's intention. Although he stresses that Lerski's project, unlike others, was inclusive rather than exclusive, Ebner situates him with regard to attempts to create a photographic *Bildatlas* (or as Walter Benjamin labeled Sander's book, *ein Übungsatlas*, a practice atlas) such as those constructed by Lendavi-Dircksen, August Sander, or even Günther or Clauß or Ruppin.[64] But Lerski's originality lay precisely in his destabilization of deictic belief, and he conceived the human face as a stage awaiting the photographer, who at his or her will expresses whatever quality he or she wishes upon each and every human face.[65]

There are nevertheless tensions in Lerski's (failed) project, and these bear directly on the subject of this chapter. In his communication with his intended publisher, Charles Peignot, Lerski stated his intention to focus only on "original types," as focusing on "modern Jews" would mean taking a position on the racial question of modern Jewry (*modernen Judentums*). By focusing powerfully on the *Ur*-types of the Jews, he argued, all later developments (*Entwicklungsstufen*) would become legible.[66] Thus Lerski differentiated between European Jewry, which he perceived as modern, and all the rest, which he perceived as closer to more primordial Jews. This statement undermines both his expressed photographic philosophy, which puts the weight on the photographer rather than on visible properties or hidden essences of the photographed, as well as his egalitarian social philosophy. There are further complications. First of all, his abstention from photographing European Jews in this context is connected to the situation in Europe, not Palestine. Second, he did in fact photograph European Jews. Third, and most importantly, the quoted statement stands in opposition to the actual photographic language that he employed, which was always universal.

Lerski photographed several Yemenite individuals in 1931 (fig. 5.12). He extended to them his mirror technique on glass plates of 30 cm on 40 cm, that is, 1:1. Producing extreme modernistic images, their most essential marker of difference—skin color—was removed. The color of the skin of the Yemenite man Lerski photographed on the roof of his Tel Aviv building, which served as his studio, leaves no trace of the dark skin conveyed by other photographs of Yemenites. The elderly man's face fills almost the entire frame—standard in Lerski's practice. The man's hair and beard are blond. The blemished spectacles reflect the light falling on the face from the left.[67] Apart from the man's glasses, there are no markers of social context. The man's round, modern glasses distance Lerski's portrait from the practices that differentiated between peoples of "culture" and peoples of "nature," between "modern" and "primitive," between "us" and "them," between "white" and "nonwhite." This individual is subject to the same photographic manipulations of light to which Lerski subjects all his subjects. This is far more a record of a photographer applying light to the face of an individual human being than "a photograph of a Yemenite Jew."

Lerski's application of his universal photographic language to Yemenite Jews stands for a different, universal anthropology. In coupling manipulation of light and humanistic conceptions of man, then, Lerski marks the outer

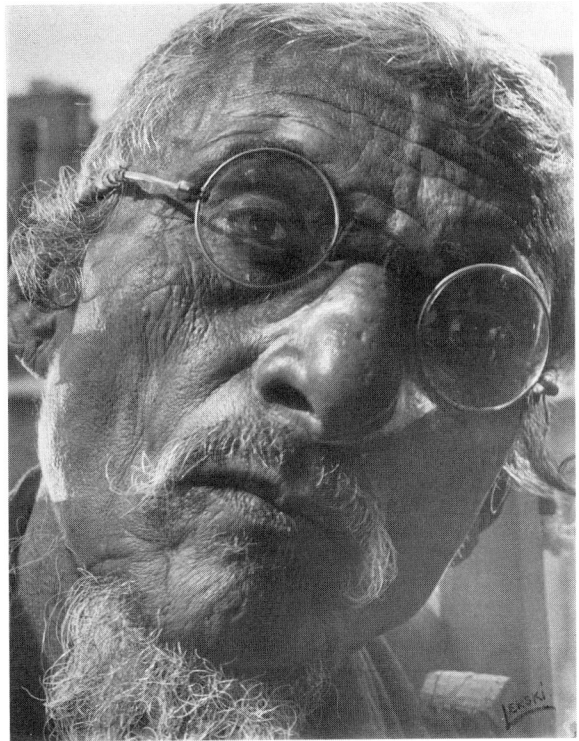

FIGURE 5.12. Yemenite man photographed by Lerski on his Tel-Aviv roof studio in the early 1930s. Helmar Lerski, *Der Mensch: Mein Bruder*, ed. Anneliese Lerski (Dresden: Kunst, 1958).

boundaries of racial photography as well as the failure built into any attempt to draw a clear, permanent line between science, art, and politics.

*

In this chapter I documented the shift of racial photography of Jews from Europe to Palestine after the end of World War I. While I show that photographic practices and ideas of race were to a great extent scientifically derivative, by paying special attention to "small tensions," I argued that with the transfer of these ideas and practices to Palestine, a new context was generated. For now they no longer bore primarily on questions of representation of Jewish variation but partook in the emergence of new and intricate asymmetric social relations and social structures. Nevertheless, I also showed that at no moment can this Palestinian history really be separated from its European and in particular

German context. True, "race" never acquired the political, legal, cultural, or scientific prominence it did in Germany. Involving processes of classification and representation, there was no equivalent—internal or external—to the antisemitic discourse on Jews. But all the persons studied in this chapter were educated in Germany and deeply tied to German culture, science, and scholarship, and even when they immigrated to Palestine permanently or temporarily, they continued to publish in German, write for German audiences, and most importantly, engage with the German cultural, scientific, and shifting political scene. From this latter perspective racial photography in Palestine never ceases to be a "parallel history."

INTRODUCTION

1. I draw on Lorraine Daston's notion of historical epistemology in Lorraine Daston, "'Baconian Facts': Academic Civility and the Prehistory of Objectivity," *Annals of Scholarship* 8 (1991): 337–64, and Lorraine Daston and Peter Galison, "The Image of Objectivity," *Representations* 40 (1992): 81–128. I also draw on Ian Hacking, "Historical Ontology," in *Historical Ontology* (Cambridge, MA: Harvard University Press, 2002), 1–26. See also Bruno Latour, "Practical Metaphysics," in *Reassembling the Social: An Introduction to Actor-Network-Theory* (New York: Oxford University Press, 2005); Friedrich A. Kittler, *Discourse Networks 1800/1900*, trans. Michael Metteer and Chris Cullens (Stanford, CA: Stanford University Press, 1990), which points to close correlations between technological forms and cultural discourses.

2. Steven Shapin and Simon Schaffer, *Leviathan and the Air-Pump: Hobbes, Boyle and the Experimental Life* (Princeton, NJ: Princeton University Press, 1985), 342.

3. Hike Thode-Arora, "Herbeigeholte Ferne: Völkerschauen als Vorläfer exotisierender Abenteuerfilme," in *Triviale Tropen: Exotische Reise- und Abenteuerfilme aus Deutschland, 1919–1939*, ed. Jörg Schöning (Munich: Text und Kritik, 1997), 25, 29. See also Britta Lange, *Echt. Unecht. Lebensecht. Menschenbilder im Umlauf* (Berlin: Kadmos, 2006), 153–88. Apart from forms of production of authenticity, Lange discusses the issue of serialization of ethnographic dummies achieved through their dissemination throughout European museums (169). I will be compelled to deal with serialization in chap. 4.

4. Thode-Arora, "Herbeigeholte Ferne," 30.

5. Georges Didi-Huberman, *Invention of Hysteria: Charcot and the Photographic Iconography of the Salpetriere*, trans. Alisa Hartz (Cambridge, MA: MIT Press, 2003).

6. The classical historical work in this area remains Jeffrey Herf, *Reactionary Modernism: Technology, Culture and Politics in Weimar and the Third Reich* (Cambridge: Cambridge University Press, 1984).

7. Dirk Rupnow and I discuss this in more detail in another context. See Amos Morris-Reich and Dirk Rupnow, eds., *Ideas of Race in the History of the Humanities* (London: Palgrave, forthcoming).

8. An example of a disputed historical case is that of Wilhelm Schalmayer. See Sheila Faith Weiss, *Race Hygiene and National Efficacy: The Eugenics of Wilhelm Schalmayer* (Berkeley: University of California Press, 1987), 38–39, 47–48, 141–42; Richard Weikart, *From Darwin to Hitler: Evolutionary Ethics, Eugenics, and Racism in Germany* (New York: Palgrave, 2004), 177.

9. Ann Laura Stoler, *Carnal Knowledge and Imperial Power: Race and the Intimate in Colonial Rule* (Berkeley: University of California Press, 2002), 42.

10. Ian Hacking, "The Looping Effects of Human Kinds," in *Causal Cognition: A Multidisciplinary Debate*, ed. Dan Sperber, David Premack, and Ann Premack (Oxford: Oxford University Press), 351–83.

11. Sander L. Gilman, *Making the Body Beautiful: A Cultural History of Aesthetic Surgery* (Princeton, NJ: Princeton University Press, 1999), 3–41.

12. Cf. Jon Darius, *Beyond Vision* (Oxford: Oxford University Press, 1984); Ann Thomas, ed., *Beauty of Another Order: Photography in Science* (New Haven, CT: Yale University Press in association with the National Gallery of Canada, Ottawa, 1997); Corey Keller, ed., *Brought to Light: Photography and the Invisible, 1840–1900* (New Haven, CT: Yale University Press, 2008); Peter Geimer, "Picturing the Black Box: On Blanks in Nineteenth Century Paintings and Photographs," *Science in Context* 17, no. 4 (2004): 467–501; "Noise or Nature? Photography of the Invisible around 1900," in *Shifting Boundaries of the Real: Making the Invisible Visible*, ed. Helga Nowotny and Martina Weiss (Zurich: Hochschulverlag, 2000), 119–36.

13. Jonathan Crary, *Techniques of the Observer: On Vision and Modernity in the Nineteenth Century* (Cambridge, MA: MIT Press, 1992). Crary defines subjective vision as "a vision that had been taken out of the incorporeal relations of the camera obscura and relocated in the human body" (16).

14. Jonathan Crary, *Suspensions of Perception: Attention, Spectacle, and Modern Culture* (Cambridge, MA: MIT Press, 1999), 215.

15. Crary, *Suspensions of Perception*, 294. On Renaissance perspective, see Erwin Panofsky, *Perspective as Symbolic Form* (New York: Zone, 1993), 30–31.

16. Frederic J. Schwartz, *Blind Spots: Critical Theory and the History of Art in Twentieth-Century Germany* (New Haven, CT: Yale University Press, 2005), 77.

17. Crary, *Suspensions of Perception*, 290.

18. Ibid., 294.

19. For Benjamin's oft-quoted contention of "Reception in a state of distraction" as "symptomatic of profound changes in apperception," see Walther Benjamin, "The Work of Art in the Age of Mechanical Reproduction," in *Illuminations,* ed. Hannah Arendt, trans. Harry Zohn (New York: Schocken, 1969), 240–41. Chapters 2, 3, and 4 apply Benjamin's contention to the academic or scientific domain. They demonstrate that authors developed strategies of presentation, demonstration, and illustration of visual materials to meet the challenge of reception in a state of distraction.

20. One famous example is sensory experience after the development of the train: the effect of rapid movement, power, and speed on visual perception (the subject of one of Laszlo Moholy-Nagy's first photographic series). Schwartz, *Blind Spots*, 47.

21. On the impressionists and cubists in this context, see Stephen Kern, *The Culture of Time and Space 1880–1910* (Cambridge, MA: Harvard University Press, 1983), 21–22, 140–41, 160–61, and 7–8, 22–23, 117, 195–96.

22. Schwartz, *Blind Spots*, 60.

23. Quoted without reference in Schwartz, *Blind Spots*, 65.

24. Laszlo Moholy-Nagy, "Fotografie, die objektive Sehform unserer Zeit," *Teleohr* (1936): 120–22. On Moholy-Nagy in the wider context of interwar photography, see Bernd Stiegler, "Objektives Sehen und subjektiver Blick: Zur Theorie der Fotografie in den zwanziger Jahren," in *Mediengebrauch und Erfahrungswandel: Beiträge zur Kommunikationsgeschichte*, ed. Detlev Schöttker (Göttingen: Vandenhoeck und Ruprecht, 2003), 157–169. In the following chapters I show that elements of Moholy-Nagy's understanding of photography (such as transforming of perception through photography, photography as a weapon, and a specifically photographic language independent from that of art) were put into practice by racial writers.

25. Schwartz, *Blind Spots*, 162.

26. See Rolf Tiedemann, "Dialectics at a Standstill," in *Walter Benjamin: The Arcades Project*, ed. R. Tiedemann, trans. Howard Eiland and Devin McLaughlin (Cambridge, MA: Harvard University Press, 1999), 940.

27. Schwartz, *Blind Spots*, 178–81.

28. Quoted in Ibid., 199.

29. Ibid., 217.

30. Wolfgang Kabatek, *Imagerie des Anderen im Weimarer Kino* (Bielefeld: Transcript, 2003), 163.

31. Ibid.; Béla Balázs, *Der sichtbare Mensch; oder, Die Kultur des Films* (Vienna: Deutsche-Österreichischer Verlag, 1924), 46.

32. Balázs, *Der sichtbare Mensch*, 71, 76.

33. Erwin Panofsky, "Style and Medium in the Moving Pictures," *Transition* (1937): 124–25: "In a movie theater . . . the spectator has a fixed seat, but only physically. . . . Aesthetically, he is in permanent motion, as his eye identifies itself with the lens of the camera which permanently shifts in distance and direction. And the space presented to the spectator is as movable as the spectator is himself. Not only do solid bodies move in space, but space itself moves, changing, turning, dissolving and recrystallizing."

34. Crary, *Suspensions of Perception*, 11–12.

35. Within philosophy, already in 1883 Alois Riehl differentiated between genuine philosophical questions and matters of worldview (*Weltanschauung*). The philosophical stances of Cassirer and Heidegger expressed in the 1929 Davos debate could be classified along similar lines. Indeed, after the controversy, Rudolf Carnap devoted himself to a careful reading of *Being and Time* and concluded that Heidegger's metaphysical ruminations "are not consistent with logic and the scientific mode of thinking"; Carnap quoted in Peter E. Gordon, *Continental Divide: Heidegger, Cassirer, Davos* (Cambridge, MA: Harvard University Press, 2010), 99. It is important for understanding the cultural atmosphere of the time that, according to common perception, Heidegger emerged as victor from this debate.

36. See J. P. Spiro, "Nordic vs. anti-Nordic: the Galton Society and the American Anthropological Association," *Patterns of Prejudice*, 36, no. 1 (2002): 35–48.

37. After the war, in an ironic twist, leading German racial scholars attempted to rewrite their country's history by appropriating the terms *pseudoscience* and *pseudogenetics* to distinguish their work from that of Nazi ideologists. Benoit Massin has pointed out the incongruity between this West German semantic strategy and the historical record; see Benoit Massin,

"Anthropologie und Humangenetik im Nationalsozialismus; oder, Wie schreiben deutsche Wissenschaftler ihre eigene Wissenschaftsgeschichte?," in *Wissenschaftlicher Rassismus: Analysen einer Kontinuität in den Human- Naturwissenschaften*, ed. Heidrun Kaupen-Haas and Christian Saller (Frankfurt: Campus, 1999), 23.

38. On the history of the term and the concept, see Dirk Rupnow, Veronika Lipphardt, Jens Thiel, and Christina Wessely, eds., *Pseudowissenschaft: Konzeptionen von Nichtwissenschaftlichkeit in der Wissenschaftsgeschichte* (Frankfurt: Suhrkamp, 2008).

39. On this tension, see James Elkins, *The Object Stares Back: On the Nature of Seeing* (San Diego, CA: Harcourt, 1997), 101–4.

40. Benedict Anderson, *Imagined Communities: Reflections on the Origin and Spread of Nationalism* (London: Verso, 1991). On photography in this respect, see James Ryan, *Picturing Empire: Photography and the Visualization of the British Empire* (London: Reaktion, 1997).

41. Anderson, *Imagined Communities*, 184: "The particular always stood as a provisional representative of a series, and was to be handled in this light. This is why the colonial state imagined a Chinese series before any Chinese, and a nationalist series before the appearance of any nationalists."

42. Mary Carruthers, *The Book of Memory: A Study of Memory in Medieval Culture* (Cambridge: Cambridge University Press, 1990), 52; Richard Kearney, *The Wake of Imagination: Ideas of Creativity in Western Culture* (London: Hutchinson, 1988), 106–13; Harry A. Wolfson, "The Internal Senses," *Harvard Theological Review* 28 (1935): 69–133: "the construction out of images of things existent, new composite images of things non-existent, or the breaking up of images of things existent into images of things non-existent" (92).

43. Carruthers, *Book of Memory*, 53.

44. The Encyclopedists, such as Voltaire, contrasted "active" imagination (crucial for mathematical or poetic innovation) with "passive" imagination (such as that of pregnant women, who were conceived of as an instrument of error and passion). See Lorraine Daston, "Nature by Design," in *Picturing Science, Producing Art*, ed. Caroline A. Jones and Peter Galison (New York: Routledge, 1998), 247.

45. Immanuel Kant, *Critique of Judgment*, trans. James Creed Meredith (Oxford: Clarendon, 1952), 120–21. See Kearney, *Wake of Imagination*, 167–77.

46. Immanuel Kant, *Critique of Pure Reason*, trans. N. K. Smith (London: Macmillan, 1933), 112: "Synthesis in general is the mere result of the power of the imagination, a blind but indispensable function of the soul, without which we should have no knowledge whatsoever, but the existence of which we are scarce ever conscious."

47. Richard Kearney, *Poetics of Imagining Modern to Post-Modern* (New York: Fordham University Press, 1998), 53: "Imagination is the foundationless foundation of our 'knowledge of all things.' It is the blind spot of truth which enables us to see things as identifiable objects without itself being seen. It is the invisible source of our vision: that which makes the world possible."

48. Kant, *Critique of Pure Reason*, 15, 28.

49. Kearney, *Poetics of Imagining*, 6.

50. For a detailed historical and philosophical, account see Gordon, *Continental Divide*: on Heidegger's interpretation of imagination (*Einbildung*) in Kant's first *Critique* as independent of reason and as relative to time, see 122, 130; on the productive power of the imagination as that

which connects thought and intuition, see 145; on imagination as either productive or reproductive of an object, see 166.

51. Pierre Bourdieu, *The Political Ontology of Martin Heidegger* (Stanford, CA: Stanford University Press, 1991).

52. Kearney, *Poetics of Imagining*, 3.

53. Cornelius Castoriadis, *The Imaginary Institution of Society*, trans. Kathleen Blamey (Cambridge: Polity, 1985), 146–47.

54. Kearney, *Poetics of Imagining*, 19.

55. Elkins, *Object Stares Back*, 229.

56. The two principal studies of anthropology and photography are Thomas Theye, ed., *Der Geraubte Schatten: Photographie als ethnographisches Dokument* (Munich: Münchner Stadtmuseum, 1989), and Elizabeth Edwards, ed., *Anthropology and Photography, 1860–1920* (New Haven, CT: Yale University Press, 1994).

57. Charles Sanders Peirce, "Logic as Semiotic: The Theory of Signs," in *Philosophical Writings of Perice*, ed. Justus Buchler (London: Routledge, 1955), 98–119, esp. 101–12; Roland Barthes, *Camera Lucida: Reflections on Photography*, trans. Richard Howard (New York: Hill and Wang, 1981); Susan Sontag, *On Photography* (New York: Farrar, Straus and Giroux, 1977); Daston and Galison, "Image of Objectivity." For an account of how photography gained scientific authority in psychiatry, see Sander L. Gilman, *Hugh W. Diamond and the Rise of Psychiatric Photography* (New York: Brunner/Mazel, 1976). For an account of how photography gained scientific and legal authority, see Tal Golan, *Laws of Men and Laws of Nature: The History of Scientific Expert Testimony in England and America* (Cambridge, MA: Harvard University Press, 2004).

58. For a study of photography in the context of one such murderous experiment in deep cold, see Peter Thompson's film *Universal Hotel/Universal Citizen* (1987).

59. Klaus Kreimeier, "Mechanik, Waffen und Haudegen überall Expeditionsfilme: Das bewaffnete Auge des Ethnographen," in *Triviale Tropen: Exotische Reise- und Abenteuerfilme aus Deutschland, 1919–1939*, ed. Jörg Schöning (Munich: Text und Kritik, 1997), 53.

60. Kabatek, *Imagerie des Anderen*, 9.

61. Ibid., 37; Helmuth Plessner, *Mit anderen Augen: Aspekte einen philosophischen Anthropologie* (Stuttgart: Reclam, 1980), 170.

62. Kabatek, *Imagerie des Anderen*, 104–5.

63. Richard Dyer, *White* (London: Routledge, 1997), 1: "Other people are raced, we are just people. . . . This assumption that white people are just people, which is not far off saying that whites are people whereas other colors are something else, is endemic to white culture."

64. Kabatek, *Imagerie des Anderen*, 122.

65. Anton Kaes and Martin Jay, eds., *The Weimar Republic Sourcebook* (Berkeley: University of California Press, 1994), 641–43.

66. Kabatek, *Imagerie des Anderen*, 15. For a detailed analysis of the use of photographs in books and the close relationship between photography and positivist philosophy in the first decades following the invention of photography, see Carol Armstrong, *Scenes in a Library: Reading the Photograph in the Book, 1843–1875* (Cambridge, MA: MIT Press, 1998).

67. Hayden White, "The Public Relevance of Historical Studies: Reply to Dirk Moses," *History and Theory* 44 (2005): 337.

68. For a discussion of the notion of evidence in related contexts, see James Chandler, Arnold I. Davidson, and Harry Harootunian, eds., *Questions of Evidence: Proof, Practice, and Persuasion across the Disciplines* (Chicago: University of Chicago Press, 1994).

CHAPTER ONE

1. All three approaches mentioned in this section are commonly used today, but in the digital age, the distinction between the production of an image and of a number has become blurred. Martin maybe rejects Galton's photographic method but shares his statistical preparatory step of the creation of group average. In their statistics (but not in their photography), Galton and Martin (as well as Franz Boas, who is not discussed here) were interested not only in the average but in the variance—the relative scale of variability between individuals.

2. Elisabeth Edwards, ed., *Anthropology and Photography* (New Haven, CT: Yale University Press, 1994).

3. Carl Victor and Friedrich Wilhelm Dammann, *Ethnological Photographic Gallery of the Various Races of Men* (London: Trübner, 1876). As Michael Banton notes for nineteenth-century racial typologies, "the notion of type was a convenient one because it was not tied to any particular classificatory level in zoology"; hence, it could be employed "without having to establish just what that type was": Michael Banton, *The Idea of Race* (London: Tavistock, 1977), 31.

4. During the Dreyfus affair, Bertillon was invited to identify the handwriting of the "Bordereau"—a document that had been delivered torn in small pieces by a discharged concierge of the German military attaché von Schwarzkoppen. Bertillon constructed a theory that the "Bordereau" was written partly in Dreyfus's handwriting, partly in a disguised hand, and partly in characters traced from original documents. Even after Esterhazy's confession and Dreyfus's vindication, Bertillon adhered to this view. When Colonel Picquart showed a sample of Esterhazy's handwriting to Bertillon without disclosing to him the identity of the writer, Bertillon immediately pronounced its identity with that of the "Bordereau." On learning that the handwriting was not that of Dreyfus, Bertillon became enraged and declared that the Jews had succeeded in making an exact copy of Dreyfus's handwriting. Henry T. F. Rhodes, "Alphonse Bertillon," in *In the Tracks of Crime*, ed. Henry T. F. Rhodes (London: Turnstile, 1952), 73–76. See also Henry T. F. Rhodes, *Alphonse Bertillon: Father of Scientific Detection* (London: Harrap, 1956), 161–87. Earlier, Bertillon relates to the "convex noses of the Jewish type;" Alphonse Bertillon, *Instructions for Taking Descriptions for the Identification of Criminals and Others, by Means of Anthropometric Indications*, trans. Gallus Muller (Chicago: American Bertillon Prison Bureau, 1889), 68.

5. Francis Galton, "Photographic Composites," *Photographic News* 29 (April 17, 1885): 243.

6. Bertillon's system was eventually found to be flawed: despite detailed instructions, two different officers could obtain different numbers; unlike fingerprinting, two individuals could be identified as the same person. In 1903, the system was discredited when Will West was arrested in Kansas and found with anthropometrics to be a previously arrested man, which was contradicted by fingerprinting. Rhodes, *Alphonse Bertillon*, 98.

7. See Alan Sekula, "The Body and the Archive," in *The Contest of Meaning: Critical Histories of Photography*, ed. R. Bolton (Cambridge, MA: MIT Press), 356. Quetelet also developed

a model of racial classification from objects of art: a hierarchy of head types, with presumably upright Caucasian rows approaching a lost ideal more closely than did the apelike brows of Africans. Quetelet's aesthetic ambition was to compare his project to Dürer's studies of human bodily proportion. Ian Hacking has suggested that the rise of social statistics in the mid-nineteenth century was crucial to the replacement of strictly mechanistic theories of causality by a more probabilistic paradigm reference. Sekula argues that Quetelet's determinism was based on iron laws of chance, a development that ultimately led to indeterminism (ibid., 354). Quetelet admitted that the average man was a statistical fiction that lived within the abstract configuration of the binomial distribution. Quetelet defined the central portion of the curve, the large number of measurements clustered around the mean, as a zone of normality. Divergent measurements tended toward darker regions of monstrosity and biosocial pathology (ibid., 355).

8. Sekula, "Body and the Archive," 360.

9. On the history of the frontal and profile photographic angles in the daguerreotypes of slaves ordered by Harvard scientist Louis Agassiz in 1850, see Lisa Gail Collins, "Historic Retrievals: Confronting Visual Evidence and the Documentation of Truth," *Chicago Art Journal* 8, no. 1 (1998): 5–17, and Shawn Michelle Smith, *Photography on the Color Line: W.E.B. Du Bois, Race, and Visual Culture* (Durham, NC: Duke University Press, 2004), 46–47.

10. Josh Ellenbogen, "Photography and the Imperceptible: Bertillon, Galton, Marey" (PhD diss., University of Chicago, 2005), 1:21.

11. With regard to classification of eye color, see Bertillon, *Instructions for Taking Descriptions*, 54.

12. Alphonse Bertillon, *La photographie judiciaire* (Paris: Gauthier-Villars, 1890), 25, quoted in Ellenbogen, "Photography and the Imperceptible," 1:23–24. Ellenbogen argues that "Bertillon's system meant to produce a version of the object with only the most tenuous links to its visual original" (1:50).

13. Ellenbogen, "Photography and the Imperceptible," 1:40 goes as far as stating that Bertillon studies "visible objects by means of data without visual existence."

14. Sekula, "Body and the Archive," 360; Ellenbogen, "Photography and the Imperceptible," 1:65.

15. Quoted in Ellenbogen, "Photography and the Imperceptible," 1:74–75. "The system does not base itself upon human beings' everyday visual or tactile perception of the world, but instead on a set of procedures that transpose the elements of such perception into data with a manageable form" (ibid., 91). According to newspaper accounts, a course of study for police agents was established in 1889 to "learn how to see." Police officers met four times a week over three months. Bertillon sometimes led the courses himself (ibid., 118). Bertillon repeatedly stated that "The eye only sees in each thing that for which it looks, and it only looks for that of which it already has an idea"; Bertillon, *Photographie judiciaire*, 67.

16. One issue touched on whether to focus on moments of rest or movement. Bertillon believed moments of rest show their objects in a characteristic way; Ellenbogen, "Photography and the Imperceptible," 1:154.

17. Sekula, "Body and the Archive," 360.

18. Ibid.

19. Ellenbogen, "Photography and the Imperceptible," 1:39–40, 294.

20. Ibid., 1:20, 32.

21. On the biographical context of the development between 1877 and 1875, see Nicholas Wright Gillham, *A Life of Sir Francis Galton: From African Exploration to the Birth of Eugenics* (Oxford: Oxford University Press, 2001), 215–20. Gillham stresses the significance of spirituality in this respect (216).

22. Ibid., 217.

23. David Green, "Veins of Resemblance: Photography and Eugenics," *Oxford Art Journal* 7, no. 2 (1984): 3.

24. Ibid., 4.

25. "The number of races of mankind that have been entirely destroyed under the pressure of the requirements of an oncoming civilization, reads us a terrible lesson. Probably in no former period of the world has the destruction of the races of any animal whatever been effected over such wide areas and with such startling rapidity as has been the cases of the savage man. . . . the human denizens of vast regions have been entirely swept away in the short space of three centuries, less by the pressure of a stronger race than through the influence of a civilization they were incapable of supporting"; quoted without reference, C. P. Blacker, "Appendix I," *Eugenics: Galton and After* (London: Duckworth, 1952), 326.

26. Gerd Gigerenzer et al., eds., *The Empire of Chance: How Probability Changed Science and Everyday Life* (Cambridge: Cambridge University Press, 1989), 53–59. Michael Bulmer, *Francis Galton: Pioneer of Heredity and Biometry* (Baltimore: Johns Hopkins University Press, 2003), 181.

27. On the differences between terminologies, see Stephen M. Stigler and W. H. Kruskal, "Normative Terminology," in *Statistics on the Table: The History of Statistical Concepts and Methods* (Cambridge, MA: Harvard University Press, 1999), 403–32.

28. Bulmer, *Francis Galton,* 170.

29. See John Herschel, "Quetelet on Probabilities," *Edinburgh Review* 92 (1850): 23. For discussion, see Theodore M. Porter, *The Rise of Statistical Thinking: 1820–1900* (Princeton, NJ: Princeton University Press, 1986), 120–21.

30. Stephen M. Stigler, *The History of Statistics: The Measurement of Uncertainty before 1900* (Cambridge, MA: Harvard University Press, 1986), 267.

31. Bulmer, *Francis Galton,* 180–81.

32. Porter, *Rise of Statistical Thinking,* 128–46.

33. Ibid., 138.

34. Bulmer, *Francis Galton,* 181.

35. Ibid., 198.

36. Ibid., 276–77; Stigler, *History of Statistics,* 281–83.

37. Michael Banton, "Galton's Conception of Race in Historical Perspective," in *Sir Francis Galton, FRS: The Legacy of His Ideas,* ed. Milo Keynes (London: Macmillan, 1993), 170.

38. The self-evidence of "races" emerges from discussion of other topics such as the statistical study of height: "It clearly would not be proper to combine the heights of men belonging to two dissimilar races, in the expectation that the compound results would be governed the same constants"; Francis Galton, *Hereditary Genius: An Inquiry into Its Laws and Consequences* (New York: Macmillan, 1880), 29.

39. Galton applies the law of deviation in a chapter he devotes to the different worth of races in *Hereditary Genius*, 336-50.

40. Francis Galton, *Inquiries into Human Faculty* (London: Macmillan, 1883), 10.

41. Francis Galton, "Discontinuity in Evolution," *Mind* (1894): 362-63.

42. See Green, "Veins of Resemblance," 11n25; Francis Galton, "Address to the Department of Anthropology, Section H," *British Association Report* (1877): 94-100; reprinted in *Nature* 16 (1877): 344-47.

43. Green, "Veins of Resemblance," 11-12.

44. This question interfaced with the semiotic contrast between the objectivity of the mechanically produced photograph and that of drawing produced by a trained draftsman (although by the third part of the nineteenth century, the drive to automaticity affected both sides). Lorraine Daston and Peter Galison, *Objectivity* (New York: Zone, 2007), 164.

45. Ibid., 169.

46. Sekula, "Body and the Archive," 368.

47. Galton presumed that the typical was found at the center and that the idiosyncratic could be relegated to the periphery of the photograph. But, Sekula notes, blurring occurs over the entire surface of the image (ibid.). Ellenbogen argues that Galton was appropriating Cameron's photographic/artistic blurring (Ellenbogen, "Photography and the Imperceptible," 256-65).

48. Sekula, "Body and the Archive," 368n69.

49. Francis Galton, "Photographic Chronicles from Childhood to Age," *Fortnightly Review* 181 (1882): 26-31.

50. Sekula, "Body and the Archive," 367.

51. Pierre Saurisse, "Portraits composites: la photographie des types physionomiques à la fin du XIX siècle," *Histoire de L'Art* 37-38 (1997): 70.

52. Daston and Galison, *Objectivity*, 168.

53. Francis Galton, "Composite Portraits," *Journal of the Anthropological Insitute* 8 (1878): 137; *Inquiries into Human Faculty and Its Development* (London: Macmillan, 1883), 345-46.

54. In *The Critique of Aesthetical Judgment*, Kant states, "the Imagination can, in all probability, actually though unconsciously let one image glide into another, and thus by the concurrence of several of the same kind come by an average, which serves as the common measure of all. Every one has seen a thousand full-grown men. Now if you wish to judge of the normal size, estimating it by means of comparison, the imagination allows a great number of images (perhaps the whole thousand) to fall on one another. If I am allowed here the analogy of optical presentations, it is the space where most of them are combined and inside the contour, where the place is illuminated with the most vivid colours, that the *average size* is cognizable; which, both in height and breadth, is equally far removed from the extreme bounds of the greatest and smallest stature. And this is the stature of a beautiful man"; Immanuel Kant, *Kant's Critique of Judgement*, trans. J. H. Bernard, 2nd ed. (London: Macmillan, 1914), 87-88; See also Sekula, "Body and the Archive," 338n38.

55. Sekula, "Body and the Archive," 370.

56. Daston and Galison, *Objectivity*, 167-68n71. Galton's method of image making, Daston and Galison observe, is poised between two modes of observation: it aimed for an ideal type that lay "behind" any single individual. But Galton's face machine proceeded toward that ideal

not with what he and others come to see as subjective idealization but with the quasi-automated procedures of mechanical objectivity (ibid., 169).

57. Ibid., 171. The composite method became Galton's lasting contribution to scientific photography. Galton complemented this method with less famous scientific inventions, such as "analytical photography," which, by superimposing positive and negative images, isolated unshared elements; Sekula, "Body and the Archive," 368. In 1884 Galton published the *Record of Family Faculties*, a long questionnaire for distribution to the medical profession, and *Life History Album*, to be bought by parents on the birth of a child, including a schedule of measurements and observations on the child's development for a continuous photographic record; Green, *Francis Galton*, 12–13. Galton's interest in photography never ceased up until his death in 1911, but the years between 1877 and 1884 were the most intensive period of his experimentation with photography; Green, *Francis Galton*, 13–14).

58. Galton, quoted in Karl Pearson, *The Life, Letters and Labours of Francis Galton* (Cambridge: Cambridge University Press, 1924), 2:292.

59. Gary Alan Fine, "Joseph Jacobs: A Sociological Folklorist," *Folklore* 98, no. 2 (1987): 183–93.

60. See "Eugenics and the Jew: Sir Francis Galton," *Jewish Chronicle*, July 29, 1910. Pearson praised this specific composite: "We all know the Jewish boy and Galton's portraiture brings him before us in a way that only a great work of art could equal—scarcely excel for the artist would only idealise from *one* model"; quoted in Sekula, "Body and the Archive," 370n79) Green notes that apart from the composite portraits of Jews, Galton did not use this method to analyze racial types, and only one other instance of such use is known; Alice C. Fletcher, "Composite Portraits of the American Indians," *Science* 170 (1886).

61. Galton, "Composite Portraits," 98. See also Porter, *Rise of Statistical Thinking*, 140.

62. "Eugenics and the Jew"; "Photographic Composites," *Photographer News* 29, no. 1389 (April 17, 1885).

63. Jacobs, quoted in Pearson, *Life, Letters and Labours*, 2:293.

64. Jacobs wrote, "here we have something . . . more spiritual than a spirit. . . . The composite face must represent this Jewish forefather. In these Jewish composites we have the nearest representation we can hope to possess of the lad Samuel as he ministered before the Ark, or the youthful David when he tended his father's sheep"; Joseph Jacobs, "The Jewish Type, and Galton's Composite Photographs," *Photographer News* 29, no. 1390 (April 24, 1885).

65. One of the central aims of this study is to contextualize the "reactionary paradigm" within contemporary versions and practices of science and scholarship. In this framework, a note on the connection between Galton's composite photography and Wittgenstein's notion of "family resemblance," a subject elaborated on by Carlo Ginzburg as well as Lorraine Daston and Peter Galison, is in place. In the posthumously published *Philosophical Investigations*, Wittgenstein criticized the idea that concepts can be picked out by a set of necessary and sufficient conditions—precisely the problem Galton's composite photographic method came to answer. Already in 1929 or 1930, Wittgenstein tied his concept of "family resemblances" directly to Galton's composite photography; Daston and Galison, *Objectivity*, 336 and n43. Daston and Galison show how a certain notion of "immediate knowledge" within science was modeled on "race": the metaphor for immediate knowledge in various scientific branches was the grasping

of "race." Wittgenstein's notion of "family resemblance," an idea that now enjoys wide circulation in analytical philosophy, according to which concepts or games share something essential that cannot, nonetheless, be formulated in terms of necessary and sufficient conditions derived, therefore, from Galton's work on race. I elaborate on the significance of Wittgenstein's notion of family resemblance in the context of racial photography in chap. 3.

66. Pearson, *Life, Letters and Labours*, 2:290 (in n. 1 on that page Pearson lists numerous instances of attempts, most of them scientifically meaningless in his view, to employ the composite method). For anthropological purposes, photographs of the head needed to be taken full face and in exact profile and, if possible, also from above. Galton suggested an arrangement of three mirrors surrounding the subject to be photographed so as to reflect the three different views of the head. Galton, though, did not make use of the technique himself, and only later did this kind of anthropometric photography secure much stricter scientific standards, particularly through the work of the anthropologist Rudolf Martin. For the employment of Galton's method in France, see Saurisse, "Portraits composites," 73.

67. Rudolf Martin, "Über einige neure Instrumente und Hilfsmittel für den anthropologischen Unterricht," *Correspondenz-Blatt der Deutschen anthropologischen Gesellschaft* 34 (1903): 127–32.

68. Martin improved a *Haartafel* (hair table) and a *Hauttafel* (skin table), which were employed by Eugen Fischer and Felix von Luschan. Martin notes that his eye table is an improvement over Broca's insufficient one and over Bertillon's, which was found to be too complicated. See Hildegard Hugentobler-Schwanger, "Der Anthropologe Rudolf Martin (1864–1925)" (PhD diss., University of Zurich, 1990), 11, 53; Rudolf Martin " Über einige neure Instrumente," 131.

69. Uwe Hoßfeld, *Geschichte der biologischen Anthropologie in Deutschland: Von den Anfängen bis in die Nachkriegszeit* (Stuttgart: Franz Steiner, 2005), 182. See also Gerfried Ziegelmayer, "100 Jahre Anthropologie in München," *Würzburger medizinhistorische Mitteilungen* 5 (1987): 255. Martin's students in Zürich included Bruno Oetteking, Theodor Mollison, Otto Schlaginhaufen, Jan Czekonowski, and Adolph H. Schultz; in Munich they included Walter Scheidt, Wilhelm Gieseler, and Karl Saller—a genealogy, according to Hoßfeld, that held up to 1978. For a comparison of definitions of anthropology in editions of the *Lehrbuch*, see Hoßfeld, *Geschichte*, 34, 41–43. For a more comprehensive treatment than afforded here, see also my "Anthropology, Standardisation and Measurement: Rudolf Martin and Anthropometric Photography," *British Journal for the History of Science* 46, no. 3 (2013): 487–516.

70. Rainer Knußman, *Vergleichende Biologie des Menschen: Ein Lehrbuch der Anthropologie und Humangenetik* (Frankfurt: Fischer, 1980). This book was reprinted in 1986 and 1999. On tensions in the use of photography here, see my "After the Fact: 'Jews' in German Physical Anthropology, 1945–1992," in *Race, Color, Identity: Rethinking Discourses about "Jews" in the Twenty-First Century*, ed. Efraim Sicher (Oxford: Berghahn, 2013), 217–33.

71. See Ziegelmayer, "100 Jahre Anthropologie," 256. Martin's motto was "tolerance is the first step to inner freedom." See his son's description: Kurt Martin, "Rudolf Martin und die Kunst," *Anthropologischer Anzeiger* 27, no. 2 (1965): 246–51 (Martin's motto quoted on p. 251). Hugentobler-Schwanger, "Der Anthropologe Rudolf Martin," 49, 82.

72. Hoßfeld, *Geschichte*, 307.

73. Ibid., 228.

74. On Martin's earlier work concerning race, see Hugentobler-Schwanger, "Der Anthropologe Rudolf Martin," 10, 11.

75. Ibid., 87.

76. Rudolf Martin, *Anthropologie als Wissenschaft und Lehrfach: Eine akademische Antrittsrede* (Jena: Fischer, 1901), 19.

77. Ibid., 17. Similarly to Boas, Martin criticizes as "dilettante" and unscientific (*unwissenschaftlich*) certain studies of skulls and denies that nationality can ever be deduced from skull shape.

78. Cf. Martin "Über einige neure Instrumente," 131.

79. Rudolf Martin and Karl Saller, *Lehrbuch der Anthropologie: In systematischer Darstellung mit besonderer Berücksichtigung der Anthropologischen Methoden* (Stuttgart: Fischer, 1957), 1:168.

80. Karl Saller, "Rudolf Martin," *Münchener medizinische Wochenschrift* 7 (August 1925): 1343.

81. Martin, *Anthropologie als Wissenschaft*, 14–15. See also Martin, "Über einige neure Instrumente," 127–32.

82. Martin, *Anthropologie als Wissenschaft*, 24.

83. Ibid.

84. Ibid., 15.

85. Martin developed or improved a host of measurement instruments, which he defined and described with a photograph or a drawing. Martin "Über einige neure," 127–32.

86. Martin, *Anthropologie als Wissenschaft*, 6–11.

87. Hugentobler-Schwanger, "Der Anthropologe Rudolf Martin," 51.

88. On the differentiation between the two, see Rudolf Martin, *Lehrbuch der Anthropologie in systematischer Darstellung* (Jena: Fischer, 1914), 2; "Körperbedeckung und Schmück," in *Die Inlandstämme der Malayischen Halbinsel: Wissenschaftliche Ergebnisse einer Reise durch die Vereinigten malayischen Staaten* (Jena: Fischer, 1905).

89. Ibid., 680–720.

90. Theodore Mollison, "Die Verwendung der Photographie für die Messung der Körperproportionen des Menschen," *Archiv für Anthropologie* 37 (1910): 305–21.

91. Ibid., 314.

92. Frank Spencer, "Some Notes on the Attempt to Apply Photography to Anthropometry during the Second Half of the Nineteenth Century," in *Anthropology and Photography, 1860–1920*, ed. Elizabeth Edwards (New Haven, CT: Yale University Press, 1992), 103.

93. Mollison, "Die Verwendung der Photographie," 317–18.

94. Ibid., 321.

95. Martin, *Lehrbuch*, 34.

96. Ibid.

97. Spencer, "Some Notes," 99–106. See also Andrew D. Evans, "Capturing Race: Anthropology in German and Austrian Prisoner-of-War Camps during World War I," in *Colonialist Photography: Imag(in)ing Race and Place*, ed. Eleanor M. Hight and Gary D. Sampson (London: Routledge, 2002), 228.

98. Martin, *Lehrbuch*, 34.

99. Ibid.

100. Ibid., 35.

101. Ibid., 35–36.

102. Alphonse Bertillon and Arthur Chervin, *Anthropologie métrique: Conseils pratiques aux missionnaires scientifiques sur la manière de mesurer, de photographier et de décrire des sujets vivants et des pièces anatomiques* (Paris: Imprimerie Nationale, 1909), 67.

103. Martin, *Lehrbuch*, 36.

104. Ibid., 39.

105. Ibid., 42.

106. Evans, "Capturing Race."

107. Evans, "Capturing Race," 226, 229, 336. The format of profile and frontal views forced the body of the prisoner into a prearranged position, its agency taken away, based on methods from criminal photography.

108. Ibid., 250. The racialization of Jews occurred primarily through classification, that is, by viewing Jews as a separate category. See Rudolf Pöch, "Bericht über die von Wiener Anthropologischen Gesellschaft in den K.u.K. Kriegsgefangenenlagern veranlaßten Studien," *Mitteilungen der Anthropologischen Gesellschaft Wien* 48 (1918): 146–61 (see esp. Bergjuden, 146; Karaite, 148; a Jew from Petrowkow, 149; a Jew from Kiev, 150; and the similarity between Jews and Gypsies). In the second section of the article, Pöch deals with Martin's photographic method directly, particularly with the issue of control.

109. Evans, "Capturing Race," 235, 247. Evans shows that the exchange between Luschan and Struck negotiated the representation of typicality in drawings in contrast to photographs.

110. The fiercest controversy was between Martin's Zurich student Otto Schlaginhaufen and R. Neuhauß, touching on questions of authorship. Most interesting, from the perspective of this study, is Schlaginhaufen's discussion of the limitations of the photographic method. Schlaginhaufen emphasized that photography is merely a tool for reproduction; he objected to expectations that photography could not possibly satisfy. Sometimes the observation of a photograph was better than the living being, but the belief that photographic observation could replace living observation verged on the mystical, in his view. He fiercely opposed comparison of types from photographs as "subjective." He referred, in this context, to Eugen Fischer's scientific use of photographs to classify racial bastards where no written genealogy is available. See Otto Schlaginhaufen, "Die Stellung der Photographie in der anthropologischen Methodik und die Pygmänfrage in Neuguinea," *Zeitschrift für Ethnologie* 47 (1915): 53–58; R. Neuhauß, "Die Pygmänfrage in Neuguinea," *Zeitschrift für Ethnologie* 46 (1914): 753–54.

111. Evans, "Capturing Race," 231.

112. Cf. Schlaghinhaufen, "Die Stellung der Photographie," 53–58. Rudolf Martin, "Anthropologische Untersuchungen an Kriegsgefangene," *Die Umschau* 19 (1915). Pöch emphasized the advantages of the war situation and the prisoners at hand in the scientific observation; see "Anthropologische Studien Kriegsgefangene," *Die Umschau* 20 (1916): 988–91. In his response, Martin emphasized the need for control to enable comparison of materials collected by distinct teams; "Anthropologische Studien Kriegsgefangenen," 1027.

113. Arguing in the Boasian vein, Martin emphasized that all present peoples are racially heterogeneous. To prove this point, he entered a lengthy discussion of the racial history of European peoples. Hugentobler-Schwanger, "Der Anthropologe Rudolf Martin," 72.

114. Rudolf Martin, "Anthropometrie," in *Handbuch der Sozialen Hygiene und Gesundheits-fürsorge*, ed. A. Gottstein, A. Schlossmann, and A. Teleky (Berlin: Lehmann, 1925): 256–301.

115. Ibid., 294.

116. Ibid., 297–98. Martin develops equipment to move the camera around the photographed person in order to have photographs taken from several fixed angles without the subject having to move.

117. Ibid., 300–301.

118. Martin, *Anthropologie als Wissenschaft*, 27.

119. In one place, however, while discussing photographs for "technical purposes of instruction," Martin mentions a failed attempt he made to collaborate with Orell Füssli to create a poster of race photographs "to be hung in the classroom," for which, Martin confessed, he had taken photographs himself. To obtain the highest degree of similarity, the black-and-white photographs were painted by W. von Steiner of Zurich. Here Martin required that the specimen represent all typical traits (hair color and form, facial shape). Martin "Über einige neure Instrumente," 132. During the World War I prisoners of war study, anthropologists directly confronted the question of typicality. In their view, this was an intuitive decision of the anthropologist based on preexisting categories of classification. In Evan's interpretation, the photographic project created categories of types. Evans, "Capturing Race," 233.

120. Martin and Saller, *Lehrbuch der Anthropologie*, 1:148. In this volume, Saller added short notes on the ethical use of photography. He included Mollison as one of the founders of the photographic method together with Pöch and Martin (150).

121. Ibid., 1:110–20: "In fact, it is hardly possible to force races living in proximity into a rigid scheme." (118). Race is certainly not described as a social construction, but its plasticity difficulties in empirical classifications are emphasized. Note in this context references to Boas (111, 118).

122. Sekula, "Body and the Archive," 373–74.

123. See Anne Laura Stoler, *Race and the Education of Desire: Foucault's History of Sexuality and the Colonial Order of Things* (Durham, NC: Duke University Press, 1995), 134, 205–206. Foucault identifies knowledge based on the invisible as a turn of the nineteenth century invention, where things would be grouped "by the vigour that is hidden down below, in the depths"; Michel Foucault, *The Order of Things: An Archeology of the Human Sciences* (London: Tavistock, 1970), 251.

124. George Mosse, *Toward the Final Solution: A History of European Racism* (Madison: University of Wisconsin Press, 1978).

125. The most extensive treatment of Stratz is afforded by Michael Hau in *The Cult of Health and Beauty in Germany: A Social History, 1890–1930* (Chicago: University of Chicago Press, 2003). Hau notes that Stratz's books on the feminine body made lavish use of photographs of nude females, thus appealing to the male voyeur at least as much as to female readers (40). He also stresses that this use of photographs made the books expensive, a fact that limited the readership to respectable, wealthy, and educated *Bürger* (40).

126. Carl Heinrich Stratz, *Die Rassenschönheit des Weibes* (Stuttgart: F. Enke, 1901).

127. Carl Heinrich Stratz, *Die Rassenschönheit des Weibes*, 5th ed. (Stuttgart: Enke, 1904), 7, quoted in Hau, *Cult of Health and Beauty*, 86 (analysis of Stratz's racial typology on 87–88).

128. Hau, *Cult of Health and Beauty*, 69.

129. Stoler, *Race and the Education of Desire*, 184.

130. Carl Heinrich Stratz, *Die Frauen auf Java: Eine gynäkologische Studie* (Stuttgart: Enke, 1897); Stoler, *Race and the Education of Desire*, 187.

131. Hau, *Cult of Health and Beauty*, 85. Stratz viewed himself as particularly indebted to Johannes Ranke, the Munich liberal anthropologist (96), who in 1918 was succeeded by Rudolf Martin. When Stratz, who believed that white women most closely approached the ideal of beauty, came across members of other races that approached the ideal, he assumed the presence of white blood in their ancestry (88).

132. Stratz, *Was sind Juden? Eine ethnographisch-anthropologische Studie* (Leipzig: G. Freytag 1903), 134.

133. On the medicalized study of the Jewish gaze, see Sander Gilman, *The Jew's Body* (London: Routledge, 1991), 68–73; *Freud, Race, and Gender* (Princeton, NJ: Princeton University Press, 1995), 72–75. The argument that this notion of the gaze could only be documented through photographs is further corroborated by Charles Darwin's use of photographs and actors in his 1872 *The Expression of Emotion in Man and Animals* (London: Penguin, 2009).

134. Stratz, *Was sind Juden?*, 22–23.

135. For a detailed biography, see Todd M. Endelmann, "Anglo-Jewish Scientists and the Science of Race," *Jewish Social Studies* 11 (2004): 52–92. See also Raphael Falk, "Three Zionist Men of Science: Between Nature and Nurture," in *Jews and Sciences in German Contexts: Case Studies from the 19th and 20th Centuries*, ed. Ulrich Charpa and Ute Deichmann (Tübingen: Mohr Siebeck, 2007), 137–39.

136. Redcliffe N. Salaman, "The Inheritance of Colour and Other Characters in the Potato," *Journal of Genetics* 1, no. 1 (1910): 7–46.

137. Ibid., 22, 23, 26.

138. Ibid., 22.

139. Ibid., 14.

140. Ibid., 17.

141. Redcliffe N. Salaman, "Heredity and the Jew," *Journal of Genetics* 1 (1911): 273–92.

142. Ibid., 278.

143. Ibid., 279.

144. Ibid., 280.

145. Ibid., 287

146. Ibid., 287, 288.

147. Ibid., 286.

148. On Fischer's significance in the shift from anthropological models to human genetic ones and in reorienting research from the study of racial types to that of racial mixture, see Veronika Lipphardt, *Biologie der Juden: Jüdische Wissenschaftler über "Rasse" und Vererbung 1900–1935* (Göttingen: Vandenhoeck und Ruprecht, 2008), 131.

149. Eugen Fischer, *Die Rehobother Bastards und das Bastardierungsproblem beim Menschen: Anthropologische und ethnographische Studien am Rehobother Bastardvolk in Deutsch-Südwest-Afrika* (1912; Graz: Akademische Druck, 1961).

150. Ibid., 58.

151. In a lecture published in 1914, Fischer opens by stating that the most burning problems

in human biology pertain to racial mixing. Eugen Fischer, *Das Problem der Rassenkreuzung beim Menschen* (Freiburg: Spezer und Kaerner, 1914), 1. In this lecture Fischer refers to Salaman several times (6, 13, 20).

152. Fischer, *Die Rehobother Bastards*, 168.

153. Ibid., 57–58.

154. Fischer ends his article calling for the adaptation of a methodology of anthropological study of families. Fischer, *Das Problem der Rassenkreuzung*, 30.

155. Fischer, *Die Rehobother Bastards*, 318.

156. My "Album of the Extinct Race: Eugen Fischer and Photography," which is a chapter in the book I am currently writing, affords a comprehensive history of Fischer's use of photography throughout his career.

157. Christopher S. Wood, "Introduction," in *The Vienna School Reader: Politics and Art Historical Method in the 1930s*, ed. Christopher S. Wood (New York: Zone, 2000), 10, 12, 26.

158. Alois Riegl, "The Place of the Vapheio Cups in the History of Art," trans. Tawney Becker, in *The Vienna School Reader: Politics and Art Historical Method in the 1930s*, ed. Christopher S. Wood (New York: Zone, 2000): 105–29. Numerous other references to race in similar contexts can be found in Riegl's *Historical Grammar of the Visual Arts*, trans. Jaqueline E. Jung (New York: Zone, 2004); for a short comment on photography, see 360.

159. Riegl, "Vapheio Cups," 106.

160. Ibid., 111.

161. Ibid., 117.

162. The discourse about the close connection between art and a specific (German) way of seeing was not confined to Wölfflin or Simmel. Oskar Hagen, the founder of the department of art history at the University of Wisconsin, published *Deutsches Sehen: Gestaltungsfragen der deutsches Kunst* (Munich: Piper, 1920). In 1923, a pupil of Wölfflin, Kurt Gerstenberg, developed the idea of *Optik*, or "optical zones," that arise from the reaction to the terrain and qualities of light. Kurt Gerstenberg, *Ideen zu einer Kunstgeographie Europas*, Bibliothek der Kunstgeschichte 48/49 (Leipzig: Seemann, 1923). Otto Pächt developed similar ideas in publications in 1933: "Gestlatungsprinzipien der westlichen Malerei des 15. Jahrhunderts," *Kunstwissenschaftliche Forschungen* 2 (1933): 75–100. See Thomas DaCosta Kaufmann, *Toward a Geography of Art* (Chicago: Chicago University Press, 2004), 75, 78.

163. Connections between contemporary art history and racial or national ideas were widespread. In the German language alone the list includes Alois Riegl, Otto Pächt, Wilhelm Pinder, Hans Sedlmayer, and Wilhelm Worringer, to name just some of the most prominent writers. Claire Farago, " 'Vision Itself Has Its History': 'Race,' Nation, and Renaissance Art History," in *Reframing the Renaissance: Visual Culture in Europe and Latin America* (New Haven, CT: Yale University Press, 1995), 67–88. Kaufmann, *Geography of Art*, 43–67. Particularly interesting in this respect is Kaufmann's discussion of Karl Frankl, a Halle professor of Jewish descent, whose book was published on the eve of the war in Czechoslovakia and who "might have seen his demise in a death camp" had he not emigrated to Princeton's Institute of Advanced Studies (83–85).

164. For a comprehensive account of Simmel and art, see Barbar Smitmans-Vajda, "Die Bedeutung der Bildenden Kunst in der Philosophie Georg Simmels" (PhD diss., University of Tübingen, 1994).

165. I wish to stress the complex role of individuals of Jewish descent in the development of "German Ideology" through interpretations of Rembrandt. Carl Neumann (a German of Jewish descent), a Heidelberg historian of art, wrote in 1906 the first biography of Rembrandt. In the 1926 revised edition of the book he compares Simmel's book and Langbehn's antisemitic *Rembrandt als Erzieher*. Langbehn, according to Neumann, was not a disciplined mind but a good "listener" who heard several slogans and brought them together. His book won enormous success, and the idea of Rembrandt's "German mission" became taken for granted. Its value, according to Neumann, is not as a history of art book but in its ideology. Similarly, Simmel, according to Neumann, is maybe a brilliant philosopher, but his discussion lacks real facts. Neumann connected this tendency with World War I: "The terrible events of recent history have bred indifference to reality and fact. So much that we thought real has become a dream and melted into air. This indifference to or even contempt for facts and data makes it easy to flee from truth rather than to come to terms with it" (38); Carl Neumann, *Rembrandt* (Munich: F. Meiner, 1924), 1:27-38. In 1927, Aby Warburg, distinguished historian (also Jewish German) drafted a letter to Carl Neumann in which he attempted to amplify the "'Teutonic" aspect of Rembrandt: "The classical heritage functions as a memory for Europe's mentality. The Germanic [is] oblivious of the self, [possessed] by restless drive for freedom in the infinite, and the restless submission to established classical forms"; quoted in Charlotte Schoell-Glass, *Aby Warburg and Anti-Semitism: Political Perspectives on Images and Culture* (Detroit, MI: Wayne State University Press, 2008), 136, 138. Neumann criticizes Simmel for using Rembrandt as an "experiment field" to test his "philosophical system" of Rembrandt as the purest example of the spirit of German art (*germanischen Kunstgeistes*), to which the book is devoted. Neumann rejects the opposition of the Roman vs. German culture as populist and Simmel's quest for Rembrandt's "essence" as ahistorical. Note that in 1921 Erwin Panofsky delivered a lecture to the Akademie für die Wissenschaft des Judentums in which he addressed Rembrandt's portrayal of Jews in sixteenth-century Amsterdam. Panofsky argued that Jews were represented neither as types nor as individuals but as essence (*Wesen*). Rembrandt depicted their metaphysical substance in the kind of painting that Spinoza sought philosophically: "the Greatest *German* painter at the end of his life represented the world as the greatest *Jewish* philosophy conceived of it"; Erwin Panofsky, "Rembrandt und das Judentum," *Jahrbuch der Hamburger Kunstsammlungen* 28 (1974): 106 (italics added).

166. Georg Simmel, *Rembrandt: An Essay in the Philosophy of Art*, ed. and trans. Helmut Staubman and Alan Scott (New York: Routledge, 2005), 17. Svetlana Alpers describes Simmel's book as "disappointing"; Svetlana Alpers, *Rembrandt's Enterprise: The Studio and the Market* (Chicago: Chicago University Press, 1995), 153n53.

167. Simmel, *Rembrandt*, 17. There is a sense of ambivalence in Simmel's statement, as it lends itself to either a racial interpretation or a cultural one (precisely of the kind suggested by Anne Stoler in *Race and the Education of Desire*).

168. Georg Simmel, "How Is Society Possible?," in *Georg Simmel on Individuality and Social Forms*, ed. and trans. Donald N. Levine (Chicago: University of Chicago Press, 1971), 11 (italics added). Simmel's discussion involves considerations that are not unlike the theoretical foundations of impressionism. Involving subjective and objective sensations, impressionists sought to study the essentials of vision before the epistemological moment of recognition. The moment of recognition is conflated with the pollution of language, which unconsciously fills

the blanks. Cf. Richard Shiff, *Cézanne and the End of Impressionism: A Study of the Theory, Technique, and Critical Evaluation of Modern Art* (Chicago: University of Chicago Press, 1984). The question of the potential politicization of perceptual theory incessantly resurfaces in the following chapters.

169. Simmel, *Rembrandt*, 22. Alpers ends her book *The Art of Describing* by comparing the entire tradition of Dutch painting, culminating with Vermeer and Rembrandt. Rembrandt heightened his antirealism by applying thick layers of paint, thereby undermining the picture as a transparent representation of a real referent. Svetlana Alpers, *The Art of Describing* (Chicago: University of Chicago Press, 1983), 222–28.

170. Simmel, *Rembrandt*, 22.

171. Ibid., 39: "with total immediacy and in a completely realistic sense, the painting, whose impression concentrates continuous movement via some means, is closer to reality than . . . is the snapshot." Similarly, Simmel views the moving picture as less real than the work of art, where the imagination of the viewer completes the movement of the "before and after the represented moment" (38). In Rembrandt, movement is not a quality of the object, according to Simmel, but "a quality of a particular viewing" (39). The work of art, therefore, "offers much more 'truth' than does the photographic snapshot."

172. Simmel, *Rembrandt*, 142.

173. Ibid., 143–44, 151.

174. *Rembrandt*, 150, 83 (italics added). The translators note the similarity between Simmel's antirealism and Oscar Wilde's *Intentions: The Decay of Lying* (1891), which states that "life imitates art far more than the opposite" (87), and Walter Pater's (Wilde's teacher) antirealism in *The Renaissance: Studies in Art and Poetry* (1873; Oxford: Oxford University Press, 1986), which also opposes southern and northern (Teutonic) in the first chapter of the book (171n13).

175. Simmel, *Rembrandt*, 43

176. Ibid., 44.

177. Ibid., 58.

178. Ibid., 104: *schlechtrassig jüdisch*, translated as "petit bourgeois, poorly bred Jewish."

179. Ibid., 62.

180. Ibid.

181. Ibid., 66.

182. Ibid., 98–100.

183. Georg Simmel, "Die ästhetische Bedeutung des Gesichts" (1901) and "Ästhtik des Por-träts" (1905), in *Aufsätze und Abhandlungen 1901–1908 Band I* (Frankfurt: Suhrkamp, 1995), 36–42, 321–32; "Der Ausdruck des Seelischen" in *Rembrandt: Eine Kunstphilosophischer Versuch* (Leipzig: K. Wolff, 1916); "Das Problem des Porträts" (1918), in *Aufsätze und Abhandlungen 1909–1918 Band II* (Frankfurt: Suhrkamp, 1995), 370–81.

184. Simmel enters into a discussion of the relationship between the different parts of the face (the accord between the nose and the lips and forehead—the way they justify each other) of the kind found later in Günther and Clauß. Smitmans-Vajda, "Die Bedeutung der Bildenden Kunst," 156.

185. Smitmans-Vajda, "Die Bedeutung der Bildenden Kunst," 157–58.

186. Ibid., 160.

187. Ibid., 157–58.

188. Ibid., 150.

189. Ibid., 26.

190. Ibid., 44.

191. Heinrich Wölfflin, *Principles of Art History: The Problem of the Development of Style in Later Art*, trans. M. D. Hottinger (New York: Dover 1932).

192. Wölfflin, *Principles of Art History*, 230.

193. Ibid.: "the gradual depreciation of line as the path of vision and guide of the eye" (14).

194. Ibid., 15.

195. Ibid., 18.

196. Martin Warnke, "On Heinrich Wölfflin," *Representations* 27 (1989): 173–75. For an opposing interpretation of Wölfflin in this respect, see the first chapter of Paul Jaskot, *The Nazi Perpetrator: Postwar German Art and the Politics of the Right* (Minneapolis: University of Minnesota Press, 2012).

197. Warnke, "On Heinrich Wölfflin," 176. To Germans, according to Warnke, art and culture were synonymous with political and state will. Amid a total appropriation of culture by politics, Wölfflin insisted on the autonomy and discrete organizational integrity of optical culture (177).

198. Warnke dismisses the significance of "race" as "a type of natural law to which Wölfflin pays tribute, and which is at the same time the most precarious concession that he was ever willing to offer to the general spirits of the times" (Warnke, "On Heinrich Wölfflin," 181).

199. Wölfflin's interest in racial and national character in art increased in the 1920s; *Italien und das deutsche Formgefühl* (Munich: Bruckmann, 1931); see also his 1940 *Gedanken zur Kunstgeschichte: Gedrucktes und Ungedrucktes* (Basel: Schwabe 1947), 109–31.

200. Wölfflin, *Principles of Art History*, viii.

201. Ibid., 6, 11.

202. Ibid., 106; "Northern beauty is not a beauty of the self-contained and self-limited, but of the boundless and infinite" (148; see also 147); on the difference between Italian and German bells, 183; on Rembrandt as the product of Nordic imagination, 183; "There exists in the north from the outset a general feeling for the immersion of the detail in the whole, the feeling that every entity can have sense and significance only in connection with others, with the whole world" (183); on Dürer as "an exclusively Nordic possibility," (184); the North subordinates the individual to the whole (194)—such are significant: "It is ultimately the contrast of seeing in detail and seeing as a whole" (155).

203. Ibid., 194, 235.

204. Wölfflin, *Principles of Art History*, sees the methodological question of abstraction of "pure" styles from "mixed" expressions as central to his project: "It is certainly no easy task to reveal this inward visual development, because the representation possibilities of an epoch are never shown in abstract purity but, as is natural, are always bound to a certain expressional content" (12). Günther and Clauß transported Wölfflin's notion of style from art to race, where they faced similar questions.

205. Ibid., 156.

206. Ibid., 229 (italics added). Indeed, according to Wölfflin, "The more clearly the opposition of the types is recognized, the more interesting does the history of the transition become"

(99). Specifically on the painterly in this context, see Simmel, *Rembrandt*, 138, 141, and Wölfflin, *Principles of Art History*, on the painterly and movement (27–28) and as running "in the veins of the Germanic race" (67).

CHAPTER TWO

1. Cf. John M. Efron, *Defenders of the Race: Jewish Doctors and the Race Science in Fin-de-Siècle Europe* (New Haven, CT: Yale University Press, 1994): "Race scientists could see almost anything they wished in the objects of their gaze" (92), and "Race science was highly subjective and prone to fanciful aesthetic judgments" (93).

2. Adding considerable ambivalence to the following, Stefan Ihrig has recently shown that a politicized discourse about Armenians commenced in late nineteenth and early twentieth century Germany and Austria, closely echoing the antiphilo-Semitic discourse. Armenians were ascribed the same characteristics as Jews: their superior acumen for and cunning in business; their being parasites in the Turkish national body; and their inability to build (their own) state. See Stephan Ihrig, *Justifying Genocide: Germany and the Armenians from Bismarck to Hitler* (Cambridge, MA: Harvard University Press, forthcoming). Considering the dynamics of imagination in politics and science, it would seem that the "familiar" and close, i.e., central European Jewry and images thereof, were used to make sense of the Armenians and probably not the other way around.

3. For his biography, see Anja Laukötter, *Von der "Kultur" zur "Rasse"—vom Objekt zum Körper? Völkerkundemuseen und ihre Wissenschaften zu Beginn des 20. Jahrhunderts* (Bielefeld: Transcript, 2007), and Christine Stelzig, *Afrika am Museum für Völkerkunde zu Berlin 1873–1919* (Herbolzheim: Centaurus, 2004).

4. On the discussion of Jewish racial purity, see Efron, *Defenders of the Race*, 20–26. This discussion included Richard Andree, Karl Asmund Rudolphi, Carl Vogt, Friedrich Maurer, Ludwig Stieda, and Julius Kollmann.

5. Michael Hagener, "Mikro-Anthropologie und Photographie: Gustav Fritschs Haarspaltereien und Klassifizierung der Rassen," in *Ordnungen der Sichtbarkeit: Fotographie in Wissenschaft, Kunst und Technologie*, ed. Peter Geimer (Frankfurt: Suhrkamp, 2002), 254.

6. He was a keen photographer (he was once awarded a prize for amateur photography). Apart from some brief instructions for traveling anthropologists, he never discussed the status of photographs, their technical aspects, or criteria for successful photographs. Felix von Luschan, *Anleitung zu wissenschaftlichen Beobachtungen auf dem Gebiete der Anthropologie, Ethnographie und Urgeschichte* (Leipzig: Jänicke 1914), 6, 108. Von Luschan instructs anthropologists to take photographs of the head from direct and profile angles but does not discuss control. He refers to Martin's *Lehrbuch* as in the making and highly expected (5).

7. See Andrew Zimmerman, *Anthropology and Antihumanism in Imperial Germany* (Chicago: University of Chicago Press, 2001), 30–35. See also the important discussion of clothes in the colonial context (260), the tactics used to achieve the cooperation of colonial subjects, von Luschan's policies with regard to the acquisition of anthropological materials (including skulls and brains; 245), and the relationship of these to the revolt of the Maji Maji in the colony under German rule (156–58).

8. Zimmerman, *Anthropology and Antihumanism*, 33.

9. Rather than interpret von Luschan's statements as essentially inconsistent, it is more helpful to acknowledge his ideas, including those pertaining to his photographic practices, as part of a multifaceted scientific structure. W. E. B. DuBois pointed to some of von Luschan's contradictions, and American historian John David Smith has elaborated on these in his "W.E.B. Du Bois, Felix von Luschan, and Racial Reform at the *Fin de Siècle,*" *Amerikastudien: A Quarterly* 47, no. 1 (2002): 23–38.

10. Felix von Luschan, "Sammlung Baessler, Schädel von Polynesischen Inseln," in *Veröffentlichungen aus dem Königlichen Museum für Völkerkunde* (1907): 251.

11. Felix von Luschan, "The Early Inhabitants of Western Asia," *Journal of the Royal Anthropological Institute of Great Britain and Ireland* 41 (1911): 237; "Sammlung Baessler," 254; see also "Hamitische Typen," in *Die Sprachen der Hamiten,* ed. Carl Meinhof (Hamburg: L. Friedrichsen, 1912), 247.

12. Felix von Luschan, "Anthropological View of Race," in *Inter-Racial Problems: Papers from the First Universal Races Congress Held in London in 1911,* ed. Gustav Spiller (New York, 1911), 22. For antiracist statements see von Luschan's "Anthropological View of Race," Laukötter provides examples of racist comments von Luschan made in his unpublished notes, such as those directed at individuals who refused to undergo anthropometric measurements; Laukötter, *Von der "Kultur" zur "Rasse,"* 78.

13. Felix von Luschan, "Die Anthropologische Stellung der Juden", *Correspondenz-Blatt der deutschen Gesellschaft für Anthropologie, Ethnologie und Urgeschichte* 23, no. 10 (1892): 94. Von Luschan's first connection between Jews and the Hittite type was made even earlier, in "Die Tachtadschz und andere Überreste der alten Bevoelkerung Lykiens," *Archiv für Anthropologie* 19, no. 1/2 (1890): 31–53. For a discussion of von Luschan in the Jewish context, see Efron, *Defenders of the Race,* 137, 140. Efron's recapitulation of von Luschan's views, however, is not entirely accurate.

14. Von Luschan, "Die Anthropologische Stellung der Juden," 95.

15. Ibid.

16. Ibid., 96.

17. The Hittite texts were not yet deciphered, but von Luschan assumed that they belonged neither to the Indo-European nor the Semitic language group. But in 1902 Norwegian Assyriologist Jorgen Alexander Knudston identified the language on some clay tablets as Indo-European. His results were largely ignored by the scholarly world. In 1915 Czech philologist Bedřich Hrozný published an article in which he proved that Hittite belonged to the Indo-European branch of languages. Despite his claim that he differentiated between language and race, philological and anthropological classification, von Luschan (as well as other prominent scholars such as Eduard Meyer) found the thesis that "the Hittites, whose extreme short heads and extreme prominent noses we know from so many contemporary representations, had initially spoken an European language" almost grotesque. Friedrich Hrozný, "Die Sprache der Hethiter: Ihr Bau und ihre Zugehörigkeit zum indogermanischen Sprachstamm: Ein Entzifferungsversuch," *Boghazkoi-Studien* 1, nos. 1, 2 (Leipzig: Hinrichs, 1916–1917); Felix von Luschan, *Völker, Rassen, Sprachen* (Berlin: Welt 1922), 121; Felix Wiedemann, "What Have the Jews to Do with the Hittites? On Archeological and Anthropological Cartographies of the (Ancient) Near East at the Turn of the 20th Century" (unpublished manuscript).

18. On von Luschan's sojourns in Anatolia and Syria, see Ralf-B. Wartke, "Felix von Luschan

und die Ausgrabungen in Sendschirli," in *Felix von Luschan (1854–1924): Leben und Wirken eines Universalgelehrten*, ed. Peter Ruggendorfer and Hubert D. Szemethy (Vienna: Böhlau, 2009), 307–20. See also Wiedemann, "What Have the Jews to Do with the Hittites?," 9–10; Jörg Klinger, *Die Hethiter: Geschichte—Gesellschaft—Kultur* (Munich: Beck, 2007), 7–31; and Itamar Singer, *The Hittites and Their Civilization* [in Hebrew] (Jerusalem: Bialik Institute, 2009). Von Luschan, "Die Anthropologische Stellung der Juden," 99.

19. Von Luschan, "Die Anthropologische Stellung der Juden," 99.

20. Ibid., 100. There are certain similarities between von Luschan and the leader of German anthropology, Rudolf Virchow. Virchow was an open opponent of (political) antisemitism. His anthropometric studies of six million German school children, however, included Jews, so to speak, by way of exclusion: the total number of children measured is accompanied by the statement "darunter Juden" (Jews among them), and he listed Jews in a separate column. See Zimmerman, *Anthropology and Antihumanism*, 135–46. Von Luschan's student Elias Auerbach responded to his argument and rejected the idea that Jews were a mixed race. In particular, Auerbach focused on opposing the Hittite theory. On the particulars of this controversy see Veronika Lipphardt, *Biologie der Juden: Jüdische Wissenschaftler über "Rasse" und Vererbung, 1900–1935* (Berlin: Vandenhoeck und Ruprecht, 2008), 66ff. For a detailed discussion of Auerbach, see Efron, *Defenders of the Race*, 127–41.

21. Von Luschan, "Zur physischen Anthropologie der Juden," *Zeitschrift für Demographie und Statistik der Juden* 1, no. 1 (1905): 3.

22. Ibid.

23. Carl Meinhof, *Die Sprachen der Hamiten* (Hamburg: L. Friedrichsen, 1912); Felix von Luschan, *The Racial Affinities of the Hottentots* (London: Spottiswoode, 1907), 5, 8.

24. Von Luschan, *Racial Affinities*, 5.

25. See Margaret Olin, "Touching Photographs: Roland Barthes's Mistaken Identification," *Representations* 80, no. 1 (2002): 99–118, esp. 114–15, and *Touching Photographs* (Chicago: University of Chicago Press, 2011), 6, 10.

26. On the index in Peirce's philosophy, see Thomas Loyd Short, *Peirce's Theory of Signs* (Cambridge: Cambridge University Press, 2007). On the index in the development of economics, see Robert W. Dimand, "The Quest for an Ideal Index: Irving Fisher and the Making of the Index Numbers," in *The Economic Mind in America: Essays in the History of American Economics*, ed. Macolm Rutherford (New York Routledge, 1998), 128–44.

27. Felix von Luschan, "Zusammenhänge und Konvergenz," *Mitteilungen der Anthropologischen Gesellschaft in Wien* 48 (1918): 1–117; Hermann Struck and Felix von Luschan, *Kriegsgefangene: Ein Beitrag zur Völlkerkunde im Weltkriege* (Berlin: Reimer, 1916), 47–50.

28. Von Luschan, "Hamitische Typen," 247.

29. Von Luschan, "Zusammenhänge und Konvergenz," 44.

30. Margaret Olin, "Jews among the Peoples: Visual Archives in German Prison Camps during the Great War," in *Doing Anthropology in Wartime and War Zones: World War I and the Cultural Sciences in Europe*, ed. Reinhard Johler, Christian Marchetti, and Monique Scheer (Bielefeld: Transcript, 2010), 255–77.

31. Olin, "Jews among the Peoples," 267–68.

32. Struck and von Luschan, *Kriegsgefangene*, 2.

33. Bea Schröttner, "Hermann Struck during the First World War," in *Hermann Struck (1876–1944)* [in Hebrew], ed. Ruthi Ofek and Chana Schütz (Tefen, Israel: Open Museum, 2007) 142.

34. In the major 1907 Berlin exhibition of Jewish artists, Struck (and Lilien) were excluded because of their Zionist convictions, which were seen as undermining the Berlin Jewish community. See Chana Schütz, "Hermann Struck in Berlin," in *Hermann Struck (1876–1944)* [in Hebrew], ed. Ruthi Ofek and Chana Schütz (Tefen, Israel: Open Museum, 2007), 38. For a brief account of Struck's life, see the memoirs of his brother-in-law Henry Pachter, *Weimar Etudes* (New York: Columbia University Press, 1982), 203–7. K. Schwartz, "Hermann Struck," in *Hermann Struck: The Man and the Artist* [in Hebrew], ed. Itzhak Man (Tel Aviv: Dvir, 1954), 36.

35. On the one hand Struck viewed Jews as strangers in Germany. See his letter to his brother quoted in Schröttner, "Hermann Struck during the First World War," 129. On the other hand, when he was off guard, he identified as a German; ibid., 142.

36. Ibid., 125 and 129, respectively.

37. Ibid., 123–24.

38. See Noah Isenberg, ed., "Preface to the English Edition," in Arnold Zweig and Hermann Stuck, *The Face of East European Jewry*, ed. and trans. Noah Isenberg (Berkeley: University of California Press, 2004), xvii. See also Jane Rusel, *Hermann Struck (1876–1944): Das Leben und graphische Werk eines jüdischen Künstlers* (Frankfurt: Lang, 1997), 99–100. As Rusel shows, Struck's intensive work on eastern European Jews was published on numerous occasions in various newspapers already during the war (171–83). On the cooperation between Struck and Zweig in the context of the German Jews and the *Ostjuden*, see Steven Aschheim, *Brothers and Strangers: The East European Jew in German Jewish Consciousness, 1800–1923* (Madison: University of Wisconsin Press, 1982), 184–214. On the book, see Leslie Morris, "Reading the Face of the Other: Arnold Zweig's and Hermann Struck's *Das ostjüdische Antlitz*," in *The Imperialist Imagination: German Colonialism and Its Legacy*, ed. Sara Friedrichsmeyer, Sara Lennox, and Susanne Zantop (Ann Arbor: University of Michigan Press), 189–203. Aby Warburg, too, started collecting images of the Jews encountered by the advancing German army as well as episodes of antisemitic violence; see Matthew Rampley, "Aby Warburg: *Kulturwissenschaft*, Judaism, and the Politics of Identity," *Oxford Art Journal* 33, no. 3 (2010): 328.

39. See Paul Fechter, "The Jewish Struck in the Eyes of a Non-Jew," in *Hermann Struck: The Man and the Artist* [in Hebrew], ed. Itzhak Man (Tel Aviv: Dvir, 1954), 78. Hermann Struck, "The Way to the Land of Israel," in ibid., 90–91, 116–117. Compare the 2007 review of the Tefen Museum catalog, which notes that one is today shocked by the POW book and by Struck's cooperation with a race expert and with the hagiographic 1954 collection, which saw no problem with Struck's drawings of "Jewish types of the Jewish race"; Yeshiahu Wolfsberg, "The Artist Jew—The Jew Artist," in ibid., 18; Smadar Shefi, "Hermann Struck: Here and There in Europe and the Land of Israel," *Ha'aretz*, December 20, 2007 [in Hebrew], accessed April 10, 2015, http://www.haaretz.co.il/gallery/art/1.1466553.

40. *Antlitz der Zeit: 60 Fotos Deutscher Menschen* (Munich: Transmare, 1929).

41. Schröttner, "Hermann Struck during the First World War," 147.

42. Struck and von Luschan, *Kriegsgefangene*.

43. Ibid., 1–2, 92–93.

44. Cf. ibid., 8–9, 14–15.

45. Andrew Evans, *Anthropology at War: World War I and the Science of Race in Germany* (Chicago: University of Chicago Press, 2010), 179. With regard to Pöch's team, Evans shows that while Russians were studied and racialized as enemies of the Central powers, Russian soldiers of German descent, such as Volga Germans, were not (184).

46. Ibid., 156.

47. Struck to von Luschan, March 14, 1916, file H. Struck, NL von Luschan, quoted and translated in Evans, *Anthropology at War*, 164: "Your remarks in relation to the Negro-Type were thoroughly correct, and I immediately did his hair. He now has received the very pretty, frizzy Negro hair, and I believe that he will please you. In addition, on the Russian that you already eliminated some time ago, I have enlarged the skull and ear. I will present these new prints to you after your return, and then bow very happily to your dictum."

48. Evans, *Anthropology at War*, 161.

49. Struck also altered his portrait of Sigmund Freud following a short exchange with him in 1914. Struck, then, viewed making adjustments and adaptations part of his working process.

50. Lorraine Daston and Peter Galison, *Objectivity* (New York: Zone, 2007), 309–63.

51. On the penetration of photography into the cognitive style of Freud and his contemporaries, how it anticipated the meeting with referents, see Mary Bergstein, *Mirrors of Memory: Freud, Photography, and the History of Art* (Ithaca, NY: Cornell University Press, 2010), and her earlier "Lonely Aphrodites: On the Documentary Photography of Sculpture," *Art Bulletin* 74, no. 3 (1992):475–98.

52. See Ruthi Ofek, "The Books of Hermann Struck," in *Hermann Struck: 1876–1944* [in Hebrew], ed. Ruthi Ofek and Chana Schütz (Tefen, Israel: Open Museum, 2007), 207; James Henry Breasted, *Ancient Times: A History of the Early World: An Introduction to the Study of the Ancient History and Career of the Early Man* (Boston: Ginn, 1916), 240.

53. Hans Ulrich Gumbrecht, *Production of Presence: What Meaning Cannot Convey* (Stanford, CA: Stanford University Press, 2004), 8 ,11.

54. The notion of the archive has been the subject of growing scholarly reflection from the mid 1990s onward. Following Foucault, and even more so Derrida, the contingent, decentered, unstable, and partial structure of the archive as a generator of signification have been well established and need not be repeated here. The standard dictionary definition relates to a place where documents and other materials of public or historical interest are preserved. Not all writers distinguish between archives as repositories of documents, manuscripts, and images and libraries as repositories of published books, journals, and other media. In the current context I mean the former. In this excursus I make a rather specific argument, not found in the literature on the archive, pertaining to the discussion of race, photography, and the imagination. In the context of existing literature on the archive, I would like to note briefly only a few points. Following the recognition of the archive as a power structure, questions arise concerning its legitimacy as well as the truth claims of archival material. What the archive contains is already a reconstruction—a recording of history from a particular perspective. This is true with regard to racial photography in more than one sense. On a lower, "empirical" register, however, this is also true with regard not only to what goes into the archive in the first place but also—as will be shown—with regard to what is eradicated or removed later. Influenced by Michel Foucault, *The*

Archeology of Knowledge, trans. A. M. Sheridan Smith (London: Routledge, 2002), the discussion of the 1990s was to a great extent fired by Jacques Derrida's seminal *Archive Fever: A Freudian Impression* (Chicago: University of Chicago Press, 1998). For an overview of the discussion, see Marlene Manoff, "Theories of the Archive from across the Disciplines," *Libraries and the Academy* 4, no. 1 (2004): 9–25; key publications include two issues of the *Journal for the History of the Human Sciences* that appeared in 1998 and 1999, and articles and books by English historian Carolyn Steedman, in particular, *Dust: The Archive and Cultural Memory* (New Brunswick, NJ: Rutgers University Press, 2001), and "Something She Called a Fever: Michelet, Derrida, and Dust," *American Historical Review* 106, no. 4 (2001): 1159–80. Ann Stoler offers intriguing observations about practical epistemological considerations involved in conducting work in archives in similar contexts to those discussed here. See especially the first two chapters of her *Along the Archival Grain: Epistemic Anxieties and Colonial Common Sense* (Princeton, NJ: Princeton University Press, 2010). Rather than interpret archival materials along postcolonial conventions, she calls for "a commitment to a less assured and perhaps more humble stance— to explore the grain with care and read along it first" (50). Methodologically, the closest attempt to what this excursus attempts is found in Harriet Bradley, "The Seductions of the Archive: Voices Lost and Found," *History of the Human Sciences* 12, no. 2 (1999): 107–22. Richard Hobbs affords a fascinating reading of Christian Boltanski with regard to photography, the archive, and the Nazi past, interpreting some of Boltanski's artistic projects as undermining the authenticity of the types of archival evidence and in particular photography; see Richard Hobbs, "Boltanski's Visual Archive," *History of the Human Science* 11, no. 4 (1998): 121–40.

55. Stelzig, *Afrika*, 106.

56. See the letters in the Photothek Inventar der Anthropologischen Abt, Nr. 1, Nr. 18.128, Des Naturhistorischen Museums Wien.

57. Andreas Mayer, "Museale Inszenierungen von anthropologischen Fiktionen: 'Rasse' und 'Menschheit' im Wiener Naturhistorischen Museum nach 1945," in *Repräsentationsformen in den biologischen Wissenschaften*, ed. Armin Geus, Thomas Junker, Hans-Jörg Rheinberger, Christa Riedl-Dorn, and Michael Weingarten (Berlin: VWB, 1999), 73–87.

58. G. Fritsch, "Vergleichende Betrachtungen über die ältesten ägyptischen Darstellungen von Volkstypen," in *Abdruck aus der Naturwissenschaftlichen Wochenschrift* (Jena: Gustav Fischer, 1904), 695–96.

59. Cf. Paul Buberl, *Die Griechisch-Ägyptischen Mumienbildniesse der Sammlung Th. Graf* (Vienna: Krystallverlag, 1922), 15, where, on the basis of explicit realistic assumptions, he determines the possibility that some of the portrayed were rich Jews from Philadelphia.

60. On Fishberg in the American context, see Eric L. Goldstein, *The Price of Whiteness: Jews, Race, and American Identity* (Princeton, NJ: Princeton University Press, 2006), 111–15. On Fishberg in the inner-Jewish political context, see Mitchell B. Hart, "Racial Science, Social Science, and the Politics of Jewish Assimilation," *Isis* 90, no. 2 (1999): 289–95, and chap. 6 of Allan Kraut, *Silent Travelers: Germs, Genes, and the Immigrant Menace* (New York: Basic, 1994). On Fishberg's use of photography, see Mitchell B. Hart, *Social Science and the Politics of Modern Jewish Identity* (Stanford, CA: Stanford University Press, 2000), 180–81. See also Anne C. Rose, " 'Race' Speech—'Culture' Speech—'Soul' Speech: The Brief Career of Social-Science Language in American Religion during the Fascist Era," *Religion and American Culture: A Journal*

of Interpretation 14, no. 1 (2004): 86 (italics added). Contrary to Rose's position, I argue that only in retrospect can Fishberg's argument be read as counterracial or his use of photographs as counterevidence.

61. Maurice Fishberg, "Physical Anthropology of the Jews I. The Cephalic Index," *American Anthropologist* 4 (1902): 684.

62. Several years later Fishberg criticized Radosavljevich's critique of Franz Boas's work in the field of physical anthropology. See Maurice Fishberg, "Remarks on Radosavljevich Critical Contribution to 'School Anthropology,'" *American Anthropologist*, n.s., 14 (1912): 131–41. In 1936 G. M. Morant and Otto Samson criticized Fishberg's technique, method, and erroneous and mistaken conclusions; see G. M. Morant and Otto Samson, "An Examination of Investigations by Dr. Maurice Fishberg and Professor Franz Boas Dealing with Measurements of Jews in New York," *Biometrika* 38, no. 1/2 (1936): 1–31.

63. Fishberg, "Physical Anthropology of the Jews I," 706.

64. Maurice Fishberg, "Physical Anthropology of the Jews II: Pigmentation," *American Anthropologist*, n.s., 5 (1903): 89.

65. Ibid.

66. Ibid., 94, 106.

67. Ibid., 103.

68. Maurice Fishberg, *The Jews: A Study of Race and Environment* (New York: Walter Scott, 1911).

69. Fishberg, *Rassenmerkmale der Juden: Einfürung in ihre Anthropologie* (Munich: E. Reinhardt, 1913), 216–25.

70. Ibid., 228, 231.

71. "A Study of the Jewish Race; Dr. Maurice Fishberg Seemingly Denies Their Claim to be a Particular People," *New York Times*, February 12, 1911.

72. Ibid.

73. Fishberg, *Rassenmerkmale der Juden*, 192, 193.

74. Sigmund Feist, *Stammeskunde der Juden: Die jüdischen Stämme der Erde in alter und neuer Zeit: Historisch-anthropologische Skizzen* (Leipzig: J. C. Hinrichs, 1925). A collection of recently published letters written to him by 77 of his pupils during their service in the German military during the Great War has brought him back to public attention. Sabine Hank and Hermann Simon, eds., *Feldpostbriefe jüdischer Soldaten 1914–1918* (Berlin: Hentrich und Hentrich, 2002).

75. Feist, *Stammeskunde der Juden*, 13n1.

76. Feist's 1916 criticism of Ludwig Wilser, quoted by Bernard Mees, *The Science of the Swastika* (Budapest: Central European University Press, 2008), 176; Feist's daughter is quoted on 276.

77. Sigmund Feist, *Europa im Lichte der Vorgeschichte und die Ergebnisse der vergleichenden indogermanischen Sprachwissenschaft: Ein Beitrag zur Frage nach den Ursitzen der Indogermanen* (Berlin: Weidmannsche, 1910), 47–54. On the racial subject Feist bases himself on von Luschan and Friedrich Ratzel.

78. Sigmund Feist, *Germanen und Kelten in der antiken Überlieferung* (Halle: Max Niemeyer, 1927).

79. Mees, *Science of the Swastika*, 177.

80. Ibid.

81. Feist, *Stammeskunde der Juden*, 25.

82. Ibid.

83. Ibid., 27.

84. Ibid.

85. Ibid., 28.

86. Ibid., 177–78.

87. Ibid., 178, 181n1.

88. Ibid., 173, 177.

89. Gunther R. Kress and Theo van Leeuwen, *Reading Images: Grammar of Visual Design* (London: Routledge, 1996), 181.

90. See Deborah Poole, *Vision, Race, and Modernity: A Visual Economy of the Andean Image World* (Princeton, NJ: Princeton University Press, 1997), 159.

91. Franz Boas, "Are the Jews a Race?," *World Tomorrow*, January 1923; reprinted as "The Jews," in *Race and Democratic Society* (New York: Biblo and Tannen, 1969), 39–42.

92. Feist, *Stammeskunde der Juden*, 176.

93. Ibid., 70. Somewhat more ambivalent is Feist's direction to his reader to observe the difference between light- and dark-skinned Jews in India (pictures 38 and 43), or on p. 123, where Feist refers to skin color and refers the readers to picture 73 in table 32.

94. My use of terms here draws on Aleida Assmann in *Einführung in die Kulturwissenschaft: Grundbegriffe, Themen, Fragestellungen* (Munich: Beck, 1998), 224–28, 232.

95. I am drawing on Hans-Jörg Rheinberger, *An Epistemology of the Concrete: Twentieth-Century Histories of Life* (Durham, NC: Duke University Press, 2010), 30.

96. Daston and Galison, *Objectivity*, 55–113.

97. Ibid., 318.

CHAPTER THREE

1. Michael Hau, *The Cult of Health and Beauty in Germany* (Chicago: University of Chicago Press, 2003), 153. The most comprehensive treatment of Günther remains Hans-Jürgen Lutzhöft, *Der nordische Gedanke in Deutschland, 1920–1940* (Stuttgart: Klett, 1971). See also the more recent Peter Schwandt, *Hans F.K. Günther: Porträt, Entwicklung und Wirken des rassistisch-nordischen Denkens* (Saarbrücken: VDM, 2008).

2. Günther, *Rassenkunde des deutschen Volkes* (Munich: Lehmann, 1937), 255; *Der nordische Gedanke* (Munich: Lehmann, 1927), 41.

3. Günther Kress and Theo van Leeuwen, *Reading Images: The Grammar of Visual Design* (New York: Routledge, 2006), 18.

4. Kress and van Leeuwen, *Reading Images*, 26.

5. Out of probably different motivations, Moholy-Nagy formulated a similar observation. Laszlo Moholy-Nagy, "Fotografie, die objektive Sehform unserer Zeit," *Teleohr* (1936): 122.

6. Rudolf Arnheim, *The Power of the Center: A Study of Composition in Visual Arts* (Berkeley: University of California Press, 1982).

7. Hans F. K. Günther, *Rassenkunde des jüdischen Volkes* (Munich: Lehmann, 1930), 231.

8. Günther, *Rassenkunde des jüdischen Volkes*, 214–215, 217, 218: height, chest, flat feet, eye form, or ears.

9. Ibid., 8, quotation from 215 (and 252).

10. Note Günther's semantic slippage from Salaman's "Anglo-Jewish" to "Jews" and from "Gentiles" to "English." Hence, while criticizing Salaman for treating the two elements as race rather than as racial mixtures, he in fact removes the "Anglo" from Jews, making the racial opposition even more direct and rendering the "Anglo" mere paratypical; Günther, *Rassenkunde des jüdischen Volkes*, 289.

11. Ibid., 287.

12. Ibid., 286.

13. Günther is not the first to reject Galton while reproducing his images. Already in 1911 Maurice Fishberg does the same. Unlike Günther, who rejects Galton's method, Fishberg rejects the series and its interpretation by Jacobs on empirical grounds, noting that "A glance at the composite portraits herewith reproduced shows that very few of the above characteristics are discernible, notwithstanding the fact that the boys were carefully selected as such who look more Jewish than the average." See Maurice Fishberg, *The Jews: A Study of Race and Environment* (New York: Walter Scott, 1911), 104 (the German translation of this book appeared in 1913).

14. Günther, *Rassenkunde des jüdischen Volkes*, 209–11. Maurice Fishberg and Jack Jacobs, "Anthropological Types," in *The Jewish Encyclopedia* (New York: Funk and Wagnalls, 1904), 12:291–95.

15. Günther, *Rassenkunde des jüdischen Volkes*, 211.

16. Ibid.

17. Alison Scott-Baumann, *Ricoeur and the Hermeneutics of Suspicion* (New York: Continuum, 2009).

18. Bruno Latour, *Reassembling the Social: An Introduction to Actor-Network-Analysis* (Oxford: Oxford University Press, 2005), 63–86.

19. Rebecca Mae Salokar and Mary L. Volcansek, eds., *Women in Law: A Bio-Bibliographical Sourcebook* (Westport, CT: Greenwood, 1996), 31–33.

20. Ann Laura Stoler, *Race and the Education of Desire: Foucault's History of Sexuality and the Colonial Order of Things* (Durham, NC: Duke University Press, 1995); *Along the Archival Grain: Epistemic Anxieties and Colonial Common Sense* (Princeton, NJ: Princeton University Press, 2010); Sander L. Gilman, *The Jew's Body* (London: Routledge, 1991); Shawn Michelle Smith, *Photography on the Color Line: W.E.B. Du Bois, Race, and Visual Culture* (Durham, NC: Duke University Press, 2004).

21. See Amos Morris-Reich, "Race, Ideas, and Ideals: A Comparison of Franz Boas and Hans F. K. Günther," *History of European Ideas* 32, no. 3 (2006): 313–32.

22. Already at the time, scientists questioned this interpretation of Mendel, and in the late 1920s the synthetic theory of Mendel and Darwin was formulated. Günther's interpretation, however, was certainly within the scientific purview of the time.

23. Kress and van Leeuwen, *Reading Images*, 50. Günther had no interest in individuals as such: when he incorporated photographs of famous individuals such as Albert Einstein or Theodor Herzl, the aim was to identify them racially.

24. Kress and van Leeuwen, *Reading Images*, 50, 161.

25. Ibid., 161.

26. "You watch a painting; a friend of yours points out a feature you had not noticed; you are thus *made to see* something. Who is seeing it? You, of course. And yet, wouldn't you freely acknowledge that you would not have seen it *without* your friend. So who has seen the delicate feature? Is it you or your friend? The question is absurd. Who would be silly enough to deduct from the total sum of action the influence of pointing something out?" Latour, *Reassembling the Social*, 237.

27. Ludwig Wittgenstein writes "I wanted to put this picture before your eyes, and your *acceptance* of this picture consists in your being inclined to regard a given case differently; that is, to compare it with *this* series of pictures. I have changed your *way of seeing*." Ludwig Wittgenstein, *Zettel*, ed. G. E. M. Anscombe and G. H. von Wright, trans. G. E. M. Anscombe (Los Angeles: University of California Press, 1970), 82e.

28. Cristina Grasseni, "Video and Ethnographic Knowledge: Skilled Vision in the Practice of Breeding," in *Working Images: Visual Research and Representation in Ethnography*, ed. Sarah Pink et al. (New York: Routledge, 2004), 15–30; "Good Looking: Learning to be a Cattle Breeder," in *Skilled Vision: Between Apprenticeship and Standards*, ed. Cristina Grasseni (New York: Berghahn, 2007), 47–66; *Developing Skill, Developing Vision* (New York: Berghahn, 2009).

29. Quoted in David Lodge, *The Art of Fiction* (New York: Viking 1992), 53.

30. Hans Sedlmayr, "Bruegel's *Macchia*," trans. Frederic J. Schwartz, *The Vienna School Reader: Politics and Art Historical Method in the 1930s*, ed. Christopher S. Wood (New York: Zone, 2000), 323–76. In his 1931 "Toward a Rigorous Study of Art," trans. Mia Fineman, ibid., 133–79, Sedlmayr emphasizes the relativity of seeing that I addressed in chap. 1 (145) as affecting the observer's sense of spatial boundary (146, 149).

31. Sedlmayr, "Bruegel's *Macchia*," 336.

32. Ibid., 339.

33. Ibid., 340. Sedlmayr's analysis of Bruegel is not of our concern in the current context. It is important, however, that Sedlmayr claims that Bruegel's "people [characters portrayed in his pictures] look uncanny and strange because, in their essence, they themselves are problematic hybrid and alien (if perhaps not downright "bad") creatures" (351).

34. Ibid., 354.

35. Kress and van Leeuwen, *Reading Images*, 181.

36. Günther included this photograph already in a different series, in the first version of the appendix on the Jews in *Rassenkunde des deutschen Volkes* (Munich: Lehmann, 1928), 404.

37. On structuralism, see Jonathan Culler, *Structuralist Poetics: Structuralism, Linguistics, and the Study of Literature* (London: Routledge and Kegan Paul, 2002).

38. Stoler, *Race and the Education of Desire*.

39. Eugen Fischer and Hans F. K. Günther, *Deutsche Köpfe nordischer Rasse* (Munich: Lehmann, 1927).

40. See in this context C. L. Schleich, "Juedische Rassenkoepfe," *Ost und West* 4 (1906): 227–40. While unquestionably racializing Jews, this essay, which comprises a text and numerous images of mostly bearded Jews, indicates a form of Jewish beauty. In the context of Günther's

dependence on photography it is interesting to note that in 1906 Schleich still criticizes photography as inferior to expressions that only artists can capture.

41. Günther's photographic series accord with Desrosieres orientation toward statistics as "making things that hold." But it diverges from it in the sense that Günther on occasion does not strive to remove the variable from the invariable. Desrosieres, *The Politics of Large Numbers: A History of Statistical Reasoning*, trans. Camille Naish (Cambridge, MA: Harvard University Press, 1998), 9–12.

42. See Amos Morris-Reich, "Method, Project, and the Racial Characteristics of Jews: A Comparison of Franz Boas and Hans F. K. Günther," *Jewish Social Studies* 13, no. 1 (2006): 136–69.

43. Günther, *Rassenkunde des jüdischen Volkes*, 77.

44. Ibid.

45. See David K. Naugle, *Worldview: The History of a Concept* (Grand Rapids, MI: Eerdmans, 2002).

46. Cf. Friedrich Nietzsche, "On Truth and Lie in an Extra-Moral Sense," *The Portable Nietzsche*, ed. and trans. Walter Kaufmann (New York: Penguin, 1982), 46–47. The notion of *Weltanschauung* was criticized by Husserl and Heidegger, but there is no question that Wittgenstein offers its most powerful and thoroughgoing deconstruction; Naugle, *Worldview*, 148–67. See also Judith Genova, *Wittgenstein: A Way of Seeing* (New York: Routledge, 1995), 50.

47. Houston Stewart Chamberlain, *Foundations of the Nineteenth Century*, trans. John Lees, with an introduction by Lord Redesdale (London: J. Lane, 1911), 241–42.

48. Houston Stewart Chamberlain, *Immanuel Kant* (London: John Lane, 1914), 17–18; Geoffrey G. Field, *Evangelist of Race: The Germanic Vision of Houston Stewart Chamberlain* (New York: Columbia University Press, 1981), 281–91; Chaimberlain, *Foundations*, 235–246. In Chamberlain's short 1905 *Aryan World-View* (Munich: F. Bruckmann, 1938), he does not discuss "seeing." On the morphological version of biology more generally see Lynn K. Nyhart, *Biology Takes Form: Animal Morphology and the German Universities, 1800–1900* (Chicago: University of Chicago Press, 1995).

49. Quoted in G. Field, *Evangelist of Race*, 284.

50. Eugen Fischer and Hans F. K. Günther, *Deutsche Köpfe nordischer Rasse* (Munich: Lehmann, 1927), 9–16.

51. Günther, *Rassenkunde des jüdischen Volkes*, 217–18. Günther refers to John Beddoe, "On the Physical Characteristics of the Jews," *Transactions of the Ethnological Society of London*, n.s. (1861): 22, and on William Ripley, "Über die Anthropologie der Juden," *Globus* 76 (1889): 21. Before turning to the discussion of the Jewish ear, Günther rejects Reinach's "Lamarckist" interpretation of the Jewish gaze as the result of suffering and persecution.

52. Hans F. K. Günther, *Ritter, Tod, und Teufel: Der heldische Gedanke*, 2nd ed. (Munich: Lehmann, 1924), 166–67.

53. Jürgen Kocka, "Asymmetrical Historical Comparison: The Case of German *Sonderweg*," *History and Theory* 38, no. 1 (1999): 40–50.

54. Ludwig Wittgenstein, *The Blue and Brown Books: Preliminary Studies for the Philosophical Investigations* (Oxford: Blackwell, 1958), 17; Hans Sluga, "Family Resemblance," *Grazer Philosophische Studien* 71 (2006): 8.

55. Wittgenstein's full quote reads, "And to make you see as clearly as possible what I take to be the subject matter of Ethics I will put before you a number of more or less synonymous expressions . . . and by enumerating them I want to produce the same sort of effect which Galton produced when he took a number of photos of different faces on the same photographic plate in order to get the picture of the typical features they all had in common. And as by showing you such a collective photo I could make you see what is the typical—say—Chinese face; so if you look through the row of synonyms which I will put before you, you, I hope, be able to see the characteristic features they all have in common." Quoted from Daston and Galison, *Objectivity*, 338; originally in Ludwig Wittgenstein, aphorism 67, "A Lecture on Ethics," *Philosophical Review* 74 (1965): 4–5. See Carlo Ginzburg, "Family Resemblances and Family Trees: Two Cognitive Metaphors," *Critical Inquiry* 30 (2004): 537–56. Before the line of thought that developed into the "family resemblance" idea, Wittgenstein reflected on racial difference in his remarks on James Frazer's *Golden Bough: A Study in Magic and Religion* (1890) in Ludwig Wittgenstein, *Bermerkungen über Frazers Golden Bough/ Remarks on Frazer's Golden Bough*, ed. Rush Rhees, trans. A. C. Miles (Redford, England: Brynmill, 1979).

56. Wittgenstein, *Blue and Brown Books*, 18.

57. Daston and Galison, *Objectivity*, 335–37, 475n44.

58. Ludwig Wittgenstein, *Philosophical Investigations* (Oxford: Blackwell, 1953), sec. 66 (italics in original). On the notion of family resemblance, see Sluga, "Family Resemblance," 1–21; H. Wennerberg, "The Concept of Family Resemblance in Wittgenstein's Later Philosophy," *Theoria* 33 (1967): 107–32; R. K. Gupta, "Wittgenstein's Theory of 'Family Resemblance,' in His Philosophical Investigations (Secs. 65–80)" *Philosophia Naturalis* 12 (1970): 282–86.

59. Wittgenstein, *Philosophical Investigations*, sec. 65–67); quoted in Sluga, "Family Resemblance," 1.

60. Sluga, "Family Resemblance," 14. Sluga concludes that "family" in Wittgenstein's notion is far more concerned with kinship than with similarity (16); "language" is far more a similarity term (17).

61. Wittgenstein, *Philosophical Investigations*, sec. 66; Sluga, "Family Resemblance," 14.

62. Siegfried Kracauer, *From Caligari to Hitler: A Psychological History of the German Film* (1947; facsimile repr. Princeton, NJ: Princeton University Press, 1966).

63. Kracauer, *From Caligari to Hitler*, analyzes the powerful use of maps in propaganda films, with their moving arrows and lines, as "resembling physical processes" that "show how all known materials are broken up, penetrated, pushed back and eaten away by the new one, thus demonstrating its absolute superiority" (279) as well as the panning of the camera "rising and diving," thus "deepening the spectator's conviction of complete control of the field." Günther's racial theory clearly backed that sense of racial superiority and dominance.

64. Ibid., 279.

65. Ibid., 259.

66. Günther, *Rassenkunde des deutschen Volkes*, 21, 26.

67. Hans F. K. Günther, *Kleine Rassenkunde des deutschen Volkes* (Munich: Lehmann, 1933), 140.

68. Primitive also means simply old. In *Rassenkunde des jüdischen Volkes*, he introduces far more images of individuals with a wild impression (middle and bottom rows of 72). Even more

suggestive is his reproduction of El Faiyûm mummy portraits (middle row of 73, 76, 102), those most impressive portraits some of which date back to the first century BC, imply (although they are not of Jews) the Jews' old racial age. A frequent theme in contemporary racial literature was the "young" age of the Nordic races as compared with the "old" age of the Eastern races. By pointing to affinities between contemporary Jews and primitive types, the primitiveness of the former is implied.

69. Quoted in Field, *Evangelist of Race*, 193.

70. Kracauer, *From Caligieri to Hitler*, 63.

71. Richard T. Gray, *About Face: German Physiognomic Thought from Lavater to Auschwitz* (Detroit, MI: Wayne State University Press, 2004), 219–72.

CHAPTER FOUR

1. Ludwig Ferdinand Clauß, *Rasse und Charakter: Das lebendige Antlitz* (Frankfurt: Diesterweg, 1938), 8. Such expressions are ubiquitous throughout Clauß's work; cf. *Die Seele des Andern: Wege zum Verstehen im Abend- und Morgenlande* (Baden Baden: Grimm, 1958), 28; *Fremde Schönheit: Eine Betrachtung seelischer Stilgesetze* (Heidelberg: Kampmann, 1928), 36.

2. Peter Weingart, *Doppel-Leben: Ludwig Ferdinand Clauß: Zwischen Rassenforschung und Widerstand* (Frankfurt: Campus, 1995). For an apologetic account of Clauß as undermining National Socialist race ideology, written by a student of his, see Reinhard Walz, "Ludwig Ferdinand Clauß zum 70. Geburtstag: Die Entstehung einer Psychologie der Psyche," *Jahrbuch für Psychologie, Psychotherapie und medizinische Anthropologie* 9 (1962): 149–65. For a criticism of these earlier accounts and an assessment of Clauß's influence on the educational system, see Matthias Schwerendt, *"Trau keinem Fuchs auf grüner Heid, und keinem Jud bei seinem Eid": Antisemitismus in nationalsozialistischen Schulbüchern und Unterrichtsmaterialien* (Berlin: Metropol, 2009), 146–54.

3. Clauß's Yad Vashem file reveals that the process by which his title was revoked was as shallow as the basis on which it was first awarded.

4. See chap. 18 and 19 in Heather Pringle, *The Master Plan: Himmler's Scholars and the Holocaust* (New York: Hyperion, 2006). To add to the incomprehensibility of this story, Pringle notes that after the war, Beger was employed by Margaret Lande in her paper business.

5. The main alternative to Clauß's insistence on rationality in the discourse on race in the humanities is the mystical approach, most notably that of Rosenberg, who insists that race is spontaneously recognizable, "a mystical synthesis, an activity of soul," "which cannot be explained rationally." Alfred Rosenberg, *Race and Race History and Other Essays by Alfred Rosenberg*, ed. Robert Pois (New York: Harper and Row, 1970), 84.

6. Clauß introduced the notions of *Mitleben* and *Stil* in his attempt to integrate Husserl's phenomenology with the principle of fundamental racial difference. Petermann, extending Clauß's fundamentalist perspective, attempted to define racial souls in terms of *Ideenforschung* and *Urbildern des Seienden*, in which both body and soul flow from a "mutual Gestalt-idea"; Bruno Petermann, *Das Problem der Rassenseele: Vorlesungen zur Grundlegung einer allgemeinen Rassenpsychologie* (Leipzig: Barth, 1935), 63. Eduard Ortner, focusing on a different difficulty, attempted to correlate Clauß's racial types with six corresponding biological types categorized

according to their formal relationship to the environment. In the process he differentiates between average type (*Durchschnittstyp*) and ideal type (*Idealtyp*), the latter being obtained through the ideational procedure of abstracting an essence from its partial, empirical appearance. Eduard Ortner, *Biologische Typen des Menschen und ihr Verhältnis zu Rasse und Wert* (Leipzig: Thieme, 1937), 9, 45. For a devastating review of this book, see M. F. Martin, *American Journal of Psychology* 51, no. 1 (1938): 190–91.

7. Ludwig Ferdinand Clauß, *Grundfragen der Rassenseelenkunde, Rasse und Seele: Eine Einführung in den Sinn der leiblichen Gestalt* (Berlin: Buechergilde Gutenberg, 1933), 121.

8. Ibid., 87.

9. On Jews as "desert people," cf. Werner Sombart, *The Jews and Modern Capitalism* (Boston: Dutton, 1913). Strictly speaking, according to Clauß, the desert is not a livable landscape. In a postwar article devoted to the "desert," published in the years following the establishment of Israel, Clauß condemns the Zionist project in racial terms: out of yearning for their desert essence, the Jews returned to their home desert, but betraying their law they de-desertified (*entwüsten*) the land, transforming it into a flourishing garden. Ludwig Ferdinand Clauß, "Die Kraft der Wüste," *Jahrbuch für Psychologie, Psychotherapie und medizinische Anthropologie* 10, no. 3/4 (1963): 250.

10. Ludwig Ferdinand Clauß, *Rasse und Seele: Eine Einführung in den Sinn der leiblichen Gestalt* (Munich: Gutenberg, 1937), 11.

11. Ibid., photographs 53, 54, and 55 on pages 68, 69.

12. Ibid., photograph 53 on p. 68.

13. Ibid., photographs 56–61 on pp. 70–71.

14. Ludwig Ferdinand Clauß, "Der semitische Mensch," *Rasse: Monatsschrift der Nordischen Bewegung* 1, no. 4/5 (1934): 170.

15. Clauß, *Rasse und Seele: Eine Einführung in den Sinn der leiblichen Gestalt*, 68, 74, 75.

16. Ibid., 58, 66, 72, 74.

17. Ibid., photographs 62–64 on p. 56.

18. But Clauß's generalizations are not fundamentally different from much of the art history that is contemporary with him. Margaret Olin discusses Karl Schnaase's (1798–1875) trope of the Jewish gaze as a failure to focus because of its constant motion and Martin Buber's adaptation of the same idea based on contemporary perception theory. Margaret Olin, *The Nation without Art: Examining Modern Discourses on Jewish Art* (Lincoln: University of Nebraska Press, 2001), 14 and 121, respectively. This may be the source of Wilhelm Worringer's 1909 generalizations on Northern art in *Abstraction and Empathy: Contributions to the Psychology of Style*, trans. Michael Bullock (Cleveland: Meridian, 1967), 106–115.

19. Clauß, *Rasse und Seele: Eine Einführung in den Sinn der leiblichen Gestalt*, photographs 62–69 on pp. 78–81.

20. Asceticism, intellectuality, and financial interest can be seen as stemming from this type. This duality is also the source of this type's mistrust of others (ibid., 84). Materialism, too, stems from this same duality (ibid., 98).

21. Erich Brauer (addressed in the next chapter) notes in his diary (March 1929) that he visited Clauß in his home and that reviews of Clauß's book by Grünwald, J. Hamburger, and Heinz Caspari had appeared: "Clauß's descriptions are maybe the deepest and most beautiful

in literature on the Jews ever to appear." He then discusses the photographs specifically: "Whoever looks at the photographs closely no longer thinks of Jewish race or Jewish type in the ethnological sense. Clauß depicts the Jews as good and nice. Clauß is above parties and hatreds. Only seeks the truth. . . . His work should be a new sign in science." V1943, notebook 83, pp. 167 and 178–179, National Library of Israel. I wish to thank Orit Abuhav for referring me to this entry.

22. Clauß, *Rasse und Seele: Eine Einführung in den Sinn der leiblichen Gestalt*, 99; *Rasse und Charakter*, 104.

23. Ludwig Ferdinand Clauß, "Woran erkennt man den Juden?," *Wir und die Welt* 11 (1940): 449–61.

24. Ibid., 449. After World War II, Clauß states—with regard to horses rather than Jews—that their original idea (*Urbild*) is independent of their existence. Even if they were to become extinct, the *idea* of the "horse" would persist. See Ludwig Ferdinand Clauß, "Menschsein auf Türkisch," *Jahrbuch für Psychologie, Psychotherapie und medizinische Anthropologie* 16 (1968): 96. The eyes of the Germans were opened to race in 1918, according to Clauß (in an earlier article), revealing to them that some people, though they spoke German and enjoyed German citizenship, had faces, thoughts, and feelings that were un-German. See Ludwig Ferdinand Clauß, "Ziel und Gegenwart: Gedanken zur deutschen italienischen Rassenbewegung," *Wir und die Welt* 19 (1939): 39.

25. Clauß, "Woran erkennt man den Juden?," 450.

26. Ibid., 452–53.

27. Ibid., 453–54.

28. Ibid., 453. Throughout his writing, Clauß often differentiates between race and character, the latter being an individual rather than a racial feature. Here, however, he suggests that the character of the Jews is an inherited character (Erbcharakter) arising from their ghetto breeding (Ghetto-Züchtung). The ghetto gaze is constituted from within, but it can evaporate (454).

29. Cf. Clauß, *Rasse und Charakter*, 104; *Rasse und Seele: Eine Einführung in den Sinn der leiblichen Gestalt*, 110–11; "Stile der Wahrhaftigkeit und des Lügens," *Rasse* 2, no. 11 (1935): 426.

30. Adolf Hitler, *Mein Kampf*, trans. Ralph Manheim (Boston: Mariner, 1943), 302.

31. Clauß, "Der semitische Mensch," 176.

32. Ludwig Ferdinand Clauß, "Rasse im Raum," *Wir und die Welt* 1 (1940): 6. Richard Walther Darré developed the theory of race and space; see his *Das Bauerntum als Lebensquell der nordischen Rasse* (Munich: Lehmann, 1928). See also Anna Bramwell, *Blood and Soil: Richard Walther Darré and Hitler's "Green Party"* (Abbotsbrook, Bourne End, Buckinghamshire: Kensal, 1985). On the history of *Lebensraum*, see Woodruff D. Smith, *The Ideological Origins of Nazi Imperialism* (New York: Oxford University Press, 1985), 83–111. Clauß's interpretation reflects Husserl's notion of *Lebenswelt* (*Gesammelte Werke*, ed. Marly Biemel, vol. 4, *Ideen zu einer reinen Phänomenologie und phänomenologishen Philosophie, Second Book* [The Hague: Nijhoff, 1952], sec. 50), according to which person and environment are inseparable. "We call a set of stylistically structured psychic surroundings a landscape"; Clauß, *Rasse und Seele: Ein Einführung in die Gegenwart* (Munich: Lehmann, 1926), 33. Sections of Clauß's description of the French garden read much like Panofsky's famous discussion of the English garden. See Ludwig Ferdinand Clauß, "Geist romanisch gesehen," *Wir und die Welt* 8 (1941): 317; Erwin Panofsky, "The Ideological Antecedents of the Rolls-Royce Radiator," *Three Essays on Style* (Cambridge, MA: MIT Press, 1997), 129–68.

33. In a short essay, "Philosophy of the Landscape" (1913), Simmel contrasts landscape and nature. The latter has no borders, while the former is a historically modern construction which transcends its individual components (478), unified by an atmosphere (*Stimmung*) in which subject and object merge; Georg Simmel, "Philosophie der Landschaft," *Aufsätze und Abhandlungen 1909–1919 Band I* (Frankfurt: Suhrkamp, 2001), 471–82.

34. Clauß, "Rasse im Raum," 7.

35. Ibid.

36. Ibid., 9–12.

37. Ibid., 9.

38. Ibid., 10.

39. See also Clauß, *Die Seele des Andern*, 120, 160–164, 174–176, as well as Ludwig Ferdinand Clauß, "Die innere Landschaft: Gedanken zu einer Philosophie der Rasse," *Rasse* 3, no. 12 (1936): 457–61. Rembrandt serves as a central example in this publication (459).

40. Clauß, "Rasse im Raum," 8.

41. Ibid.

42. Hannah Arendt, *Lectures on Kant's Political Philosophy*, ed. Ronald Beiner (Chicago: University of Chicago Press, 1992), 79–85.

43. Ibid., 80.

44. Ibid., 81n6.

45. Ibid., 82.

46. Ibid., 83.

47. In Ferrara's reformulation of Arendt, examples "provide us with holistic images that remain concrete"; Alessandro Ferrara, *The Force of the Example: Explorations in the Paradigm of Judgment* (New York: Columbia University Press, 2008), 53.

48. Clauß, *Rasse und Seele: Eine Einführung in den Sinn der leiblichen Gestalt*, 76, 77

49. Ibid., 78.

50. Ibid., 80.

51. Arendt, *Lectures*, 80 (italics in the original, square brackets added). The crucial difference between Clauß and Arendt, as we shall see in the discussion that follows, pertains to the ontological status of examples as embedded in foundationalist (Clauß) or nonfoundationalist (Arendt) principles.

52. Ferrara, *Force of the Example*, 27.

53. Ibid., 22.

54. Defining race as gestalt, Clauß connects race to the problem of the perception of complexes, that is, to the problem of the recognition of sameness in difference (e.g., the same melody in two different keys) through specific types of complex networks of acts of presentation (perception, memory, and imagination). See Kevin Mulligan and Barry Smith, "Mach and Ehrenfels: The Foundations of Gestalt Theory," in *Foundations of Gestalt Theory*, ed. Barry Smith and Kevin Mulligan (Munich: Philosophia, 1988), 124–57, esp. 125–29.

55. Mitchell G. Ash, *Gestalt Psychology in German Culture 1890–1967: Holism and the Quest for Objectivity* (Cambridge: Cambridge University Press, 1998), 52.

56. Ibid., 120.

57. Ibid., 186. Gestalt theorists were important in initiating the great interest in seeing, expressed in Wolfgang Metzger's 1936 *Gesetze des Sehens* (*Laws of Seeing*; Ash, *Gestalt Psychology*,

347) and in his article "Laws of Seeing—Applied." As Ash shows, there were several concepts of gestalt in Germany. Clauß did not belong to the Berlin group, either institutionally or socially.

58. Examples focused on the spontaneity of knowledge, as with the identification of color. Ehrenfels stated, "Descartes pointed out that all the wise men in the world could not define the color 'white'; but I need only to open my eyes to see it and it is the same with race" (quoted in Ash, *Gestalt Psychology*, 90). This spontaneous experience is central to Clauß: he repeats the example with the color white as well. See Clauß, *Rasse und Seele: Eine Einführung in den Sinn der leiblichen Gestalt*, 28.

59. On the notion of thought experiments and their history, see R. Arthur, "On Thought Experiments as A Priori Science," *International Studies in the Philosophy of Science* 13, no. 3 (1999): 215–29, and, more recently, Ulrich Kühne, *Die Methode des Gedankenexperiments* (Frankfurt: Suhrkamp, 2005).

60. Ash, *Gestalt Psychology*, 179–180. Among many other examples, one could mention Koffka and Kenkel's motion experiments (reproduced in ibid., 140) or Friedrich Schumann's experiments concerning subjective groupings (reproduced in ibid., 92).

61. See Ludwig Ferdinand Clauß, *Rasse ist Gestalt* (Munich: Lehmann, 1937), 14–21; *Rassenseele und Einzelmensch: Lichtbildervortrag* (Munich: Lehmann, 1938), 32–35; "Grundfragen der Rassen-Psychologie," *Wir und die Welt* 2 (1941): 252–253; *Die Seele des Andern*, 108–115.

62. Clauß, *Die Seele des Andern*, 108.

63. Ibid., 108–9.

64. Ibid., 109 (italics in original).

65. Ibid., 145.

66. Tania Munz, "An Ethology of the Scientific Cinéaste: Karl von Frisch, Konrad Lorenz, and Animal Behavior on Film" (unpublished manuscript), 11; in German as "Eine Ethologie des wissenschaftlichen Cineasten: Karl von Frisch, Konrad Lorenz und das Verhalten der Tiere im Film," *Montage/av* 14, no. 2 (2005): 52–68.

67. On specimens in the history of science, see Lorraine Daston, "Type Specimens and Scientific Memory," *Critical Inquiry* 31 (2004): 153–82.

68. Hans Rainer Sepp, ed., *Edmund Husserl und die phänomenologische Bewegung: Zeugnisse in Text und Bild* (Munich: Alber, 1988). Clauß is mentioned only twice, briefly (15, 270). In the short biographies of Husserl's students he is described as a *Völkerpsychologe*, and titles with the word *race* are omitted from the selective list of publications (425). Similarly, Spiegelberg's account minimizes Clauß's role with respect to Husserl (and is wrong about his relationship to Nazism): "There would be no point in listing here in similar fashion the names of all the other figures who during the twenties had minor roles in the field between Husserl and Heidegger, eventually all to move on along their own lines. Thus Ludwig Ferdinand Clauß (1892–1973), who later became interested in questions of racial psychology (without joining the Nazis) was for some time an unofficial assistant to Husserl." Herbert Spiegelberg, ed., *The Phenomenological Movement: A Historical Introduction* (Hague: Springer, 1976), 249.

69. Ludwig Ferdinand Clauß, "Das Verstehen des sprachlichen Kunstwerks: Ein Streifzug durch Grundfragen der verstehenden Wissenschaften," in *Festschrift Edmund Husserl zum 70. Geburtstag gewidmet* (Halle: Niemeyer, 1929), 53–70. This essay is described as "charming and witty" in a review in *Philosophical Review* 39, no. 6 (November 1930): 625–30. Cf. "Nordische

Kampfgesinnung," *Wir und die Welt* 12 (1941): 475. Elsewhere, the racial aspect is explicit: Clauß discusses the story of Gunnar, from Nordic mythology, who knowingly sacrificed his own life, and Clauß recalls a philosophy professor (who could have been Husserl) who questioned the rationality of Gunnar's act; from the utter misunderstanding of German values, Clauß deduced foreign Jewish blood (476).

70. For a comprehensive analysis of Clauß in the context of phenomenology, see Richard T. Gray, *About Face: German Physiognomic Thought from Lavater to Auschwitz* (Detroit, MI: Wayne State University, 2004), 273–332. Per Leo's recent analysis of Clauß emphasizes his affinity rather to Ludwig Klages's graphology and to the tradition of characterology; see Per Leo, *Der Wille zum Wesen: Weltanschauungskultur, charakterologisches Denken und Judenfeindschaft in Deutschland 1890–1940* (Berlin: Matthes und Seitz, 2013), 452–78.

71. Husserl, *Ideen II*, sec. 125, 132–33.

72. Ibid., sec. 21, 95. Worringer, in *Abstraction and Empathy*, uses the theory as promoted by Theodor Lipps (1903), which is probably Husserl's source as well. Robert Vischer (1873) is usually cited as the originator of this widely disseminated idea. Directly or indirectly, and aside from Clauß, Husserl's conception of empathy influenced some of the most important phenomenological work of the twentieth century: Edith Stein extended Husserl's account in *On the Problem of Empathy* (1917): empathy consists in imaginatively putting oneself in the place of another I, reproducing in one's own imagination the form of the other's experience. Heidegger developed his interpretation as being-with (*Mitsein*) in 1927; Sartre developed "the look of the other" (1943); Merleau-Ponty drew on it in *Phenomenology of Perception* (1945).

73. Clauß, *Rasse und Seele: Eine Einführung in den Sinn der leiblichen Gestalt*, 117. Clauß further articulates mimesis and *Mitleben* in postwar articles: Ludwig Ferdinand Clauß, "Mimesis und Mimema: Art und Schwierigkeit des methodischen Mitlebens," *Jahrbuch für Psychologie und Psychotherapie* 2, no. 3 (1953): 268.

74. Clauß meticulously studies what he calls "racial blockages." In *Rasse und Seele: Eine Einführung in den Sinn der leiblichen Gestalt* he studies the following modalities: discrepancies between individual and type (84), the inversion and radicalization of types of style (98–99), life in a world that is foreign to style (109), stylistic mixtures (126), blockages caused by the discrepancy between body and soul (143–46), and beauty as relative to style (147).

75. Frederic J. Schwartz, "Cathedrals and Shoes: Concepts of Style in Wölfflin and Adorno," *New German Critique* 76 (1999): 4.

76. Clauß, *Rasse und Seele: Ein Einführung in die Gegenwart*), 20–21 (translated in Gray, *About Face*, 307). Clauß conceives of style in aesthetic categories, reflected in a tradition from Karl Philipp Moritz through Kant and beyond, where the work of art is "beautiful" when its interacting parts conform to one single coherent style or obey one single "law." Clauß, *Rasse und Seele: Ein Einführung in die Gegenwart*, 24; "Mimesis und Mimema," 269.

77. See Ernst H. Gombrich, "Style," *The International Encyclopedia of the Social Sciences* Vol. 15, 360. Within the history of art, Clauß is aligned with Wölfflin's notion of style.

78. Clauß, *Rasse und Seele: Eine Einführung in den Sinn der leiblichen Gestalt*, 127.

79. See Barry Smith and David Woodruff Smith, "Introduction," in *The Cambridge Companion to Husserl*, ed. Barry Smith and David Woodruff Smith (Cambridge: Cambridge University Press, 1995), 32–33. On the history of *Anschauung* in physiology, see Henning Schmidgen,

"Pictures, Preparations, and Living Process: The Production of Immediate Visual Perception (*Anschauung*) in Late-19th-Century Physiology," *Journal of the History of Biology*, 37 (2004): 477–513, esp. 483–484, for the history of the notion in German philosophy and psychology.

80. Edmund Husserl, *Phantasie, Bildbewusstsein, Erinnerung* (The Hague: Springer, 1980), quoted in Julia Jansen, "On the Development of Husserl's Transcendental Phenomenology of Imagination and Its Use for Interdisciplinary Research," *Phenomenology and the Cognitive Sciences* 40 (2005): 122.

81. Husserl, *Phantasie*, 109. Husserl briefly describes a photograph of his child. The photograph is an image (*Bild*) represented in our imagination.

82. "What can be seen as an *as if*-appearance of an *actual* object is in fact also an actual appearance of a *possible object*"; Husserl, *Phantasie*, 507 (quoted in Jansen, "On the Development of Husserl's Transcendental Phenomenology," 126), 265. See also Brian Elliott, *Phenomenology and Imagination in Husserl and Heidegger* (New York: Routledge, 2005); Eva Brann, *The World of Imagination: Sum and Substance* (Boston: Rowman and Littlefield, 1991); J Sallis, "Spacing Imagination: Husserl and the Phenomenology of Imagination," in *Eros and Eris: Contributions to a Hermeneutical Phenomenology*, ed. Paul Van Tongeran et al. (Dordrecht: Springer, 1992), 201–15.

83. Edmund Husserl, *Gesammelte Werke*, ed. Karl Schuhmann, vol. 3, *Ideen zu einer reinen Phänomenologie und phänomenologishen Philosophie, First Book* (The Hague: Nijhoff, 1976), 16.

84. "The more we see the more we must be able to imagine; and the more we imagine, the more we must think we see"; Gotthold Ephraim Lessing, *Laocoon: An Essay upon the Limits of Painting and Poetry*, trans. Ellen Frothingham (New York: Dover, 2005), 16–17. In a similar line of argument, Marey, too, noted that images "appeal to the imagination rather than the senses"; see Etienne-Jules Marey, *Movement*, trans. Eric Pritchard (London: D. Appleton, 1895), 304. In the culture of the second half of the twentieth century, a negative correlation between the two is assumed; in the beginning of the twenty-first century, it is a positive one that is assumed.

85. Clauß, "Ziel und Gegenwart," 40.

86. Clauß, *Rassenforschung im täglichen Leben* (Berlin: Brehm, 1934), 38.

87. Edmund Husserl, *Cartesian Meditations: An Introduction to Phenomenology*, trans. D. Cairns (The Hague: Nijhoff, 1960), sec. 5. For an explanation of the grades of evidence according to Husserl, see Smith and Woodruff Smith, "Introduction," 34–35.

88. This speech appeared in print as the third booklet in the Writings of the Party series (Schriften der Bewegung, ed. Philip Bouler); Clauß, *Rasse ist Gestalt* (Munich: Zentralverlag NSDAP Eher, 1937).

89. Ibid., 8.

90. Immanuel Kant, *Metaphysical Foundations of Natural Science*, trans. James Ellington (Indianapolis, IN: Bobbs-Merrill, 1970), 3.

91. Ibid., 4.

92. Ibid.

93. Cf. Edmund Husserl, "Philosophy as Rigorous Science," in *Phenomenology and the Crisis of Philosophy*, ed. Q. Lauer (New York: Harper and Row, 1965), 71–167; Georg Simmel, *The Problems of the Philosophy of History: An Epistemological Essay*, trans. and. ed. Guy Oakes (New York: Free Press, 1977), 140.

94. Elsewhere, Clauß relates directly to statistics. See Clauß, "Stile der Wahrhaftigkeit,"

417; "Mimesis und Mimema," 268. For a comparison of this "law" with that of mathematics, see Bruno Petermann, *Das Problem der Rassenseele: Vorlesungen zur Grundlegung einer allgemeinen Rassenpsychologie* (Leipzig: Barth, 1935), 96.

95. Quoted in Eric Michaud, *The Cult of Art in Nazi Germany*, trans. Janet Lloyd (Stanford, CA: Stanford University Press, 2004), 127.

96. The difference between his postwar writing and his earlier writing is not theoretical or epistemological, but Clauß now stresses that intimate knowledge of the context of the photographs' production is necessary in order for it to serve as evidence (*Beleg*). For the first time, Clauß designates the photographed person as a "partner," placing the photographer and photographed on a par. Quoting Martin Buber, Clauß concludes that the camera is a tool for being "in a relationship" (Clauß, *Die Seele des Andern*, 232).

97. Ibid., 257.

98. Ibid., 158–257.

99. Clauß's claim that a staged photograph is of little value for scientific observation (259) places him squarely within one strain of scientific photography, whereas in medical photography the assumption grew that to emphasize the important features, a thorough staging was necessary. See Martin Kemp, *Seen Unseen: Art, Science, and Intuition from Leonardo to the Hubble Telescope* (Oxford: Oxford University Press, 2006), 274.

100. See Berthold Hinz, *Art in the Third Reich* (New York: Pantheon, 1979), 173.

101. Quoted in Michaud, *Cult of Art*, 129–30.

102. Technologically, Clauß's method was dependent on the reduction of shutter speeds and the improvements in the sensitivity of the plates that had already occurred in the beginning of the twentieth century. See Marta Braun, "The Expanded Present: Photographing Movement," in *Beauty of Another Order: Photography in Science*, ed. Ann Thomas (New Haven, CT: Yale University Press, 1997), 150–84, and Mary Ann Doane, *The Emergence of Cinematic Time: Modernity, Contingency, the Archive* (Cambridge, MA: Harvard University Press, 2002). While Clauß's discourse about photography is that of mechanical reproduction, he claims at the same time to produce a reality unavailable to the eye. This latter claim needs to be situated in the context of the controversy between Lorraine Daston and Peter Galison and Joel Snyder and Josh Ellenbogen on Marey's photographic method: Daston and Galison viewed Marey's method as an instance of the use of machines to record experimental data in a more reliable way than biased human observers could perform; Snyder and Ellenbogen claimed that Marey produced events unavailable to the eye. Lorraine Daston and Peter Galison, "The Image of Objectivity," *Representations* 40 (1992): 81–128; Joel Snyder, "Visualization and Visibility," in *Picturing Science, Producing Art*, ed. Caroline Jones and Peter Galison (New York: Routledge, 1998), 379–97; Josh Ellenbogen, "Camera and Mind," *Representations* 101 (2008): 86. Curiously, Marey, like Clauß, abstracted drawings from photographs (Marey, *Movement*, 88–89; 106). Their respective ultimate scientific ideals, however, were at odds: Marey aimed for the production of mathematical curves, Clauß for understanding (Ellenbogen, "Camera and Mind," 97). In emphasizing the inherent advantage of the camera over the eye, Clauß joined a long line of writers, including Charles Darwin. By 1959, however, Clauß's method had become technologically archaic. Ethologists now studied animal behavior with motion pictures; cultural anthropologists still employed the camera but also recognized the social mediation it performed.

103. See Hinz, *Art in the Third Reich*, 176.

104. Michaud, *Cult of Art*, 255.

105. Clauß, *Rasse und Seele: Eine Einführung in den Sinn der leiblichen Gestalt*, 166.

106. Ibid., 175.

107. Clauß, *Die Seele des Andern*, 147, 145.

108. What I mean to indicate is the joining together of Clauß's linguistic signifiers and visual signs as well as their union with wider patterns that were specifically National Socialist, in which the visual image of the bearers of the flame was intended to express a direct lineage from and continuity with classical knowledge and traditions; the signifier and the visual image were, respectively, a metaphor and its concretization and materialization. Such visual expressions were omnipresent during the Nazi period, and these signifiers did not altogether disappear afterwards. Joseph Beuys described his vocation to become an artist as being inspired by a picture of a Lehmbruck sculpture: "And in the picture I saw a . . . flame, and I heard the words: protect the flame"; quoted in Hans Belting, *The Germans and Their Art: A Troublesome Relationship*, trans. Scott Kleager (New Haven, CT: Yale University Press, 1998), 82.

CHAPTER FIVE

1. While not discussing photography, this is the framework that Etan Bloom develops with regard to Arthur Ruppin. See Etan Bloom, *Arthur Ruppin and the Production of Pre-Israeli Culture* (Leiden: Brill, 2011).

2. I elaborate on this in "Jews between *Volk* and *Rasse*," in *National Races*, ed. Richard Mc-Mahon (Lincoln: University of Nebraska Press, forthcoming).

3. Yfaat Weiss, "Central European Ethnonationalism and Zionist Binationalism," *Jewish Social Studies* 11, no. 1 (2004): 93–117.

4. See Amos Morris-Reich, "Arthur Ruppin's Concept of Race," *Israel Studies* 11, no. 3 (2006): 1–30.

5. In the Hebrew edition of the 1930 *Sociology of the Jews* (Tel Aviv: Stiebel, 1934) Ruppin describes him as "my teacher" (*rabbi*); in the German edition as *meinen Lehrer*. Arthur Ruppin, *Soziologie der Juden* (Berlin: Jüdischer Verlag, 1930), 1:12.

6. Arthur Ruppin, *Die Juden der Gegenwart: Eine Sozialwissenschaftliche Studie*, 2nd ed., (Cologne: Jüdischer Verlag, 1911), 215.

7. For an analysis, see my "Argumentative Patterns and Epistemic Considerations: Responses to Anti-Semitism in the Conceptual History of Social Science," *Jewish Quarterly Review* 100, no. 3 (2010): 454–82.

8. The situation, however, was more ambivalent, as Ruppin opposed hierarchical ordering of human populations or the idea (from Gobineau, Chamberlain, or Günther) that all civilization was produced by Nordics or claims of Jewish inferiority. Ruppin, *The Jewish Fate and Future* (London: Macmillian, 1940), 233–34. His analysis of Nazi antisemitism was a classic sociological one, and he argued that the Aryan theory was scientifically unsound while nonetheless citing Günther, Clauß, and Fischer as scientific authorities and simultaneously entrenching his cooperation with opponents of determinism. See Morris-Reich, "Ruppin's Concept of Race."

9. For the internal and external contexts, see Dafna Hirsh, "Zionist Physicians, Mixed Marriage, and the Creation of a New Jewish Type," *Journal of the Royal Anthropological Institute* 15 (2009): 592–609.

10. Ruppin, *Die Juden der Gegenwart*, 218n6. Elsewhere Ruppin notes that the difference may be due to "better social situation and home care." Ruppin, *Jewish Fate and Future*, 22n1.

11. Ibid., 29.

12. Ibid., 10. Note that Ruppin implies that the monuments shown at the end of the book are of ancient Jews whereas they are in fact of types he argues the Jews resemble.

13. Ruppin, *Sociology of the Jews*, 6. Discussing the Jews' mixture, Ruppin notes that Redcliffe Salaman (discussed in chap. 1) believed that the Philistine element among modern Jews was dominant, that others who saw Bedouins were impressed by their resemblance to modern Jews, and that Felix von Luschan, while touring northern Syria, found a great likeness between the types he saw there and European Jews. All three impressions are correct because they represent the three different major types that make up the Jews. He ends this short analysis by stating that if von Luschan were to tour Arabia or southern Italy rather than Syria, he would have found the local types there to resemble Jews. Nonetheless, Ruppin determines that, from the three, von Luschan's is closest to the truth (7–8).

14. Ibid., 11.

15. Roland Barthes, "The Reality Effect," in *The Rustle of Language*, trans. Richard Howard (Oxford: Blackwell, 1986), 141–48.

16. See the Central Zionist Archives GNAR\330925.

17. This is a photograph of a young woman from frontal and profile angles that appears as illustration 102a and 102b in the German edition of the book and as GNAR\331034 and GNAR\331035 in the archive. Some of the photographs are similar to those reprinted in the book (cf. "Armenian types" or "probably Armenian orphanages" GNAR\331011—GNAR\331033 and GNAR\402873—GNAR\402888).

18. E.g., GNAR\331037 shows a person whose head is covered, almost touching the camera lens, a woman carrying a large jar, gazing, somewhat numbly, at the photographer. In the same frame there is also another woman, smiling and in the background, and several additional somewhat shocked children.

19. GNAR\330959 and GNAR\330960. The CZA identifies the woman in the photographs as Sophie Ruppin. Our attempts to find out the relationship of Sophie to Arthur led us to Yaacov Goren (Rupin's biographer), and Etan and Yereon Ruppin (Arthur's grandchildren through his son Rafael). While they could not find a Sophie in the family genealogy, Yereon speculated the photographs might be of Steffie Ruppin, daughter of Ziegfried Ruppin and thus Arthur's niece.

20. Hacking, "The Looping Effect of Kinds," 351–93.

21. Ruppin, *Die Juden der Gegenwart*, 213n1; *Sociology of the Jews*, 44.

22. Note that unlike von Luschan, Fischer, or Fishberg, Ruppin had no medical background, and unlike Goldschmidt or Struck, he had no background in art or art history. His selection, therefore, was not guided by specifically medical or artistic categories.

23. Chaim Weizmann, *Trial and Error: The Autobiography of Chaim Weizmann* (New York: Harper, 1949), 129.

24. Orit Abuhav, "The Face of Man: On the Contribution of Anthropologists Erich Brauer and Raphael Patai to the Anthropology of Jews" [in Hebrew], *Jerusalem Studies in Jewish Folklore* 20 (2003): 155–73.

25. Brauer was also marginalized because of his relationship with the renowned scholar of Jewish mysticism Gerschom Scholem. Both belonged to the same group of young Berliners

from assimilated Jewish background who showed interest in Zionism. The two were not only close friends, but in 1915–1916 they were members of the "Young Yehuda" movement. Scholem's advances to Brauer's sister, Greta, were rejected, a fact that complicated their relations in Jerusalem, where on the arrival of Brauer Scholem, Gerschom Scholem was a professor in the Hebrew University and Brauer only a temporary affiliated researcher.

26. This study was published in 1925 as *Züge aus der Religion der Herero: Ein Beitrag zur Hamitenfrage*, Institut der Völkerkunde, Erste Reihe, Ethnographie und Ethnologie (Leipzig: R. Voigtländer, 1925).

27. For detailed accounts of Brauer's status in the Hebrew University between the fields of Jewish studies and oriental studies, anthropology, and folklore, see Orit Abuhav, "The Face of Man" (unpublished manuscript); Vered Madar and Dani Schrire, "From Leipzig to Jerusalem: Erich Brauer, a Jewish Ethnographer in Search of a Field" (unpublished manuscript); and Dani Schrire, "From Tribes in Africa to the Communities of the East: Brauer and the 'Return' from Africa to Judaism" (unpublished section from the PhD diss. "Collecting the Pieces of Exile: A Critical View of Folklore Research in Israel in the 1940s–1950s" [in Hebrew], The Hebrew University, 2011).

28. Brauer, *Züge aus der Religion der Herero*, 111–17.

29. Erich Brauer, *Ethnologie der Jemenitischen Juden* (Heidelberg: Carl Winters Universitätsbuchhandlung, 1934); *The Jews of Kurdistan: An Ethnological Study* [in Hebrew], trans. and ed. Raphael Patai (Jerusalem: Israeli Institute for Folklore and Ethnology, 1947); *The Jews of Kurdistan*, ed. Raphael Patai (Detroit, MI: Wayne State University, 1993).

30. Brauer, *Ethnologie der Jemenitischen Juden*, 51–54.

31. Ibid., 54.

32. Ibid., 55.

33. Ibid., 57.

34. Brauer, *The Jews of Kurdistan: An Ethnological Study* and *The Jews of Kurdistan*.

35. Erich Brauer, "Birth Rites of the Jews of Kurdistan" [in Hebrew], *Edot: A Folklore and Ethnology Quarterly* 1 (1945–46): 64.

36. Johanes Fabian, *Time and the Other: How Anthropology Makes Its Object* (New York: Columbia University Press, 2002).

37. Examples are found in his estate at the Information Center for Jewish Art and Culture at the Israel Museum, Jerusalem. Brauer in fact mentions modern haircuts in the book on the Yemenite Jews, 54n6.

38. See Theo van Leeuwen, "Semiotics and Iconography," in *Handbook of Visual Analysis*, ed. Theo van Leeuwen and Carey Jewitt (London: Sage, 2001), 100.

39. Brauer's Promotionsurkunde (15.4.1924), Universitätsarchiv Leipzig, Phil. Fak. Prom. 9976.

40. In the last two decades Yemenite Jews have increasingly caught the interest of Israeli scholars, with some claiming that Yemenite Jews were the direct object of the racism of Zionist leaders toward them (a prominent role is ascribed to Arthur Ruppin in this respect) and describing the widespread exploitation, oppression, and suffering they endured in Palestine in the late nineteenth century and the first half of the twentieth. Bloom, *Pre-Israeli Culture*; for a comprehensive

history of the study of Yemenite Jewry, see Noah S. Gerber, *Ourselves or Our Holy Books: The Cultural Discovery of Yemenite Jewry* [in Hebrew] (Jerusalem: Yad Ben Zvi, 2013).

41. Guy Raz, *A Yemenite Portrait: Jewish Orientalism in Local Photography* [in Hebrew] (Tel Aviv: Eretz Israel Museum, 2012).

42. Richard Andree, *Zur Volkskunde der Juden* (Bielefeld: Von Velhagen und Klasnig, 1881), 230. The racial perspective and the use of photographs with regard to Yemenite Jews emerge gradually in the last decades of the nineteenth century. For instance, M. Schwarzauer's report "Die Juden in Jemen," *Monatsschrift für Geschichte und Wissenschaft des Judentums* 3, no. 1 (1854): 42–44, makes no use of either. Hermann Burckhardt's report "Die Juden in Jemen," *Ost und West* 2, no. 5 (1902): 337–42, provides photographs but makes no mention of race or physical differences (in the previous chapter we saw his photographs were a major resource for Fishberg and Feist).

43. Cf. Samuel Weissenberg, "Die autochtone Bevölkerung Palestinäs in anthropologischer Beziehung," *Zeitschrift für Demographie und Statistik der Juden* 5, no. 9 (1909): 130.

44. Samuel Weissenberg, "Das jüdische Rassenproblem," *Zeitschrift für Demographie und Statistik der Juden* 1, no. 5 (1905): 8.

45. Samuel Weissenberg, "Die jemenitischen Juden," *Palästina* 7, no. 1 (1910): 17.

46. Ibid., 17.

47. Ruppin, *Sociology of Jews*, vol. 15. Note Ruppin's differentiation between "special types" and "foreign types," Yemenite Jews belonging to the former.

48. Raz, *Yemenite Portrait*, 53. Adopting the orientalist framework led Raz to exclude from his exhibition Yemenite photographers.

49. On Lilien and Zionism, see Haim Finkelstein, "Introduction," and Mark H. Gelber, "E.M. Lilien and the Jewish Renaissance," in *E.M. Lilien in the Middle East: Etchings (1925–1908)* ([Beersheba]: Ben-Gurion University of the Negev: Avraham Baron Art Gallery, 1988).

50. Micha Bar-Am, *Painting with Light: The Photographic Aspect in the Work of E.M. Lilien* (Tel Aviv: Tel Aviv Museum of Art, Dvir, 1990), 29.

51. Ibid.

52. Bar-Am mentions emphases, distortions, and photographic framing and foreshortening; Bar-Am, *Painting with Light*, 38, 56n69. Bar-Am shows three kinds of ways that Lilien integrated photography into his art—free interpretation, transfer of photographic content by way of projection, and composite painting (based on several photographs); see 44–46.

53. Otto M. Lilien and Eva Strauss, eds., *E.M. Lilien: Briefe and seine Frau 1905–1925* (Frankfurt: Jüdischer Verlag/Athenaüm, 1985), 185.

54. Ibid., 118.

55. Ibid., 172, 120, 185. He also states that the reproductions sent to his wife do not render the horror that the originals in his possession of a massacre of Jews in Russia do (54).

56. Raz, *Yemenite Portrait*, 31e.

57. Lilien and Strauss, *E.M. Lilien: Briefe*, 37. Elsewhere he describes Jews' love for their countries as that of adopted children for their parents (138).

58. Ibid., 189, 190.

59. Bar-Am, *Painting with Light*, 56n65.

60. For comprehensive analysis of Lerski in the context of cinema (including Zionist propaganda film), see Ofer Ashkenazi, "The Symphony of a Great Heimat: Zionism as a Cure for Weimar Crisis in Lerski's *Avodah*" (unpublished manuscript).

61. For a recent analysis of Lerski, see Kathryn Alice-Steinbock, *Crisis and Classification: Photographic Portrait Typologies in Early 20th Century Germany* (PhD diss., University of Michigan, 2011), 228–94. Following the discussion of Rembrandt in the introduction of this book, note that Lerski viewed him as the father of antitypological portraiture (to which he himself belonged; 235). For Lerski's biography, see Ute Eskildsen and Jan-Christopher Horak, eds., *Helmar Lerski: Lichtbildner* (Essen: Museum Folkswang, 1982).

62. Helmar Lerski, *Köpfe des Alltags* (Berlin: Hermann Rockendorf, 1931), plates 33–42.

63. Quoted in Alice-Steinbock, *Crisis and Classification*, 228, and variations in note 30. Ebner, *Metamorphosen des Gesichts*, 3: "Every human being embodies everything; it is only a question of how the light falls." Alice-Steinbock shows that Lerski undermined the title of his "Jewish heads," "oriental types," or "Jews and Arabs" by the "photographer's domineering use of light—one which sculpts all sitters in equally theatrical appearances" (236).

64. Florian Ebner, *Metamorphosen des Gesichts: Die "Verwandlung durch Licht" von Helmar Lerksi* (Göttingen: Steidl, 2002), 53.

65. Ibid., 48.

66. Quoted in Ebner, *Metamorphosen des Gesichts*, 54.

67. The reproduction of this photograph exemplifies the kind of considerations discussed in the preface to this book. I first encountered this photograph in the widely circulated East German posthumous homage to Lerski, *Der Mensch Mein Bruder: Lichtbilder von Helmar Lerski: Text von Louis Fürnberg, Berthold Viertel u. Arnold Zweig*, ed. Anneliese Lerski (Dresden: Verlag der Kunst, 1958), 36. I found the fact that the left eye of the photographed man was not visible was essential to Lerski's modernistic photographic treatment of the subject. Preparing for the book, I received a high-quality scan of the print from Folkwang Museum in Essen (where his estate is found). When the scan arrived I realized both eyes were here completely visible. What I thought was a feature of his treatment of the subject resulted rather from the book reproduction quality.

INDEX

Page numbers in italics refer to illustrations.